Manika Gupta

Biological Pathways to Improve Pest Control in Agriculture

Anchor Academic
Publishing

Gupta, Manika: Biological Pathways to Improve Pest Control in Agriculture, Hamburg, Anchor Academic Publishing 2017

Buch-ISBN: 978-3-96067-189-3
PDF-eBook-ISBN: 978-3-96067-689-8
Druck/Herstellung: Anchor Academic Publishing, Hamburg, 2017

Bibliografische Information der Deutschen Nationalbibliothek:
Die Deutsche Nationalbibliothek verzeichnet diese Publikation in der Deutschen Nationalbibliografie; detaillierte bibliografische Daten sind im Internet über http://dnb.d-nb.de abrufbar.

Bibliographical Information of the German National Library:
The German National Library lists this publication in the German National Bibliography. Detailed bibliographic data can be found at: http://dnb.d-nb.de

All rights reserved. This publication may not be reproduced, stored in a retrieval system or transmitted, in any form or by any means, electronic, mechanical, photocopying, recording or otherwise, without the prior permission of the publishers.

Das Werk einschließlich aller seiner Teile ist urheberrechtlich geschützt. Jede Verwertung außerhalb der Grenzen des Urheberrechtsgesetzes ist ohne Zustimmung des Verlages unzulässig und strafbar. Dies gilt insbesondere für Vervielfältigungen, Übersetzungen, Mikroverfilmungen und die Einspeicherung und Bearbeitung in elektronischen Systemen.

Die Wiedergabe von Gebrauchsnamen, Handelsnamen, Warenbezeichnungen usw. in diesem Werk berechtigt auch ohne besondere Kennzeichnung nicht zu der Annahme, dass solche Namen im Sinne der Warenzeichen- und Markenschutz-Gesetzgebung als frei zu betrachten wären und daher von jedermann benutzt werden dürften.

Die Informationen in diesem Werk wurden mit Sorgfalt erarbeitet. Dennoch können Fehler nicht vollständig ausgeschlossen werden und die Diplomica Verlag GmbH, die Autoren oder Übersetzer übernehmen keine juristische Verantwortung oder irgendeine Haftung für evtl. verbliebene fehlerhafte Angaben und deren Folgen.

Alle Rechte vorbehalten

© Anchor Academic Publishing, Imprint der Diplomica Verlag GmbH
Hermannstal 119k, 22119 Hamburg
http://www.diplomica-verlag.de, Hamburg 2017
Printed in Germany

Acknowledgement

God is great and mighty to make, without whose will nothing can move in the world. Human efforts achieve success only when God smiles on them. As a human I thought of doing something in the field of my studies related to thrips population on chilli plants and biological methods to control it so that yield loss occurring to the farmers of the chilli field may be checked and at the same time harms caused by the chemicals in the pest management may be saved. Really it was due to God's grace upon me that I could complete my adventure.

God helps those who help themselves. As a novice researcher, I went to Dr. Virendra Kumar, Associate Professor, Dept. of Zoology and he agreed to guide me in the present research endeavour. Words fail to express the gratitude I owe to him. It was he who showed me the right path in the course of this research work and unreservedly afforded time whenever I needed it. His valuable suggestions and tips to carry on the research work made me reach the present stage of completion of the project.

I am also thankful to other faculty members of the Department of Zoology, D.S. College, Aligarh namely Dr. Anjana Agrawal, Dr. Sheeba, Mr. Vinay Kumar, Ms. Parul Yadav, Dr. Meera Singh, Dr. P.C. Agrawal, Dr. R.K. Agrawal, Dr. R.K. Goel, Dr. T.C. Agrawal and Dr. P.K. Gupta, who gave fruitful suggestions time to time during the progress of the present research work.

My thanks and gratitude to Dr. S.K. Goel, Head, Department of Mathematics, D.S. College, Aligarh are released spontaneously at this juncture as it was he who showed me a simple and effective method of drawing statistical conclusions for the confirmation of the stated problem.

I offer my special thanks to Dr. Mahesh Chandra, Assistant Professor, Dept. of Zoology, P.C. Bagla College, Hathras who has specialities in research programmes and works as the chief editor of several journals. It was under the benefit of his abilities freely extendable to me that the present research thesis got the current shape. He supported me by providing relevant material and technique to bring finishing touch in my thesis.

I shall be failing in my duty if I do not extend my gratitude to Dr. Sarang Savalekar [Ph.D. in Termite Management, Director – Paramount Pest Consultants Pvt. Ltd., Partner – Pest Consultants, President of Pest Management Association (India)] with whom I had frequent on-line consultations about the progress of my research thesis. His constant support, encouragement and kind faith in me always boosted me and kept me focus towards the goal. It was the help of this gentleman with a big heart that I could achieve high standard in the preparation of my thesis. Dr. Sarang Savalekar is working in Termite Management and has high aptitude for research.

I am highly thankful to Dr. Vivek Kumar, University of Florida, who provided me a number of relevant taxonomic images of thrips which I used for the beautification of my thesis. I am also thankful to Dr. Antoon Loomans, Netherlands who helped me in taxonomical identification and examination of morphological biology of natural enemies selected for our research.

I must extend my gratitude to the journals 'Bionotes', 'Nature and Environment', 'Annals of Zoology' and 'Annals of Natural Sciences' which gave space to my research papers during the progress of the present research work.

My thanks go to different libraries, like Maulana Azad Library, A.M.U., D.S. College Library, Aligarh, The British Council Library, Google Online Library and Wiley Online Library wherefrom I picked up material relevant to my research. I am also thankful to different websites freely available for visiting and downloading. I owe a great deal to IARI, New Delhi and NBAIR, Bangalore (Agricultural Institutes) which supplied to me natural enemies of thrips for getting practical results of my thesis.

I am also thankful to different farmers of the localities mentioned in the thesis, who gave permission to hold experiments in their fields. I am also thankful to Mr. Ravi Mittal who with his extraordinary computer skills made the ugly looking manuscript of the thesis into a beautiful volume as it looks now.

The last, but not the least, the support of my family should not go unnoticed and uncared for. I am highly grateful to my father Dr. Hem Prakash, Head, Dept. of English, D.S. College, Aligarh and my mother Smt. Abha Varshney who helped me financially, morally and emotionally during the progress of the research work. My younger brother Er. Akash and my younger sister Ms. Shreya Varshney also deserve my thanks as they with all naughty activities teased me time to time and again ignited my interest for the research work in ironic comments.

(Manika Gupta)

Abstract

India is the heaven of agricultural products, its vast plains contain alluvial soil with rich natural contents. Major economy of India is based on agricultural products. Green revolution in India brought high hopes for Indian farmers. Several new scientific information helped crop production to grow by leaps and bounds. The more researches, the more intricacies. Further knowledge of pests makes scientists think about several other solutions. Use of chemicals was immediately adopted to curve the population of pests and good results were also obtained. But later on, harmful effects of the pesticides were also known. It was realized later on that the regular use of chemicals in pesticides creates danger for human health. A kind of ecological disturbance is caused unnecessarily. Over the next three decades, production of food grains in India was increased at least 2 million tonnes in order to meet the food demand of the growing population. Today, further prospects of raising agricultural production in the manner as in the recent past appear to be severally constrained. Land frontiers are becoming limited and opportunities to bring additional land under cultivation are being bedimmed.

Over the last 10-15 years, *S. dorsalis* has rapidly become a major pest of chilli pepper and Solanaceous plants. It has gained prominence in several tropical regions of the world. Due to cryptic small size and thigmotactic behaviour it is not easily detectable and harms caused, it continues to affect chilli production and quality. *S. dorsalis* was known to infest a wide variety of host plants. It causes damage by extracting the contents of the cells and tissues of their host plant. This leads to necrosis of feeding in the forms of scars, leaves distortion, distortion of buds, flowers and young fruits. In order to maintain economic feasibility, it is necessary to find effective means of biological control, so as to maintain the standards of commercialization. In this study, it is suggested the biological approach is a cost-effective technology. The magnitude of the net benefits of this technology would depend on the type of input which is used in biological management package and its applications. Biological management is akin to a new technology and knowledge intensive. It requires method of perfect application before it is transferred to the farmers. Our research study indicates, some alternatives to check the growth of harmful pests for the crop of chilli. It is true that the natural control through biological methods may not kill cent percent insects harming the chilli plants. But, we have to attain that level which does not affect the growth at economic level and also gives healthy production of the fruits of chilli. We have to think about the final market quality of the crop in order to make it economically viable. For this purpose, most effective management has to be done to see the good results of biological methods through natural enemies; predators and parasitoids in the Insect Pest Management.

Contents

Chapter-1	Introduction		1-47
	1.1	Preamble	2
	1.2	Statement of the Problem	33
	1.3	Research Objectives	37
	1.4	Scope of Research Work	38
	1.5	Organization	40
	1.6	Research Hypothesis	43
Chapter-2	**Review of Literature**		**48-85**
	2.1	Historical Resume	49
Chapter-3	**Research Methodology**		**86-100**
	3.1	Research Methodology for Survey and Sampling	88
	3.2	Collection of Natural Enemies from Different Outfields	90
	3.3	Rearing of Sampled Insects	90
	3.4	Establishment of Research Laboratory	90
	3.5	Laboratory Bioassays	91
	3.6	Methods of Stock Collection	93
	3.7	Taxonomic Research Study	94
	3.8	Laboratory Processes	97
	3.9	Research Methodology for Cultivation of Chilli Plants	98
	3.10	Biological Controlled Experiments of Thrips Under Net House Conditions	99
Chapter-4	**Experimental Analysis**		**101-236**
	4.1	Population Fluctuation	104
	4.2	Sampling of Natural Enemies During Year 2014-15	108
	4.3	Biological Controlled Experiments Under Net House Conditions	124
	4.4	Results of Biological Controlled Experiments During the Year 2014-15 and 2015-16	180
	4.5	Statistical Analysis of Biological Controlled Experiments	194

	4.6	Discussion	212
	4.7	Discussion of Biological Controlled Experiments	217

Chapter-5	Observation		237-275
	5.1	Thysanopterans on Chilli Crop in District Aligarh of Western Uttar Pradesh	238
	5.2	Occurrence and Damage of Thrips on Chilli Crop	241
	5.3	Pest Investigation by the Monitoring Observation on Chilli Crop	245
	5.4	Weather Influences on Insects	248
	5.5	Taxonomy of *Scirtothrips dorsalis*	249
	5.6	Biology of *Scirtothrips dorsalis*	257
	5.7	Life Cycle of *S. dorsalis*	258
	5.8	Taxonomy and Biology of Natural Enemies of Thrips	263

Chapter-6	Summary and Conclusion		276-325
	6.1	Summary and Conclusion	277
	6.2	Recommendation	307
	6.3	Future Scope	311
	6.4	Popularization : May Reach at Gross Root Level Need	314
	6.5	General Suggestions : Through Our Entire Research	315
	6.6	Limitations : Over Restrain the Growth of Harmful Pests	319
	6.7	The "Pros and Cons" of Biological Control	322

	Bibliography	326-343

	Appendices	

List of Figures, Charts, Tables and Graphs

Total No. of	Count
Figures	74
Charts	01
Tables	28
Graphs	74

Chapter-1
Introduction

1.1 Preamble
1.2 Statement of the Problem
1.3 Research Objectives
1.4 Scope of Research Work
1.5 Organization
1.6 Research Hypothesis

Chapter-1
Introduction

1.1 Preamble

Agriculture : Extensive Background

From decades, agriculture has an extensive background. Its association with needful food crops is of vital significance. Agricultural activity is famous for its farming including forestry, marketing and distribution of crops. All these are acknowledged as a part

Figure 1.1 : Agriculture : Is the Backbone

of current agriculture status. It will be better to explain the whole system as the production and then distribution of the products among the people for its proper utilization in human needs. In India, Agriculture is a vital activity. It will be better to explain it as the backbone of our country. It also provides great opportunities in the field of self employment and labour employment.

Agriculture : Known as Source of Livelihood

The word agriculture has its origin in Latin language. Its adaptation in middle English gave its modern meaning. Two words "ager" and "culturo" went into its making. From "ager" we mean "field" and from "culturo" we understand cultivation which tends to mean "growing". At present it plays a tremendous role in the human efforts to use natural resources for the betterment of human life. It is the main source of livelihood of many people. It contributes a lot in the production of commodities; including food, fiber, forest and horticultural crops to maintain the quality of life.

Agricultural Contribution: To National Revenue

Agriculture contributes a lot in the economy of the most of the countries of the world. India is prominent country based on agriculture.

It deserves the second rank worldwide on the basis of the farm input and other life forms for food, bio-fuel, medicinal products etc. In India, agricultural production prepares the backbone of the country and place an important role in its economy. Its share in the Gross Domestic Products prepares the economic health of the country.

Figure 1.2 : Economic Health by GDP Share

Role of Agriculture

Agriculture has a broad range and acts as a main source for sustenance of human life. The sector of agriculture also provides fodder for domestic animals. Really, agriculture is the main source for several industries also - cotton industry, jute fibre industry, tobacco industry and further edible oils and non-edible industries, all use raw materials coming from agricultural products in India.

Figure 1.3 : Agriculture : The Basis of Human Survival

Economic Importance of Agriculture

We may find out economic importance of agriculture in its value towards fair and sustainable food products such as rice, wheat, potato, tomato, onion, beans, cotton, maize, millets and pulses and also chilli pepper which are present in the form of food grains. These basic and useful crops are the basis of our healthy diet.

In the developing countries agriculture is the main source of livelihood in rural areas. Today, the people of the world are conscious about eco-system and healthy-environment endangered by pollution and other damages.

Agriculture provides a fruitful ground for the sustainability of healthy environment. Still the field of agriculture plays a dominant role in the overall economic, scenario and it is more a **'way of life'** than a **'mode of business.'**

Figure 1.4 : Agriculture : Source for sustenance of human life

It is worthwhile to remember the following words of a thinker, "When it is understood that one losses joy and happiness in the attempt posses them, the essence of natural farming will be realized. The ultimate goal of farming is not the growing of crops, but the cultivation and perfection of human beings."

The Chilli: Life and Soul of Indian Dishes

Pepper, Chilli, Chile or **Chilli** belongs to the family **Solanaceae**. It is being cultivated throughout the country for thousands of years.

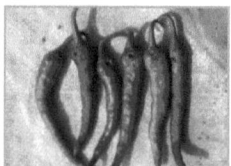

Figure 1.5 : The Chilli

Its use is still relevant despite many changes in the food habits of the people. The genus *Capsicum* has direct link to a divergent group of plants which are famous for its intense pungent taste. It is derived from the Greek word i.e. *Kapsima* means 'to bite'. In India, we have several varieties of chilli with exclusive qualities of divergent breeds.

Taxonomic Position of Chilli

Kingdom	:	Plantae
Division	:	Magnoliophyta
Class	:	Magnoliopsida
Order	:	Solanales
Family	:	Solanaceae
Genus	:	*Capsicum*
Species	:	*annuum*

World Scenario of Chilli

As per the estimations based on different researchers, it is concluded that the area of chilli production all over the world around 105 million hectare. In the total area of chilli production about 7 million tonnes of chilli is produced all over the world. The notable countries of chilli production are India, Pakistan, Indonesia, Korea, Sri Lanka and Turkey. It is used as a potent spice all over the world and in the total trade of spices in the world about 16% is covered by chilli. Its expensive use in spices is making it an impatient item of trade. Christopher Columbus, who discovered America in the last decade of 15th century was the first European to use chilli in the taste of food.

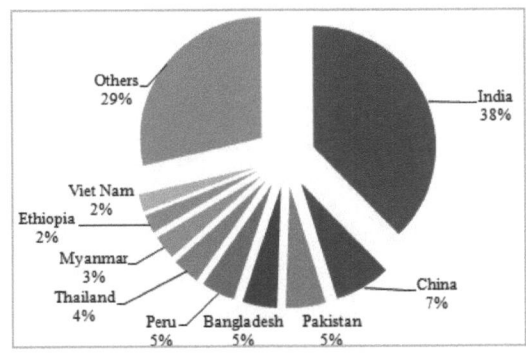

Figure 1.6 : World Scenario of Chilli

History Expansion

Chilli crop is not of Asian origin. It was brought to Asia in sixteenth century when Spanish and Portugese adventures discovered Asian countries like India and Indonesia. In 1498, chilli was brought to India by the Portuguese explorer Vasco-de-Gama. Soon, chilli became an important ingredient in Indian cuisine.

Economic Importance of Chilli with Potential Health Benefits

Chilli is an important agricultural crop. Its popularity among its increasing number of users has made it a crop of economic importance. Besides its nutritional value it is also used in medicines. Unripe fruits are green in colour, but after the process of ripening, they attain the shade of red colour. Green chillies are sometimes usefully applied for natural colours. They are also used as antioxidant compounds. Wide range spectrum of vitamins (A, C, B, E and P) is also found in this excellent crop. This crop is also rich with group of compounds like castonoids, capsaicinoids and phenolic. It is worthwhile to note that these compounds which are very essential for our immune system are largely present in fruit of *Capsicum*.

From the morphological point of view the chilli plant has dark green or purple leaves. It develops a natural crown of white colour and attains the growth upto 1.5 meters. The fruits of chilli are utilized in various ways at various stages. Fleshy pepper of large size is used and consumed as fresh vegetable. Dry fruits of chilli are used directly as well as by turning them into powder for the purpose of spices. Chilli is also liked for its aroma and sharp acidic flavour. Really it has gained wide spread popularity due to its multiple uses.

The pungency of *Capsicum annuum* is due to its active constituent *Capsaicin*. It is recognized as a powerful stimulant with no narcotic effect.

Figure 1.7 : White Colour Flower   Figure 1.8 : Large Fleshy Pepper

It is worthwhile to remember the following words of a thinker:

"Chilli is one of the great peasant foods. Its contribution has made to world cuisine. Eaten with corn bread, sweet onion, sour cream, it contains all five of the elements deemed essential by the sages of the orient, sweet, sour, salty, pungent and bitter.

Apart from traditional and culinary uses it can play a major role in therapeutic and pharmacological actions. These actions create prominent role as **Immunity Booster**.

Chilli : Acting as a Useful Agent

- They are recognized as useful agents for the benefit of human health.
- They serve as anticancer, anti-ulcer, analgesic, and anti-inflammatory, anti epileptic and anti-haemorrhoidal agents.
- Further analysis of the ingredients of *Capsicum* shows that it is helpful in the management of Burns, Psoriasis, Chronic Migraine, Heart disorders and Diabetes.
- The nutrients like Ascorbic acid and Pro-vitamin A are also found in it.

Chemical Analysis of Chilli

Chilli is an all time vegetable. It makes an integral component all over the world in preparing cuisines of better taste. However, in India and Mexico, it is largely used as a spice as well as vegetable. In preparing its chemical analysis it has been found that its dried pod yields of 100 gm contains 160 calories of energy in which we find Carbohydrates 36 gm, Protein 18 gm, Fat 16 gm, Iron 31mg, Niacin 2.5 mg, Vitamin A 640 I.U. and Vitamin C 40 mg.

Table 1.1
Nutritional Value of Chilli (*Capsicum*)

Parameters	Value (Per 100 gms.) (Chillies)
• Moisture	87.700 gm
• Protein	2.900 gm
• Fat	0.600 gm
• Minerals	1.000 gm
• Fibre	6.800 gm
• Carbohydrates	3.000 gm
• Energy	29.000 keg
• Calcium	30.000 mg
• Phosphorous	80.000 mg
• Iron	4,400 mg

Vitamins	
Carotene	175.000 µg
Thiamine	0.190 mg
Riboflavin	0.390 mg
Niacin	0.900 mg
Vitamin C	111.000 mg
Minerals & Trace Elements	
Na^+, K^+, Mg^{++}, Cu^{++}, Mn^{++}, Mo^{++}, Zn^{++}, Oxalic acid	
Caloric Value	Chilli (Dry) : 297 Chilli (Green) : 229

Chilli is called as Wonder Spice

Chilli is considered as commercial vegetable-cum-spice crops. Different varieties of Chilli are *Capsicum annuum, Capsicum frutescens, Capsicum chinense, Capsicum baccatum* and *Capsicum pubescens* etc.

Figure 1.9 : Chilli : 'A Wonder Spice'

They are cultivated for varied uses namely vegetables, pickles, spices and condiments. So, it is popularly called **'a wonder spice'**. We need it in our daily life and enjoy its taste from mild to tingling and even explosively hot.

"Chilli is one of those marvellous simple, elemental, all important and fundamental concepts that has been elaborated out of all recognition; rather like justice, or objective reality."

Important Growing States of Chilli in India

- Karnataka
- Andhra Pradesh
- Orissa
- Rajasthan
- Maharashtra
- West Bengal
- Tamil Nadu
- Uttar Pradesh

Western Uttar Pradesh : Famous for Bountiful Resources

Our area of research is in Western Uttar Pradesh, which is famous for its fertile soil, abundant water, and varied climate, cold wet and hot. In Western Uttar Pradesh agriculture is known as vital source of 'State Wealth' as far as the production of chilli is concerned. It contributes 10% of the total output of chilli in India.

Our observation areas are the villages around the district Aligarh namely, **Tappal, Jalali, Kayamganj, Sumera and Talib Nagar.** In these areas chilli is grown at a large scale. These areas contribute a lot in the chilli production of the western Uttar Pradesh. On the basis of the available data it was found that the productivity level of this crop in these areas is very low. It comes to 392 kg ha^{-1}, which is very thin in comparison to the other parts of India and still very poor on the touchstone of the international benchmark level of 5000 kg ha^{-1}. It further ignites our research to find out means to increase its productivity level by saving its damage by thrips and other harmful pests.

Growth Production of Chilli

Chilli is widely cultivated throughout the warm temperate, tropical and sub-tropical countries. Archaeologists believe that *Capsicum* was used as a food as long as 9,000 years ago. It was first used as a decorative item and later as food stuff and then was used in medicine. Chilli is known as a 'favourite item' in the use of cooking, curries, breads and appetizers. Although India is one of the biggest producers of chilli but the productivity rates and quality are not yet satisfactory. Whereas the beneath marks level of chilli is set at 5000 kg ha^{-1} in the international market. India is at 1500 kg ha^{-1} level. In order to compete in the international market effective measures are to be taken to increase its productivity and quality level.

In India

In India, the production of chilli is popular in all its corners, but there are several states which do not produce as much chilli as they consume. The production of chilli is found at large scale in those states, where climate is generally hot and dry. Its pungency level is better realized in such states of tropical areas. It is further marked on the lines of the received data that during 2012-2013, India produced around 13 lakh tons of red chilli. Chilli production for the 2014-15 crop years is estimated to be around 13 lakh tons as against 11.5 lakh ton produced during 2013-14. The total area of chilli cultivation increased to 8.06 lakh hectare from 7.48 lakh hectare.

The Chilli Ecosystem: Prone to Heavy Infestation

Vegetables play a major role in the farming of chilli crops. But such crops are heavily infected by pests during certain circumstances. Chilli is a simple crop and even ordinary farmers with lesser technical know – how can cultivate and grow it. Cosmetic damage largely affects the chilli fruits, which is directly proportional to yield loss. On the other hand, pests cause indirect damage in the form of several deformities on twigs, leaves, buds, flowers and affect the growth of the plant. High densities of pest's population cause an immediate loss in quality in a short period of time. In the studies of Lewis (1997) it has been found that the crop of chilli has been extensively damaged by insects and pests.

Figure 1.10 : The Chilli Eco-System

Anyone engaged in agricultural, horticultural or medicinal entomology needs to know something about the insects which commonly affect plants, animals or humans. The effects may be beneficial or harmful. Beneficial effects need to be improved upon whereas harmful insects need to be controlled.

Chilli is an important commercial vegetable-cum spice crop. This crop is known to harbour more than 50 insects and 2 mite pests. *Scirtothrips dorsalis* is the major constraints for higher productivity of chilli and its yields. These pests suck and attack the crop at seedling stage itself and continue till first harvest, causing severe losses upto 34%.

Insect Pest Management

Regular studies of environmental concerns have revealed varieties of insects and their behavioural patterns. It is on the basis of such studies that certain methods are evolved to curb the growth of insects. Basic foundation of insect-pest management is based on a detailed knowledge of the crop-ecosystem. Both the harmful and beneficial aspects of insect impact are recognized. Thus, entomologists have progressed in their knowledge of the chilli agro ecosystem and the related interactions.

A better side of the knowledge is related to the relative value of the beneficial insects and it is increasingly recognized.

Reduced Yield Quality By Insect and Pests

In the field of agricultural studies an insect is counted and classified as pest on the basis of the damage it causes to the crop by reducing the yield to the extent that it becomes a cause of worry to the farmers and growers of the crop.

Figure 1.11 : The Pest : Thrips

Pest

The term 'pest' is very subjective and varied accordingly to several criteria. Of course, it is a living being in the ecosystem and its presence and rapid breeding causes annoyance to crops in the way of its healthy growth. In this way, from the agricultural view point an insect which does not cause damage to crops as a whole is kept out from the list of pests animals. So many insects belong to generally accepted pest species but the individual populations are not always counted as pests and not recognized as economic pests.

From among the numerous pests, we have to recognize certain pests which cause substantial damage to the crop and on the basis of that damage we would like to call them pest species and their value is placed upon these consequences by human society.

The damage to the crop is first seen on the physiology of the plant such as twisting of leaves, discolouring of leaves and tissue, breaking on leaves and then it directly affects the quantity and quality of the fruits.

The Arthropod Complex: In Chilli Ecosystem

Large complex of Insects and their species is found in chilli ecosystems. Wherever the crop is grown a few pests are of economic significance. The large number of insects creates major pest fauna of chilli including thrips, whiteflies, aphids, mites, caterpillars etc.

Figure 1.12 : The Arthropod Complex

Chart 1.1
Classification of Insects*

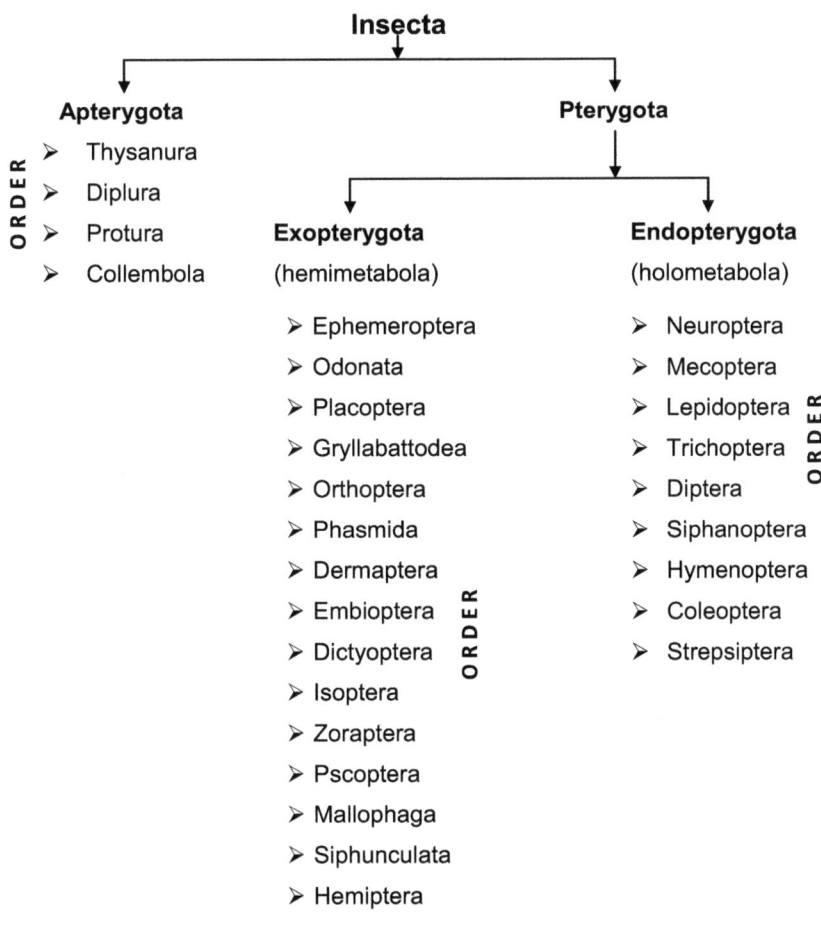

13 | Chapter 1 | Introduction

Thysanoptera: Recorded as opportunistic species

Homogenous group of insects have been recorded as the opportunistic species, they are famous for their characteristic wings having long fringe with less nervature. Thysanopteran adults are only a few mm long and they readily fly and easily migrate into and around the protected crops under green houses. Thysanopterans have peculiar piercing and sucking types mouth parts with vestigial right mandiable, protursible bladder like structure at the apex of tarsal of leg.

Figure 1.13 : Thysanopteran Thrips Adult

Up till now the two suborders 'Tubulifera' and 'Terebrantia' have been recorded under which more than 5,000 species and 8 families have been found. The families are:

- Merothripidae
- Aeolothripidae
- Heterothripidae
- Adiheterothripidae
- Thripidae
- Uzelothripidae
- Fauriellidae
- Phlaeothripidae

Thrips : Fall in order Thysanoptera

Minute insects i.e. Thrips fall in the order Thysanoptera and Family Thripidae. They show exploiting behaviour intermittently. Thrips plays an active role as a pest of several crops and causes substantial damage. Thripidae family is found throughout the world and comprise more than 200 genera.

Origin and Distribution of Thrips pest

The word 'Thrips' has been derived from a Greek word which means 'Woodworm'. Thrips also refers to "Fringed Wing Insects, Bladder footed insects, Storm flies, Thunder flies, Thunder blights and Corn lice." Due to fringed nature of wings, they come in the order Thysanoptera.

Thrips have their primary and independent status. Systematic studies on these groups have been extensive.

Biological Diversity

Large number of thrips species is considered as pests, and affects the plants in commercial value. They are widely distributed. They are most abundant in the tropics. Near about 8,000 species have been recorded all over the world. About 5,000 species have been studied with perfect details about their life history and natural habits. They spread in far and distant regions with the movement of the wind currents. Some species have also been reported upto the snow limit of the Himalayas. Under favourable conditions, many species have capability to multiply fast and their increasing population becomes the cause of irritations to the human world.

Chilli Thrips : *Scirtothrips dorsalis*

Scientific Name	:	*Scirtothrips dorsalis*
Year	:	1919
Comman Name	:	Chilli Thrips
Other Scientific Names	:	• *Anaphothrips andreae*
		• *Heliothrips minutissimus*
		• *Neophysopus fragariae*

Distribution: *S. dorsalis* is distributed in all those countries of Asia where chilli is grown. Countries like Indonesia, Bangladesh, Pakistan, Japan, Korea, China, Sri Lanka, Taiwan, Thailand and Philippines, all are affected by the infestation of *S. dorsalis*.

Origin: There are divergent views about the origin of chilli. One thing is certain that it became popular in Asian countries only after the Spanish and Portugese adventurers came here in 16th century. According to current studies, it is believed that the origin of chilli may be traced in Indonesia much before it became popular after the adventurers of Spanish and Portugese explorers. So, it is concluded that the chilli plant was present in Indonesia but its utility was not recognized in the past as much as it was recognized in the 16th century by the western explorers who encouraged its large scale production for the betterment of cuisine taste. South Asian countries became the largest producers due to favourable climate to this crop. Moreover, the better effect of its pungency is brought about by the prevailing climate conditions in South Asia. Then it became popular in Africa and other continents.

In this manner, *S. dorsalis*, the pest infecting the chilli plant is not known to be sure from which source it originated. It is found in several other crops also. On the basis of polyphagy nature they act as pestilent. From the view point of agriculture there are several hosts on which they grow as pestilent. Thrips exploit host plants and thus they become thicker and thicker and consequently change the landscape of agricultural fauna. Throughout the contiguous region of South Eastern Asia, chilli thrips has spread rapidly.

Its Hosts: There are uncertainities about the original host of *S. dorsalis*. In India, this species is found in plants of castor, pepper, cotton, tea and even in mango and peanuts. This pest is found in the new world also and has been traced in more than 100 plants of about forty families. The main wild host plants in which *S. dorsalis* is found belonging to Solanaceae and Fabaceae family. Pepper also comes within their ranges. Under the pressures of trade and globalization, thrips continued expanding through their range.

Host Damage

Thrips become serious pests as they cause wide spread damage to the host plant. Their immediate effect is seen as cosmetic infestation on the leaves. They feed on plant tissues by their mouth parts and its result is seen in the form of tissue scarification and depletion of host plant. Thrips prefers young plant tissue. They damage the host plant by their direct feeding on leaves and then on flowers and fruits. They also cause damage by transmitting viruses which aggravate contamination of the products. Thrips are phytophagous, or sometimes they show the characteristics of omnivorous herbivores. They frequently encounter an array of possible host plants and additionally, they prefer their oviposition of eggs on them.

Feeding Injury

Chilli thrips, *S. dorsalis* does not feed on mature host tissues; it feeds on young fruit calyx and leaves. Thrips feeding on delicate plant organs can create silvery scarring on leaves. This early damage to plant growth can lead to malformation of the fruits. It is the result of feeding injury caused by *S. dorsalis* on the host plant that the photosynthetic activity is also affected and the final result is significant yield loss.

Figure 1.14 : Deformities on Chilli Leaves

The infestation usually starts from seedling stage, although the severe infestation appears in the stage of vegetation. Thrips possess piercing and sucking mouthparts by which, they feeds and cause damage. They suck the epidermal cells of the host plant and cause necrosis and distortion of leaves, buds, flowers and fruits. In this manner tender leaves and buds become brittle.

Thrips are not very impressive due to its morphological nature in contrast to our better known insects. Thrips are found averaging in size at 1-2 mm. Vast majorities of thrips species are plant feeders.

Figure 1.15 : Thrips

Life History : A Simple Process

Most of the thrips attacking their host plants have a simple life history. Thrips grow at six stages, which may be enumerated like this - first step is egg. Second and third stages are larval which are non-feeding. The forth stage is prepupal. The fifth is pupate stage and the sixth and the final stage is adult through the phenomenon of metamorphosis. Under favourable environment the young attain full growth in 7-10 days.

Thrips are known as weak fliers. They do not travel long distances. But they are spread at long distances by air currents. Thrips population is mostly hidden in the ventral surface of the leaf. Nymphs as well as adults suck the cell sap mostly from the ventral surface of the leaf. As a result, the infested leaf loses its vitality and becomes curled or twisted.

The growing importance of chilli thrips compels the academic world to carry more and more researches in this field so that the major problems posed by thrips may be countered and effective solutions may be traced. Thrips infested host plant which almost failed to produce any fruits or produced very minimum number of deformed fruits.

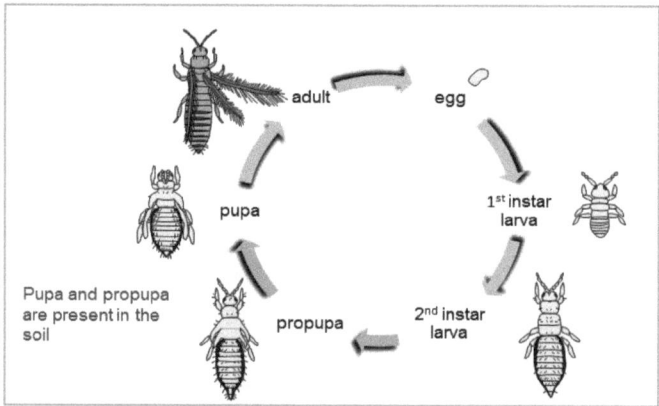

Figure 1.16 : Developmental Stages of Thrips

It is also seen that in the protected crops most of the adults are females. Males are rarely found to have full growth. Actually, males are not needed for fertilization. Thrips are gregarious in nature and the large numbers often concentrated on the same leaf or flower.

Economic Importance

Thrips are opportunist species for many scientists, agriculturists and entomologists, exploring intermittently and occurring in environment. These are mostly phytophagous and distributed universally in most ornamental crops and native plants. Phytophagous thrips mostly feed on the tender leaves and feeding areas are not restricted to leaves alone but may also extend to almost all parts of their hosts. They readily fly and can easily migrate into and around the greenhouses. Therefore, it is very difficult to control the population growth of thrips. The host range of thrips is not limited to certain few. It has its habitat in variety of host plants.

The population of thrips grows rapidly during the warmer parts of the year. In this manner, several generations of thrips overlap during the growing season. The population of thrips does not grow in the rainy season or during chilly winters with less than 4^0C temperature. Warm and dry season is the appropriate season for abundant infestation of thrips.

General Introduction of Chilli Thrips

Adult thrips	: 0.5 – 1.2mm long
Definitive Identification	: ~40-80 x magnification
Eggs	: Microscopic, generally eggs are about 0.070 mm wide and a few may be found 0.075 mm long. It is in about six to eight days that the 9 eggs of the thrips develop. Thereafter these eggs grow into two larval stages which are technically called I and II instars. The larval stage remains for six/seven days.
Prepupal Period	: Short time period approximately Twenty four hours
Pupalperiod	: 2-3 days
General Appearance – The Larvae	: Offwhite in colour

The Adults	:	The adult thrips develop into yellow and grayish colours when they are seen through microscope. Sometimes sketchy stripes are also seen on the dorsal surface adjacent to the meeting point of abdominal segments.
Life Cycle	:	Thrips shows a rapid life-cycle. The total life cycle of thrips is about 20 days and sometimes even lesser.
Duration	:	We find their rapid growth as one female oviposits more than hundred eggs at the rate of 3/4 eggs per day.
The Nymphs	:	(a) It takes two to seven days in the hatching of the nymphs at the stage of the first two instars. They carry on through pseudopodal stages. We can call them prepupa and pupa stages. (b) These developments take place on leaves or flowers as the infestation of thrips continues in the crop.
Characteristic Feature	:	The plant of chilli, above the ground is attacked by the infestation of thrips and damage occurs.
Metamorphosis	:	Thrips show a simple metamorphosis. They do not actually pupate but the pupate stages are known as pseudopodal stages because they are akin to pupal stage. Of course, it results through the process of simple metamorphosis.

Risk of Chemical Pesticides

Generally chemical pesticides are used to curb the harmful effects of insects and pests. But, the immediate gain in this process has adverse effect on the environment in the long run. Regular use of chemicals leads to insecticide resistance. Then, biodiversity is distributed by pest resurgence and pesticide residues. So the immediate gain of one genration become serious problems for the next generation, hence risk for the future.

Figure 1.17 : Agricultural practices heavily based on chemical pesticides

Pest Resurgence: Resurgence is the term which is applied in the matter of sudden increase in the number of pest population. It occurs when pests become habitual of resisting the pesticides. There is gradual decline in the treatment effect and at the later stage, rapid recovery of pests comes to the fore after initial suppression of them by chemicals.

Use of chemicals leads to several other problems in human life. It leads to residual toxicity in fruits and vegetables and pollution of water in the long run. It also causes harm to the predators and parasitoids which are beneficial to the crop in the scheme of nature. When these natural enemies of the pests are eliminated with the use of chemicals, there is always danger to the crop as the next problem of the outbreak of secondary pests also stares in our face.

On the basis of the scientific study of the applied chemicals, it is understood that the pesticides have in them toxic heavy metals, which are harmful in the long run. It is true that these chemicals have immediate effect of killing on the pests. But at the same time these chemicals also kill the natural enemies of these pests. Repeated use of chemical pesticides can benefit farmers in the pest control, but in the long run they become more harmful than even the original pests. The scheme of nature is disturbed. The gradual process of pest control with the natural enemies is checked and the rhythm of biodiversity is broken. Having looked at the immediate gains in the crop, the farmers are tempted to have the repeated use of them and this widespread use of chemicals creates adverse effects on the environment and consequently upon human health.

It was in 1960's that Rachel Carson published "Silent Spring" and tried to awaken humanity against the perils the pesticides might bring to the human world. She warned against indiscriminate use of chemicals in pest control. Now, it is realized that the continuous use of pesticides leads to altering the genetic make up of pests also and they become pesticide resistant in the long run.

It is the cause of worry that the use of pesticides in India is increasing at fast rate. It is estimated that every year there is increase of 2.5% annually from 1970 till today. Of the 96000 tonnes of pesticides produced in the country, about two thirds are used in the field of agriculture.

So, the continuous use of these pesticides in various regions of India adversely affects the plants and animal species.

Lack of Government Seriousness

Regular studies against the use of pesticides have created concern at the government levels also Central Government of India was forced to ban a number of pesticides in the field of agriculture as the harmful effects of the pesticides on human health were brought to light. But these attempts of the government remained mostly on paper. Despite the law of ban in their use, many of these pesticides are available in the market and the farmers who are tempted to use them for immediate gains can easily get them for use in agricultural crops. There is not strict enforcement of law of ban. These loopholes in the system of governance are having demonic face to annihillate the wide propensities of human health.

The government agencies are not very serious in regulating the use of pesticides. The people in the government do not pay much attention towards what is happening at the lower level where the sellers and dealers of pesticides attract the farmers far beyond the limit in the use of pesticides for immediate gains. Whatever feeble law is available for its restriction is not implemented forcefully due to corruption among the authorities. The dealers and sellers of pesticides can easily get freedom of keeping and selling even the banned pesticides. There is hardly any system of guiding the local farmers who are the actual growers and producers of the crops. The ignorant farmers very easily come under attraction of immediate gains when they are told about the large scale benefit of a particular pesticide generally possessed for sale and as canvassed by the local dealers. Further, there is no enforceable law of registration and regulations for the sale of pesticides. Every "TOM, DICK AND HARRY" having a fancy of opening sale shop of pesticides and agro products can easily begin his trade and can supply harmful pests in the farmer's active world.

Biological Control Measures: A Wide Spreading Encouraging Programme

If we study the history of pest control, we find that chemical insecticides began to be used in 1940's. These pesticides were prepared on synthetic basis and were relatively easy in use. For a very long duration these chemicals were considered safe and effective means of pest management. But, later on ill effects of these pesticides were realized. There were hazards to the applicator as well as to those people who inhaled their wasps in the process of spraying. Further, it was also realized that these pesticides also produced a few ill features in the soil and water and finally affected the wild life as a whole.

The organic standards are also brought down by synthetic chemicals. If one kind of pest is restrained by chemicals, the other kinds of pests having resistance to these synthetic chemicals began to develop. In these hazardous situations many farmers and gardeners began to explore those methods which could reduce the ill effects of the pesticides.

Biological control of pesticides is a safe alternative. This method may replace pesticides and can bring good results both ways – at economic threshold and on environmental issues. It is a safe method in which a living organism is used against another living organism in the scheme of nature where killing is the means of survival. In the biological control it is discovered which the natural enemies of the harmful pests are. In this manner, the natural enemies of thrips reduce the population of these pests in natural manner and save the crop from harmful effects. It is most successful management approach. In the simplest definition as given by International Biological Programme, biological control is "using biota to reduce biota."

Biological control is thus beneficial in both ways. It helps the farmers and does not injure environment. It is, of course, cost effective as well as environmentally safest.

Pest : It is a kind of organism which grows a living being and its multiple growth causes damage to plant.

Biological control is a method in which one living organism, which is harmful to the crop, is made food by another living organism. It is the adherence to the scheme of nature that the harmful organisms are kept a low levels, of course at **Economic Threshold Level.**

Economic Threshold

Economic threshold is related to the economic value of the crop, of course, a beneficial ratio between the cost of production and the cost gained in yield. It may be defined as the level of control of the pests at economically viable cost – to reduce the pest population so as to save economic injury to the crop.

Biological control is now gaining prominance among agricultural scientists and it is now an interesting area of entomological research. It is the safest method in both natural and man made things. It also saves eco-system from further injury. The net result of natural biological control is based on the presumption that the earth is green and plants can produce biomass enough to sustain other forms of life.

It is from the time immemorial, say some 500 million years that evolution of the eco-system is maintained by reducing the pest population in a natural way in the scheme of nature without any intervention of others in artificial way. Human knowledge of this naturally controlled eco-system led to the application of certain natural agents to control the harmful organism. On the basis of this knowledge anthropod biological control was used by using predatory ants about 300 years ago.

Having been fatigued by the synthetic chemicals in the pest management, recent pest management is further looking with hopes on biological control. It is recognized as the most sustainable method in the modern times. It is cheapest as well as safest. It is expected that biological control methods will attract about 40% agricultural area by 2050.

For the success of biological control a fair and accurate knowledge of the pest behaviour and the behaviour of their natural enemies is essential. It is on the basis of accurate knowledge of the different organisms related to the crop and the environment that the biological control can be made economically viable and with negligible risk to the eco system.

Biological control is a sub-discipline of applied ecology. To adequately practice it, one should have a firm understanding of population and behavioural ecology. Without any systematic and proper identification of insect pests, and their associations of natural enemies, biological control as a science would fail to function.

There are three basic approaches of biological control strategies : (i) Importation; sometimes called Classical Biological Control; (ii) Augmentation; (iii) Conservation.

- Finding new useful natural enemies
- Releasing additional natural enemies when those naturally present are not in adequate quantity.
- Enhancing the effectiveness of beneficial organisms of protecting them from harmful effects and providing them with necessary habitat and other resources

So, biological control of arthropods can be defined as, "The study and uses of Parasites, Predators and Pathogens for the regulation of host (pest) densities."

Importation Biological Approach

It involves the introduction of a pest's natural enemies to a new locality, where they do not occur naturally. Release of natural enemies and these can permanently establish in new areas for an effective control measures.

This is usually done by government authorities. Importation or classical biological control usually targets introduced (non-native) pests, most of which arrive here without the natural enemies.

Native pests which are not adequately controlled by existing natural enemies may also be the target of this classical approach.

"Classical Biological control is the process of finding natural enemies of invasive pest and importing and establishing these natural enemies to control the pest-reuniting old enemies."

Augmentative Approach

Augmentation involves the supplemental release of natural enemies. This approach is boosting the naturally, occurring population of natural enemies.

Augmentation of Natural Enemies – typically involves the purchase and release of natural enemies that are already present in India but may not be numerous enough to adequately control pests in a particular location. The goal of augmentative biological control is to temporarily increase the number of natural enemies, and, hence the level of biological control for the target pest.

Figure 1.18 : Augmentative Release of Natural Enemies

In this approach, a few natural enemies may be released at a critical time of the season. It is called as inoculative release. Sometimes millions may be released and it is called as inundative release.

Conservative Biological Control

The conservation of existing natural enemies in an environment is the third method of biological pest control. This control refers to the use of indigenous predators and parasitoids, usually against native pests. Natural enemies are already adopted to the habitat and to the target pest. Their conservation can be simple and cost effective. Various measures are implemented to enhance the abundance of the natural enemies and their activities. This includes the manipulation of the crop microclimate, creation of overwintering refuges, increasing the availability of alternative hosts and prey, and providing essential food resources such as flowers for adult parasitoids.

Natural Enemies: Known for their Enormous Value

The Predators, Parasites and Pathogens of pests that are used in biological control management are a large component of world's biodiversity. These natural enemies are of enormous value to the sustainable agriculture, where they can often eliminate the need for pesticide input. Natural enemies introduced to the environment are capable of sustaining themselves and they are also known for their self-perpetuating behaviour and thus they continue to reduce the pest population. Biological control is a particular method of exploiting nature against nature and to get desired effect through beneficial organism i.e. natural enemies.

Figure 1.19 : The Natural Enemies of Pest's

Predators and Parasitoids: As a Miracle

Biological control is an applied field of natural phenomenon, the strategy of bringing natural agents to feed upon the damage causing pests and thus to reduce damage to tolerable levels. This term may be used to denote one of the major ecological sources of nature.

Practically, all living species are attacked by natural enemies – parasites, predators or pathogens – which feed on them in one way or the other. Indeed, many potentially injurious pests are kept at very low levels and never reach economic pest proportions due to the effective action of naturally occurring natural enemies without deliberate intervention by man.

Figure 1.20 : A Predator   Figure 1.21 : A Parasitoid

All plant and animal species are the objects of attack by some or other natural agents for their natural habits of attaining food for the perpetuation of their lives. The impact of these natural enemies ranges from a temporary or minor effect to the death of the host or prey. Predators and parasites are animals that feed on other animals. They show the entomopathogenic nature. Effective biological control may be achieved through their utilization. **The term predation is composed of a broad base of animal → plant (herbivore) – predation supporting the complex web of animal → animal (carnivore) – predation.** The terms predator and parasite have had a long history of usage.

Predators

"A predator is a free-living organism throughout its life. It preys upon other organisms for its food." In the broad sense predation is one of the major ways of life in the animal kingdom. Many ecologists have studied on animal-prey relationships for many years. Predation is common among insects and many of our successful cases of biological control have been through predators. Many species show predation habit on both immatures and adults, but some as certain species are predaceous and feed on larvae.

Predators require food to complete their life cycle and for this purpose they attack their preys and devour thrips. They are basically chewers or suckers. Some insect orders are almost exlcusively predators. Predators share the following distinguished features :

1. They are usually larger than prey.
2. They feed on a wide range of animals and mostly they are not host specific.
3. Death to the victim is caused immediately after capture.
4. Adults or young may be predatory.
5. They are known for their activeness in capturing and eating their pest and other predators.

Parasitoids

Insect parasitoids have immature life stages and they develop by feeding on a single host and devour the host for their survival. Adult parasitoids have their independent nature and free living. On the basis of their behaviour and nature of feeding and living they may also be called parasites.

A particular life stage of a species is attacked by parasitoids in their exercise of getting food for their survival. Parasitism may occur at any trophic level of a food chain. These are organisms which lay eggs in or on the bodies of their hosts and complete their life cycle on host bodies as a result of which hosts die. They show their major characteristics as follows:
- They have special choice in choosing their hosts to feed on.
- They are keen in killing their host which are bigger in size.
- Females are more active and make search for the host.
- After the search of the host the females lay their eggs near it so that food for the larvae may be available at their growth.
- Adults have free living whereas the immature ones remain near the host.

Amblyseius cucumeris* or *Neoseilus cucumeris

Commonly called	: Predatory Mites
Distribution	: Cosmopolitan and distributed through North America, Europe, Australia, New Zealand
Common Hosts	: Among the common hosts, we may trace Aphids, small caterpillars, thrips, psyllids, mites, leafhopper nymphs etc.
Common Habitats	: They live and find growth in pastures, field margins, and habitats with herbaceious and shrubby flowering plants.
Size	
Nymphs	: 0.5 – 2mm
Adults	: 1-2mm, (i) pear shaped and tan coloured, (ii) Eggs are generally round in character and visually transparent. The size of an egg is about 0.14 cm.

Note: Females of *N. cucumeris* are undistinguishable from other species. The two species can only be separated by adult males. The eggs laid by this species can tolerate low humidity and develop rapidly.

Metamorphosis	:	Both the nymphs and adults are predatory and develop through metamorphosis.
Generations per year	:	2-3
Characteristics	:	• *A. cucumeris* is a predator and it feeds on small arthropod preys. For their survival this species consumes two to ten thrips per day. Within their life span of 30 days, one consumes more than hundred thrips. • Suppose thrips population reduces, then they continue to feed on pollen and spider mites. • At the time of application of predators ratio of 1:2 must be kept between the attacker and prey. • *Neoseiulus* moves freely at long distances and feeds on young thrips.
Monitoring	:	• 10-15x hand lens must be applied for inspection. • Mostly they are found on the underside of leaves along their veins. They are sometimes seen inside the petals of flowers. • It is a biocontrol agent widely used for the elimination of thrips, not of one species, but of several species in greenhouses through preventive, mass releases.

Macrotracheiliella nigra (Minute Pirate Bug)

Family	:	Anthocoridae
Distribution	:	The anthocoridae family, popularly known for causing more damage has minute pirate bugs which are also called flower bugs. There are many species and more than 500 are known worldwide.

Habit and Habitat	:	They are cryptic in nature and keep their habitats in galls. They are also seen in the open surface in the environment proper for their growth.
Feeding Habit	:	• They feed on the plant material. • Frequently they feed on chitnised arthropod. • Minute pirate bugs are feed on insect eggs and smaller thrips and these predators are beneficial to our purpose. • These predators are beneficial as biological control agents. • They enter a reproductive diapause in late fall.
Common Hosts	:	• Feeds on the eggs of the corn earthworm, mites and thrips. • Anthocorids are often predacious.
Size	:	Anthocorids are 1.5 – 5mm long.
Morphological Features	:	• They are oval in shape and their flat bodies are often patterned in black and white. • They are soft and elongated. • Their forehead is extended forward. • Their antennae are longer than the head and they can be properly viewed. • Their mouth is designed in such a manner that they can easily suck by piercing their mouth part. They have digestive enzymes to consume the available food.

Ceranisus menes and *Thripobius semiluteus*

Family	:	Eulophidae
Distribution	:	Cosmopolitan In about 300 generate, more than 4300 species are described.

Feeding Habit	:	• Generally larvae does not feed but in a few species it is also found feeding on plants.
• It attacks on second instar larva.
• It attacks those thrips which are generally yellow and whitish in colour. It turns black and swells around head after sucking and when the parasite wasp matures.
Diagnostic Features	:	• Gaster at base is yellow or light brown.
• *C. menes* females of this species may have either a yellow or partly brown *metasoma* and shows coloration.
• *C. menes* can record for at least 20 generations on thrips species.
• *T. semiluteus* can decline thrips population about 60% of larvae.
• Having been shipped in the pupal stage, the wasps should be held at 65^0-75^0 F attain to adult stage.
• Highly mobile adults may be found moving hither and thither in the net house in search of their preys, the thrips.
• Monitoring of the wasp growth and their reproduction must be done with keen observation on immature thrips.
• Eulophids are separable from most other chalcidoidea by the possession of only four tarsomeres. On each leg a small, straight protibial spur, and by antennae with two to four funicle segments.
Development	:	• In about three weeks, the development takes place from egg to adult, if the temperature and environment is favourable to their growth.

1.2 Statement of the Problem

In our country, the role of agriculture has an extensive background. India is densely populated and its population has been increasing every year at the rate of 1.8%. By 2020, the population of India is expected to number around 130 crores. For the increasing population, additional foodgrains will be required. It is expected that two million tonnes foodgrains per year will be needed.

In order to meet the need of foodgrains for the growing population, we require to accelerate agricultural production. But insects and pests create obstacles in the our way of getting more production in agricultural yields. According to the available data about 26% mass of the potential food production is eaten by herviborous insects.

Chilli is a popular crop in India. It is estimated that 25% of the total production of chilli in the world is produced in India. India is the largest producer of chilli and the largest consumer. Further, India is also the largest exporter of chilli. We produce about 1.1 million tonnes of chilli every year. Chilli is an all time vegetable of routine food. It is a versatile spice which is excellent source of natural colours and antioxidant compounds. Chilli is used for various food articles like spices, pickles, condiments, sauces etc. Besides chilli is the principal horticultural crop of major economic value in the country. Emerging problem due to use of insecticides is secondary pest outbreak. Management for the control of secondary pests increases the cost of plant protection.

There are big losses in the production of chilli due to proliferation of insects and pests and related diseases. About 18% of the total agricultural output is harmed by the problems caused by pests. In Western Uttar Pradesh, the crop of chilli provides a lot of profit to the ordinary farmers. Yet insect pest problems are most important constraints for chilli and its production. The chilli pest *Scirothrips dorsalis* (Chilli thrips) commonly called yellow tea thrips, castor thrips, assam thrips can create serious problems before the farmers in getting good results of healthy crop of chilli. *S. dorsalis* in chilli causes losses upto 60%. During the past years, it is recorded that the losses caused by insects and pests are increasing. Loss of about 6000 crores in agricultural production occurred due to insects and pests in the year 1983. This loss increased upto 20000 crores in 1993. Further in 1996, it was estimated about 29000 crores. This increasing loss in agriculture due to the problem of insects and pests has been reduced by effective pest management.

In the crop of chilli, thrips cause major loss. They attack the visible portion of the chilli plants. Mostly thrips prefer young leaves and gather there in heavy infestation. They damage leaves, buds, fruits of chilli, and finally the production is affected heavily. Thrips excessive feeding on the leaves turns their colour from bronze to black. When the thrips attack leaves they curl upward and the entire growth of the plant is affected leaving it dwarf. Due to heavy infestation of thrips, leaves with petioles detach from the stem, causing defoliation in infected plants and measurable result is yield loss. So, the productivity rates and quality of chilli and its products are not yet satisfactory. Thrips show cryptic nature. Due to this, they create real problems in detecting it into fresh vegetative products. In export, the fresh plant material is carried from one part of the world to another and in this process the related pests also go from place to place in the form of eggs and sometimes in the form of larvae and adults too. In this manner many species have now spread from their original natural habitats. Thrips show rapid life cycle. So translocation is easily performed and new species of thrips migrate to new areas and cause significant yield loss.

Treatments against thrips have to be taken specifically. Generally pesticides used to control the population of pests can easily destroy the insect pests but they are hazardous to biodiversity. For successful pest control regular efforts are needed at all levels. Generally it is found that the individual farmers adopt their own techniques on the basis of the propaganda on hear to hear basis. They go to pesticide dealers and obtain pesticides as suggested by the local dealer and spread this knowledge among the compatriots, the chilli producers. But in this process, further problems arise as the natural enemies of thrips are also destroyed. By regular use of pesticides the soil enrichment with nutrient decreases. The presence of natural predators and native competitive insects is eliminated. Further, the use of pesticides may control particular type of pests, but they are replaced by some other pests, and, thus, the damage perpetuates in one form or the other. In this manner, human health is adversely affected. Farmers may be benefitted temporarily but the excessive use of pesticides at regular intervals creates undesirable problems in the form of various deformities. However, with the immediate benefit in the yield began to increase and it became widespread. The overall impact and adverse effects of pesticides may give rise to environmental pollution. Regular use of pesticides will create several problems of untold description. It alters the genetic make up of the pests and several other kinds of harmful pests begin to emerge and they become resistant to the pesticides. In this manner, new varieties of pests causing more harm are activated.

In India, use of pesticides is becoming unbriddled. It has been increasing by leaps and bounds. Approximately, 96,000 tonnes of technical grade pesticides are made available in the market. In recent years, looking at the harmful effects of pesticides, there has been call from the Central Government to ban a few pesticides in the field of agriculture.

Despite this, many of these are easily available in the market and harmful effects continue to persist.

Globally, thrips have been recorded as serious pests by agriculturists, because they cause various damages. Results and experiences in various studies have indicated that in the present agricultural scenario in India and particularly in Western Uttar Pradesh around us, the use of pesticides has become a compulsory step for the farmers in the management of the agricultural crop. Although it is very difficult to control insect pests without such use, but means to control by less harmful ways are to be discovered in order to protect human beings against the probable damages related to pesticides. In the modern era people believe in economic advantages by the use of pesticides in a general way. But these economic incentives cannot continue for long and farmers will further be tempted for pesticide use. The only alternative is consumer awareness about health benefits of the agricultural products grown on pesticide free techniques. The farmers generally come under the temptations of getting lucrative yield of chilli by using pesticides in more quantity. They only take immediate profit by the use of such pesticides without having knowledge of the harms they cause. Interventions of the insecticidal applications bring down the damage of the pest but leads to the problem of pesticide residues in chilli fruits. Up till now it is a major non-tariff barrier against exportation and quality of chilli in developed countries. In this manner, the *S. dorsalis* is the most problem posing pest for farmers in India and it has to be coped up with natural enemies. Major point of concern is that the use of pesticides kills and destroys almost all kinds of insects and thrips irrespective of what may be beneficial or harmful to the host plants.

In this manner, taxonomic characterization of *S. dorsalis* species from the genus are still challenging to non-experts will regard to the biology of thrips. Expert opinions will reveal that certain species of thrips are in the form of predatory habit of attacking other harmful thrips, improper identification or misidentification of thrips can lead to the use of misapplications of pesticide practices resulting in economic waste and waste of resources as well as time.

1.3 Research Objectives

The objectives of our primary research work were in favour of ecosystem, biodiversity and useful flora and fauna. Through these objectives we obtained more and more useful yield of the chilli crop without any harmful and negative effects of pesticides. We applied the biological control experiments and methods to control the infestation of chilli thrips, *Scirtothrips dorsalis* on chilli crop. Our objectives are:

1. To find out the impacts of chilli thrips, *Scirtothrips dorsalis* and their useful natural enemies in the related regions.
2. To evaluate the infestation and damage level of thrips, *Scirtothrips dorsalis* in chilli, *Capsicum annuum*. under biological controlled conditions.
3. To study the Biology of thrips and their natural enemies by the way of exploration.
4. To evaluate the seasonal population / fluctuation of thrips.
5. To provide the best biological control management techniques for thrips, *Scirtothrips dorsalis* under controlled climatic conditions.
6. To popularize the Bio-control technology among the farmers.

1.4 Scope of Research Work

India is a big country with plenty of natural resources. It is amazingly rich in agricultural land and related means of irrigation. In India, the production of food grains are also done at a big scale. Population of India has been increasing at fast rate. It may touch the number of 1.3 billion around 2020. For this population growth, we will stand in need of additional food grains of high nutritive qualities.

Chilli is a rich crop for daily use in vegetables as well as in spices. Apart from traditional and culinary uses it can play an effective role in the therapeutic and pharmacological actions. Customary habits of the Indian people support its use in daily meals. Indian spices have attracted the peoples of the world to make their meals delicious. It is the gift of nature to the Indians that we are the largest producer of chilli as well as its largest user and also largest exporter.

But in the present context the farmers in India make use of harmful pesticides to curb the growth of insect pests harming chilli plants and their fruits. These pests have the characteristics of a detrimental common property resource. They do not recognize spatial boundaries when farmers use these pesticides for long and in this manner, adverse effect on human health and eco-system is created.

The purpose behind the present research work is to find out certain methods close to nature and environment itself with a view to the harmful pests i.e. thrips of chilli plants through biological means so that ecofriendly production of chilli may be obtained as per our hypothesis and primary results. We wish to develop the growth of pesticide free plants which attract certain beneficial insects as natural enemies, predators and parasitoids. A successful venture in this direction may provide fruitful ground to control the harmful thrips of chilli. We have favoured the application of natural enemies as the method of biological control. The presence of natural enemies may curtail damage which is caused by harmful pests. In this manner good and healthy production of chilli may be attained. The effective actions of naturally occurring natural enemies favour the environment. A consistent effort in this direction will enhance the scope of the present research work for the benefit of humanity.

It is true that the natural control through biological methods may not kill cent percent insects harming the chilli plants. But we have to attain that level which does not affect the growth at economic level and also gives healthy production of the fruits of chilli. We have to think about the final market quality of the crop in order to make it economically viable. For this purpose most effective management has to be done to see the good results of biological methods in the insect pest management. Through fruitful results of our research, farmers may be attracted towards the utilization of biological approaches in the production of chilli. In order to get better results of biological management appropriate policies should be devised by the people in governance at local level. NGOs and other social groups may also help widespreading this knowledge. It is the need of the time that all responsible persons should work in this direction. Rural unemployed and educated youths may be encouraged to establish small scale biological management through natural enemies production units at grass root level.

To sustain agriculture towards its natural mode some solutions are to be traced. The solution to reduce pesticides is present in the preference for biological management. Predators and Parasitoids may be used as natural enemies, in order to get control over the thrips pests through less harmful means over the agricultural crops, more research work is needed. Certain other methods are to be explored in favour of environment, biodiversity and other useful flora and fauna. We need to maintain the tritrophic interactions in which eating (tritrophic) relationships between several species may be traced for biological control. When the eating links pass through tropic levels, it seems to be the phenomenon of tritrophic interactions. This is a version of Predator/ Parasitoid eats herbivore, and herbivore eats plants. In order to enable the herbivore to come upto the expectation of the world's population in ecofriendly manner, it is necessary to consider the third tropic level i.e. represented by Natural Enemies. Study of the insects and their harmful effects on plants is a method to improve the management of biological control in order to dishearten the use of pesticides.

1.5 Organization

Research work in the field of Entomology with a view to discovering biological control methods in the area of pest and its management requires a well established laboratory with latest equipments and scientific tools. Department of Zoology in D.S. College, Aligarh is rich in all respects. It runs post graduate classes and also provides facilities for scientific research. Many research scholar has completed research and obtained doctoral degree from this department. All equipments and scientific tools necessitated by the research needs are easily available for the researchers in this department.

Establishment of Laboratory Net House

In order to find out the extent of infestation and damage caused by thrips on chilli plants we went to different selected outfields in Aligarh District where the crop of chilli was grown by the farmers. We took certain samples fortnightly and examined them weekly and at alternative days in the laboratory net house.

Decided net house was prepared with nylon net to grow chilli plants taken as samples from different farmer's fields. We also picked up certain samples from different nurseries and grew them within the net house in order to find out results of damage on chilli plants.

With scientific tools having discovered damage caused by thrips on the leaves of the plant. We decided to prepare a net house for biological controlled experiments. We collected certain seedling plants from different nurseries and transplanted them in the experimental net house in order to find out results over damage on chilli plants, with the help of our biological management and experiments through natural enemies.

Organization of Net House

Net house was basically naturally ventilated climate, controlled temperature, humidity, light and intensity of soil media with proper irrigation were kept suitable for the growth of sampled plants.

Products used in our research

The evaluated work was carried with the help of following research products:

I.	Micropins
II.	Spreading Board / Setting Board
III.	Rectangular and Triangular Cards
IV.	Alcohol Solutions
V.	Insect Rearing Cages
VI.	Insect Collecting Boxes
VII.	Plexiglass containers, Petri dishes
VIII.	Plastic vials
IX.	BOD (BIological Oxygen Demand) Incubator
X.	Binocular Microscope
XI.	Camera Lucida

Process

To facilitate the process, the scientific processes as defined and dictated by experts were adopted in order to find out the biological nature, level of damage and fluctuation of thrips and identification of thrips and its natural enemies in the following manner:

(i) **Process of Pinning :** By this method, we pinned the insects with the help of micropins for preserving and handling their taxonomical studies.

(ii) **Process of Spreading and Positioning :** With the help of this process we used the spreading board for collected sampled insects. Insect wings, antennae and appendages were arranged in appropriate manner.

(iii) **Process of Carding :** This process was done by rectangular and triangular cards. Some collected insects were placed on these cards and by this method, we collected some useful information related to our research survey.

(iv) **Process of Sampling :** All sampling was done with the help of Random sampling methods during our research survey. It was important to check the abnormal growth of chilli plants due to the infestation caused by thrips pests. Sampling process was also done to check the natural presence of thrips natural enemies. Tapping and Sticky card methods were applied at regular intervals.

(v) **Mounting Process :** It was an essential process for the collection sampled insects. In this process we relaxed the sampled insects and placed in alcohol media solution for preservation and taxonomical identification. Mounting of insects was done in appropriate manner through petri dishes and small glass vials. Identification of the sampled specimen was done when they were placed over a microscopic slides under laboratory conditions.

(vi) **Maceration Process :** The main objective of this process was to remove the body contents of sampled specimen. This process was done in a weak NaOH solution at appropriate period.

(vii) **Dehydration Process :** This process was used to remove excess of water from the moutant specimen for clearing. Dehydration was done during replacement of specimen from one alcohol solution to another.

(viii) **Rearing Process :** Rearing was an important aspect in favour of our scientific research. This process was done with the help of BOD (Biological Oxygen Demand) incubator at optimal temperature.

1.6 Research Hypothesis

India is the heaven of agricultural products its vast plains contain alluvial soil with rich natural contents. Major economy of India is based on agricultural products. Variety of food contents in India are gained by its own agricultural products grown in traditional ways over the centuries. During the last fifty years several scientific researches have made it possible for the Indian farmers to bring about green revolution. But still the pests harming agriculture are major concern. In order to control the damage caused by pests, easily available pesticides are used and immediate results to boost agricultural products have been gained. During the last 10-15 years, it has been realized that regular use of chemicals in pesticides is creating other harms in human health as well as in grazing animals and in this manner a kind of disturbance in ecological environment is created unwontedly. Crop protection is the basic need to sustain human life. As an assurance of crop protection, chemicals were used to get immediate results but the knowledge of biological inputs in the further step to protect the crop without lowering the yields.

Agricultural Production : Maintains Quality of Life

During the three decades food grains production has considerably increased i.e. about 2 million tonnes a year with a view to meeting the requirements of the growing population. In the past years, land areas were available and therefore agricultural production was increased by extending the area of cultivation in land, just by including uncultivated land for cultivation and by making infertile land fertile and of course by finding excess over the barren lands through scientific tools and advanced technological methods. Food production was also increased by improving seeds, chemical, fertilizers, pesticides and developed resources of water for irrigation. Today, further prospects of raising agricultural production in the manner as in the recent past are restricted by various factors. Land areas are becoming limited. Further, oppportunities to turn the uncultivated areas into agricultural land are being bedimmed. Green Revoltuion has extended several techniques to increase production and people in agriculture have adopted them at large scale. Further development in this regard is also not in sight.

Impact of Biotic and Abiotic Factors

Good and appropriate agriculture is the basis of human goals and it is one of the major human activities. These efforts of human beings have direct bearing on environment and the biological world. The urgent need in all philosophical conceptions is to use the age old experiences with latest scientific explorations. In this manner, an integrated farming system can be developed and the available resources may be applied in the most fitted manner. A systematic approach developed on the basis of the previous experiences and innovative techniques can boost agricultural productivity. In this manner, human needs may be supplied with economic viability and the quality of life may be maintained in the long run.

There are a number of biotic and abiotic factors which bring hazardous effects in agricultural production. Several kinds of diseases, weeds and several other things cause damage to agricultural production. Insect pests cause large scale damage to agricultural yields. Regular use of pesticides is becoming ineffective to control the losses on the one hand and it is further causing several other problems in the quality of production.

So, it is the need of the time to extend researches towards finding out new methods and new technologies for increasing production and maintaining its quality for better food security. During the last several years attention has been paid to seek biological control over the harmful pests.

Indian Government also became alert in this connection and extended policies to control and limit the approaches of pesticides. The Central Government of India launched fiscal policy during 1990s and it was due to this the use of pesticides in agricultural production reduced to a considerable degree. Several other technological developments were also responsible for the declining trend in the use of pesticides. People with better knowledge are trying to use other techniques away from the use of chemicals in the pest management.

Over the last 10-15 years *S. dorsalis* has rapidly become a major pest of chilli pepper and Solanaceous plants. It has gained prominence in several tropical regions of the world. Due to its cryptic small size and thigmotactic behaviour it is not easily detectable and harms caused by it continue to affect chilli production and quality. It was in the Western Hemisphere that *S. dorsalis* was first recognized as a harmful pest. But, it had already caused large scale damages to the host plants before it was recognized. It has been found out that *S. dorsalis* adversely affect the tissues of their host plant. This pest extracts the vitality of the cells of the host plant. This leads to necrosis of feeding in the forms of scars, leaves distortion, distortion of buds flowers and young fruits. Even quarantine procedures are difficult to manage and *S. dorsalis* easily slips through the net with the increased taraffic among the plants. Many entomologists expressed the opinion that chilli thrips can easily be confused in the field with several other small species of thrips. He provided a pictorial key to its identification and in this manner further research possibilities came into existence.

The scientists all over the world have started expressing worry over the extensive use of pesticides in the production of food grains. When we think of changing methods in food grains, the less use of pesticides may create several other problems like less production and infected production. Under experimental conditions lead us to think about its effect on yield either way. Alternative method to restrain problems in this regard is to find out biological control over the harmful pests. Evidences presented so far suggest that the biological management can bring substantial profits by saving crop from disadvantages. We can keep ourselves away from the harmful effects of the chemicals in food crops.

In order to maintain economic feasibility it is necessary to find effective means of biological control so as to maintain the standards of commercialization. In this Scientific research it is suggested the biological approach is an effective technology, but in the beginning it requires more cost. The use of this technology at large scale will depend on the methods developed with lower cost. It will require an integrated system. So, that large scale benefits may be achieved with average cost. Primary experiments have given promising results. Rays of hope are present as biological management can substitue chemical peticides without harming agricultural productivity.

Approaches Towards Extension System

Biological Management is a new technology, but it requires intensive knowledge. Unlike many other technologies, it requires effective implementation with the sound understanding to the workers. Specific knowledge about the targeted pest and its natural enemies is required. it can not work on the basis of guess work or approximation. We must have accurate results in laboratory before the technique is transferred to farmers. Any kind of lapses in these methods will ruin the hopes. A chain of workers and consultants is required to maintain the feasibility of biological management.

Lawful Efforts

Our research studies will indicate that there are some other alternatives to check the growth of harmful pests for the crop of chilli. But in the beginning the explored methods can not prove much fruitful if the people are not forced to adopt them. It is very difficult to side track the attention of the farmers away from their routine methods and customary habits of the use of pesticides floating around them. One of the easy alternatives for drawing the attention of the farmers towards biological control methods is through the policies of the government while imposing taxes and excise duties on inputs. It will be a lawful method to curb the use of pesticides. We could see this kind of result during early 1990s. Then, the decline in pesticide use was due to the government policy of levying heavy taxes on the producers of pesticides. But, it is also seen that there are very strong roots of pesticide industry and so it may survive even after heavy taxation. So, it is necessary that first we should switchover to certain pesticides with safer applications or we may call them biopesticides. Gradually and gradually, we may leave them and replace them with biological methods.

Another lawful measure for popularizing biological methods will be to provide subsidies on the use of biopesticides. Further, incentives may be given to the farmers using biological methods for the production of chilli.

Yet, another lawful measure will be through the government advertising agencies like radio, television and newspapers. The government may decide policies to purchase the products grown on biological methods at higher rate and may supply them to the awakened citizens by highlighting its quality and health prone culture.

Chapter 1: Introduction

Number of Figures: 21

Number of Charts: 01

Number of Tables: 01

Chapter-2
Review of Literature

2.1 Historical Resume

Chapter-2
Review of Literature

2.1 Historical Resume

Thrips have a wide range of distribution throughout the world. A very large number of species are known to occur in some parts of the India viz; South India, West Bengal, Parts of Rajasthan, Madhya Pradesh, Uttar Pradesh, North Eastern India and Himachal Pradesh. In the Biological Diversity Thysanopteran possesses remarkable structural diversity. Biological Diversity has been stimulated possibly due to fluctuations in habitat and internal environments.

i) Pre – 1900 : The first ever work on Indian Thysanoptera dates back to 1856 when Newman described two new species of Tubulifera.

ii) 1901 – 1947 : During these years the studies on Indian Thysanoptera began rather late compared to other group of insects. The contributions of Bagnall (1913-1926), Hood (1919), Moulton (1927-1929), Ramakrishna Ayyar (1925-1935), Ramakrishna Ayyar and Margabandhu (1931-1939) and Shumsher (1942-1947) prepared the foundation of thrips studies in India.

1. **Hans Larsson (2005)**

    Review : Integrated Pest Management has been applied to control the aphids and thrips in cereals. These attack the most valuable crops. Aphid and Thrips are very serious pests. Warning and Forecasting methods were used in the review of Integrated Pest Management – Programme. Pests species are specialized for cereals and their two generations provide the best possibilities to utilize different crops in the landscape. In this way, the natural enemies probably reduced the density of these harmful pests and made the production of cereals more sustainable.

2. **M. Shivaprasad, B.M. Chittapur, H.D. Mohankumar, S.A. Astaputre, M.H. Tatagar and R.K. Mesta (2010)**

Review : Disease Murda complex in chilli is a very serious disease is caused by the interaction of virus and carriers such as; mites and thrips. Generally, these disease carriers can however be controlled by the use of pesticides. Pesticide application may bring down the pest population but it leads to the problem of residues. These residues in fruits have seriously affected the level of exportation of chilli. Hence, it is imperative to produce the pesticide free chilli crop by adopting useful eco-friendly management practices such as planting date, use of inter crops and host plant resistance.

3. **Magdaline Kharbangar, S. Choudhury and S.R. Hajong (2014)**

Review : Studies were conducted during the month of June to November in the year 2009 to 2011 to check and determine the occurrence and abundance of thrips in rice field. Specimens were collected from leaf blades from seedling to panicle stages at different sites. Results indicated a total of five species of thrips beloging to 2 families i.e. Thripidae and Phlaeothripidae and showed that the mean abundance of thrips was highest in the month of July and the lowest in the month of November. It was recorded on the basis of their cryptic nature and thigmotactic behaviour.

4. **Richard L. Fery and James M. Schalk (1991)**

Review : A replicated study was conducted under greenhouse, to confirm the availability to resistance to WFT in pepper germplasm. Observations revealed the severe thrips damage in the form of poorly expanded, deformed and distorted leaves, shortened internodes and severe chlorosis. Results were predicted that the resistance to thrips in pepper appears to be due to tolerance mechanisms. So, thrips resistant cultivars could be used as a cornerstone in an insect pest management programme for greenhouse pepper production.

5. **Emilija Raspudic, Marija Ivezic, Mirjana Brmez and Stanislav Trdan (2009)**

Review : Thysanoptera is a homogenous group of insects and has characteristic wings with long fringe and very poor nervature. The samples of thrips were taken from 235 different plant species. Out of which 33 thrips species belonged to suborder Terebrantia and 14 thrips species to suborder

Tubulifera. This distribution was found out on the basis of the studies of their host plants.

6. **Ekram Atakan (2011)**

Review : Thrips is considered as a serious pest infesting a wide range of arable crops worldwide. During research *F. occidentalis* was the main thrips species and *Orius niger* was the most prevailing predatory bug of the thrips. These species were significantly and regularly distributed over the fruiting parts. Populations of both the species peaked at mid or late May, where plants had low numbers of flowers. So, this study suggests that *O.niger* could be a potential candidate for biological control of thrips in strawberry production.

7. **Vivek Kumar, Dakshina R. Seal, Garima Kakkar and Lance Osborne (2012)**

Review : Chilli thrips, *S. dorsalis* has long been a pernicious pest of cotton and various ornamental, vegetables and fruit crops. In South Florida it is known to kill newly emerged seedlings, distort leaves, scar the surface of fruits of its hosts. It acts as the vector of major plant pathogens. Experimental studies were based on the characterization of *S. dorsalis* and its infestation on fruit hosts at the commercial nursery. Results revealed that the presence of both larvae and adults were found on 9 of 11 tropical fruit species, several of which were not reported in the Global Pest and Disease.

8. **P.N. Krishna Moorthy, S. Saroja and K. Shivaramu (2013)**

Review : *Capsicum* is a popular vegetable worldwide and commonly known as bell pepper or sweet pepper. Thrips *S. dorsalis* is a major insect pest attacking this crop, sucking the sap of young leaves by By nymphs and adults, resulting in the form of upward curling of leaves. In this manner, heavy infestation results in complete defoliation. So, the neem products namely; neem soap, pulverished neem seed powder extract and essential oils of basil and mint were evaluated against chilli thrips. These evaluated products were used against the recommended insecticides and arose as best botanical alternatives for the management of chilli thrips.

9. **Vivek K. Jha, Dakshina R. Seal, David J. Schuster and Garima Kakkar (2009)**

    Review : Due to rich vegetation, Florida has always been a suitable target for the establishment of invasive flora and fauna. In Florida *Scirtothrips dorsalis* is a serious potential pest of various ornamental, vegetable and fruit crops. It is dispersing quickly all over the state. From various studies and information on various aspects it is clear that the pest's biology in relation to the host crops is needed to be develop for effective integrated pest management. In this regard the diet flight pattern and its periodicity (between temperature and abundance) in multiple cropping systems is an important factor of developing a sound management programme for *S. dorsalis*. This pattern was observed on cotton, peanut, and pepper at 2 hour intervals everyday, to find out an association among their activity, behaviour of dispersal and relation to their microclimate of the habitat. Results showed that the abundance of chilli thrips was found to be maximum on cotton and then followed by peanut and pepper. During this research the peak flight activity of chilli thrips between the range of 1000 to 1400 Eastern Standard Time was also observed and sustainable results were found out.

10. **N. Mandi and A.K. Senapati (2009)**

    Review : Chilli vegetable is a versatile spice and is used as condiment, sauce, pickles, medicine etc. Various factors are responsible for low yield of chilli. The insect pests are of prime importance, that affect the production of chilli which is varied from 60.5 to 74.3%. The pest thrips is one of the most serious pests which infected chilli from seedling stage in nursery to harvesting of crop in the field. A field experiment was conducted to evaluate the effectiveness of different conventional and eco-friendly insecticides and their use against thrips *S. dorsalis* infesting chilli was made to find out their effect. Result and observation of this investigation were most effective to minimize the thrips population but showed higher cost benefit ratio in the chilli production.

11. **Scott W. Ludwig and Carlos Bogran (2007)**

   Review : From this review, chilli thrips is an important pest in tropical and subtropical regions. They have a very broad host range. They feed on more than 150 plant species which belong to 40 plant families. Chilli thrips life cycle is similar to that of other common thrips species. In its life cycle it develops from egg to adult within 12 to 22 days. Female insert their eggs inside the plant tissues and terminal plant parts. Hatching of eggs is done within 6-8 days and it depends on the optimal temperature. They may take longer time for hatching from the egg at lower temperature. Thrips shows the behaviour of metamorphosis. Chilli thrips infestations are usually first detected in the landscape when they feed primarily on pollen and various plant tissues. Due to these changes the management programmes towards this pest are still being developed. Preliminary researches suggest that the foliar spray in the form of insecticides are effective for the control of the pest on ornamental landscape plants. The use of such insecticides is not recommended because they are not very effective against chilli thrips pest but as usual they are more damaging to the beneficial insects.

12. **Ion Oltean (2012)**

   Review : Thysanopteran species are the major pests of the protected areas, causing major damage to vegetable as well as ornamental crops by sticking and sucking plant juice. The plants that were attacked by thrips suffer as different morpho-physiological and biochemical changes. In this process the power of assimilation, transpiration and contents of Chlorophyll are reduced. Many species of thrips are considered as a vector pests. So, the grand majority is to decrease the impact of thrips based mainly on chemical insecticides in a control system.

13. **Ronald D. Oetting and Ramona J. Beshear (1991)**

   Review : Thrips management, is a major consideration to ornamental Horticulture under greenhouse conditions. Several species of thrips which present under foliage and flower are considered as a pests of greenhouse crops. These species are more difficult to be controlled, because their damage occurs more frequently on green house crops in different areas of the country. So, according to this research the potential use of natural enemies is recommended for the management of thrips under greenhouse and other environments. Entomopathogens and a hemipteran predators in the genus *Orius* occur as natural enemies – which limit the population of thrips in their respective environments. They should be considered as the potential tools against thrips and their mode of infestation.

14. Natasa Mehle and Stanislav Trdan (2012)

Review : Thrips which belong to the order Thysanoptera, are very small insects. They are the pests of commercial crops and are widespread throughout the world. Thrips grow by feeding on developing flowers or vegetables. Thrips may also serve as the vectors for plant diseases. Due to their small size and predisposition behaviour towards the enclosed places, it becomes difficult to detect them in fresh vegetation by phytosanitary inspection. Several methods known so far have been applied for the identification of thrips, but still identification from the genus is difficult and ambiguous. In the present research under review some more efforts are done on Morphometric, Molecular and biochemical methods which give the most reliable identification for identifying thrips species. Results showed that these methods may be recommended to confirm the results of modern identification methods.

15. J.E. Gonzalez – Zamora and F. Garcia-Mari (2003)

Review : Strawberries are an important crop with a production of around 2,00,000 tonnes per year. The most abundant insect pests, thrips can become a major problem in this crop. Thrips feeding may cause the russeting of the fruit receptacle around the achenes. Methods for sampling were evaluated in three commercial plots by monitoring the population densities throughout the growing season. Three sampling methods were compared during this study (i) shaking the flowers – against a mesh, (ii) visual method in which inspection of each flower with the help of 10x hand lens and (iii) turpentine method used as a repellent combined with a tunnel collector.

Results obtained were like this from the visual method the 80% adults spotted and 33% larvae. In the result it was found that turpentine procedure was most efficient as it extracted almost 100% adults, nearly 100% of second instars and 50% of first instars with no chemical sprays. So, the recommended method for routine field samplings, especially for adults is the turpentine method.

16.   Douglas A. Landis, Stephen D. Wratten and Geoff M. Gurr. (2000)

Review : Conservation biological control is the manipulation of the environment to enhance the survival, power of fecundity, longevity and behaviour of natural enemies with reference to increase their effectiveness. The unfavourable environments in many agro-ecosystems for natural enemies is due to high levels of disturbance. The goal of this research was in favour of habitat management for natural enemies in which we create a suitable ecological infrastructure within the agricultural landscape. These practices have to be done within the agricultural landscape in which spatially and temporarily favourable conditions to natural enemies may be provided. Its importance is in enhancing the natural enemy performance. So, in this regard the conservation biological control should be a keystone of all biological control efforts.

17.   M.W. Johnson (2000)

Review: From this review it is learnt that the importation of exotic biological agents; Predators, Paraitoids and Pathogens which are capable of self-replication is helpful in controlling the native pests. Biological agents like Phytophagous arthopods and pathogens were used in this research. They were released in the field and their effect on the harmful it was found that these biological agents are effective in checking the growth of the native pests.

18.   K.J. Froud and P.S. Stevens (1997)

Review : Greenhouse thrips are abundant throughout the North Island and Can also be found in some outdoors regions of South Island. It is the serious pest of citrus, avocades and a wide range of ornamentals. In this research a larval parasitoid; *Thripobius semiluteus* was applied to attack greenhouse thrips. They were released at the preimaginal developmental times of thrips. The parasitoids were release at 23^0C. Their effect was seen on the comparative life cycle of 37.7 days and 27.7 days. Results showed that *Thripobius semiluteus* has a lower net reproductive rate than greenhouse thrips (13.9 and 44.7 respectively). But on the other hand the faster generation time of this parasitoid of high intrinsic rate was marked. It can be concluded that *Thripobius semiluteus* has the potential to reduce the population levels of green house thrips under optimal conditions.

19. **Lane Greer and Steve Diver (2000)**

   Review : Thrips are tiny insects that reproduce rapidly. They are one of the most difficult pests to be controlled in greenhouses. According to this research IPM for greenhouse thrips on both vegetable and ornamental crops was done. Under IPM programmes; a monitoring, biological controls, use of biorational pesticides and some ways through insect growth regulators were discussed. So, IPM offers a sustainable approach towards the dealing of greenhouse thrips. It is a good approach to fascilitate the adoption of least toxic control measures.

20. **R.S. Giraddi, S.M. Mantur, R.K. Patil, C.P. Mallapur, K.V. Ashalatha (2012)**

   Review : *Capsicum annuum* is the most popular and highly renumerative vegetable crop. It grows in most parts of the world. In its cultivation various biotic, abiotic and physiological factors. These factors are main constraints which the farmers encounter in getting higher productivity and good quality produce. In this manner, unlike many of the field problems are peculiar to greenhouse cultivation. Insect pests, mites, thrips, whitefly, stemborer, leaf miner, aphids etc. cause serious problems under protected condition. So, the assessmental studies have been done regarding the spectrum of pest of protected capsicum.

21. **P.P. Jagtap, U.S. Shingane and K.P. Kulkarni (2012)**

Review : Chilli, *Capsicum annuum* is a widely used universal spice and it belongs to the family 'Solanaceae'. In India the nutritive value of Chilli is an excellent source of vitamins. India has immense potential to grow and export different types of chillies. During 2005-06 the productivity was 1551 kg/hac. Some important states constitute nearly 75% of total area under the production of chilli. In the present research the study was based on the economics of chilli production in India. Data analyses were pertaining to the period 2009-2010 and it was found that cost-ratio was 40541.72, Rs. 42811.07 and Rs. 53421.29 per acre. Chilli production for small, medium and large farmers respectively. From this, the appropriate extension method may be adopted to evaluate the farmers on optimum use of inputs. So, there is need to develop the labour saving practices, by using improved tools for planting and harvesting of chilli.

22. **J.M. Mari, R.B. Laghri, A.S. Mari and A.K. Shahzadi (2013)**

Review : Thrips, *Frankliniella occidentalis* is a serious pest of many crops. These crops have been grown for export and domestic markets. Thrips populations have developed rapidly in a density. In this research, it was found that *O. insidiosus* is important predator to regulate the population of thrips. It is an important predator natural enemy of thrips. Results showed that natural enemy species reached sufficient numbers to suppress the local population of thrips.

23. **Garima Kakkar, Vivek Kumar, Dakshina R. Seal, Oscar E. Liburd, Philip A. Stansly (2016)**

Review : *Thrips palmi* and *F. schultzei* are serious pests of vegetable crops in various parts of the world. They show their economic importance across the globe. Two species of phytoseiid mites were evaluated as potential predators of these thrips. A study was conducted in the laboratory, a shade house and in a commercial cucumber production field. Results showed that in a non-choice lab bioassay, both predators were preyed in equal measure on the larvae of both selected thrips placed on leaf disks. On the other hand, in a shade house mites were only recovered from leaf samples and not in flowers. It was concluded that they were only effective in controlling thrips on leaves. They can serve as an effective alternative to conventional insecticide based management of thrips in commercial field.

24. H.R. Sardana, M.N. Bhatt and Mukesh Sehgal (2012)

Review : Bell pepper fruits are likely to retain unavoidably high level of pesticide residues which may not only be hazardous to consumers but may affect the export quality as well. Numerous management strategies was developed but this was more dealt in isolation. The integration of all the pest management strategies in a participatory approach could reduce the application of harmful pesticides to great extent. Keeping in this research view, the validation of multifaceted adoptable IPM technology in bell pepper crop was carried out in a participatory approach at farmers fields to reduce the dependence on chemical pesticides on one hand and protecting the ecosystem as a whole at other hand.

25. Muhammad Rafiq Kethran, Ying Ying Sun, Shahbaz Khan, Sana Ullah Baloch, L.L. Wu, T.T. Lu, Yang Yang, Zhan Hu, Abdul Salam, Sohil Iqbal, Sakhwat Ali and Waseem Bashir (2014)

Review : Different strategies have to be involved for keeping the pest in check and stabilizing the productivity of the cropping system. Date of planting is one of the crop habitat diversification are to be looked into, to minimize the incidence of insect pests on Chilli crop. The present study was conducted to observe the activity of insect pests of chilli varieties. Observation was based on a randomized complete block design with four sowing dates i.e. January 15th January 30th February 15th and Feb. 28th were selected and the process was replicated thrice.

26. D.N.R. Reddy, Puttaswamy (1985)

Review : Observations were made of pests infesting chillies in nursery beds in Karnataka, India in May-July 1978-80. The most serious pests included the noctuids, the acridids, the gryllids, the formicids and the tarsonemid mite were observed.

27. **Levent Unlu, Ekrem Ogur, Yusuf Celik (2012)**

Review : The integrated use of all established and available control methods was done to control the harmful pests. In the process the integrated pest management was analyzed through sustainable agriculture against agricultural pests. Results of this study suggested that the use of IPM methods has a lot of beneficial effects on biodiversity, human health, environmental pollution and useful flora and fauna. In IPM it is more essential to use the sequence of cultural, mechanical, biotechnological, biological and chemical control methods. So, it was concluded that this research in favour of sustainable agriculture and can be carried out by eliminating negative effects of most pesticides.

28. **John Sivinski (2013)**

Review : Natural enemy population reproduces, more slowly or even fails to maintain themselves. The reason behind it is the seasonal absence of hosts or food in the agricultural environment. In this manner, pests can reach at the economic damage level. So, it is important to locate the predators and parasitoids by augmentative approach. This approach can help to be suppressed the pests population below economic damage. Augmentation is an opportunity to reintroduce the natural enemies to an area and also helpful in the biological control management in field situations.

29. **David R. Gillespie (1988)**

Review : The known natural predator *A. cucumeris* has a potential biological agent for various species of thrips pests. They developed and reproduced on immature stages of thrips. It required often longer time to complete their development on thrips nymphs when it prey on it. Its ability to feed on alternate hosts as well as it could survive in a greenhouse in the absence of thrips hosts. *A. cucumeris* is to survive in cold storage at 9^0C for upto 14 weeks. It was concluded that they facilitate the mass production and transportation of this useful predator.

30. **Dr. T.V. Sathe, Mithari Pranoti, Dr. S.S. Patil and A.S. Desai (2015)**

Review : Thrips shows the degree of polyphagy. Its range and easy dispersal make them is a invasive insect. They are slender body small insects with a pair of fringed wings. There are more than 8,800 species of thrips are present in the world. These insects caused damage to economic plants by sucking cell sap. It was reported that thrips are destructive pests of many agricultural, horticultural and floricultural crops plants. Damage symptoms on such crops are visualized as brownish or whitish speaks or streaks on leaves, flowers and fruits. In this manner, IPM approach will add great relevance for minimize thrips population at below the economic threshold level.

31. **Mahmut Dogramaci, Steven P. Arthurs, Jijanjun Chen, Cindy McKenzie, Fabieli Irrizary, Lance Osborne (2011)**

Review : Chilli thrips, poses an economic threat to a wide-range of ornamental and vegetable plants. In this study, the examination was based on biological control of chilli thrips with a predatory mite, *Amblyseius swirskii* and the insidious flower bug, *Orius insidiosus*. They showed that at equivalent rates *O. insidiosus* was a more effective predator of adult thrips compared with *A. swirskii*. These were not effective against with thrips larvae. Results showed that both predators were effective predators of chilli thrips on pepper and suggested that both species could be used in combination without decreased the efficacy through intraguilt predation.

32. **Anais Chailleux, Philippe Bearez Jearnnine Pizzol, Edwige Amiens-Desneux, Ricardo Ramirez-Romero and Nicolas Desneux. (2013)**

Review : The tomato leafminer *Tuta absoluta* (Lepidoptera : Gelechiidae) has recently invaded in Mediterranean countries. It is a major pest in tomato crops. Plant injury consists of mine-formation within the mesophyll by feeding larvae and thus affecting the plants photosynthetic capacity. This resulted in lower fruit yeild. *Trichogrammatid oophagous* parasitoids were shown promising potential for controlling the pest before the yield decreases. Mirid predators are commonly used for biological control of whiteflies and they also prey on *Thrips absoluta*. Results showed that these predaors do not attack *Trichogramma* adults but they may partially decrease the overall impact of parasitoids on *T. absoluta* if intraguild predation occurs. It was also demonstrate that adding of *Trichogramma* prasitoids may significantly increase the level of control of the pest over what could be attained when only the mirid predator *M. pygmaeus* is present on tomato.

33. **Steven Arthurs, Cindy L. McKenzie, Jianjun Chen, Mahmut Dogramaci, Mary Brennan, Katherine Houben and Lance Osborne (2009)**

Review : The invasive chilli thrips poses a significant risk to many food and ornamental crops. It is a highly polyphagous species, withover 100 recorded hosts from at least 40 different families. Both larvae and adults attack all above ground parts of host plants, preferring young leaves, buds and fruits. In this study the evaluation of two species of phytoseiid mites as predators of *S. dorsalis*. In leaf disc assays, gravid females of *Neoseilus cucumeris* and *A. swirskii* both fed on *S. dorsalis* at statistically similar rates. Observations of this study was in greenhouse tests with infested pepper plants, both mite species established and reduced thrips numbers significantly over 28 days following a single release (30 mites/plant). Results showed that *A. swirskii* was the more effective predator. It consistently maintaining thrips below 1 per terminal leaf, compared with up to 36 for *N. cucumeris* and 70 in control treatments.

34. **Manika Gupta and Virendra Kumar (2014)**

Review : Damage to Chilli crop, *Capsicum annuum* caused by chilli thrips, *Scirtothrips dorsalis* (Hood) (Thysanoptera : Thripidae) was noticed under net house conditions at D.S. College in District Aligarh. Severe infestation by chilli thrips on chilli crop was seen in the month of May-June 2015. We observed the adult thrips and its larvae fed on various plant parts (twig, leaf, flower and fruit) in the form of deformities. Prolonged feeding by chilli thrips reduced photosynthesis of the plant results were obtained in the form of abscission of leaves and flowers causing economic loss in agricultural yields. Thrips fluctuations recorded up to 30-35/flower at regular intervals (7-9 days and fortnightly). So, the level of damage increases up to 85-90%. There were no controlled method (biological /chemical) applied to minimize the impact (damage and infestation) of thrips on chilli crop.

35. **Manika Gupta and Virendra Kumar (2015)**

Review : A biological monitoring survey was carried out of Western Uttar Pradesh in district Aligarh on favourable growing season to provide information on infestation and abundance of thrips species on the crop of chilli *Capsicum annuum*. A total of five localities participated in our study. These localities are Tappal, Jalalli, Talib Nagar, Sumera and Kayamganj. Data was collected from these localities on the basis of infestation level caused by thrips populations. Results indicated that a total of four thrips species present such as *Scirtothrips dorsalis, Thrips tabaci, Frankiniella occidentalis* and *Frankniella schultzei* in those regions. The majority of the thrips species were recorded in the young stages (leaves, flowers and fruits) of crop development appears young ball formation, necrosis of tissues, wilted growth of flowers and fruits and finally yields loss.

36. **Manika Gupta and Virendra Kumar (2015)**

Review : Thrips pests which currently play a key role in protected cultivation (fruit vegetables/ ornamentals), causing serious damage on chilli, *Capsicum annuum* plants. The present study was conducted to determine the efficacy and optimum release rate of the parasitoid *Ceranisus menes* (Hymenoptera : Eulophidae) at experimental nethouses. Results showed that there was comparatively lower insect pests infestation on chilli yield.

37. **Dionysio Pedikis, Eleftheria Kapaxidi and Georgios Papadoulis (2008)**

Review : Solanaceous crops are susceptible to infestation by a number of insect pests. These insect pests can cause serious yield losses. An approach of Biological control is an environment friendly method that enhances the sustainability in agriculture. Biological control is based on the use of natural enemies. These natural enemies are the antagonists of the harmful pests and these natural enemies may be predators, parasitoids and pathogens. Extensive research was conducted to exploit the potential of natural enemies in biological management. Under greenhouse experiments this application proved effective and its use is steadily increasing worldwide.

38. Stephanie Williamson (1998)

Review : Farmer awareness of natural enemies is the manipulation of predatory insects by crop growers. In this review, farmer and popular understanding perceptions about natural enemies were described. Lessons drawn and limitations encountered in traditional research and extension 'top-down' educational activities were examined and some of the more promising approaches were described. In this manner, some innovative programmes using discovery-learning methods with farmers to make better use of biological control were also described. The importance of these new approaches influencing policy-makers was assessed.

39. Ali Hosseini – Gharalari, Ali Mohammadipour and Nazanin Koupi (2009)

Review : Studying insect population dynamics is one of the important steps in Integrated pest management. There are several methods for studying pest population in the field and greenhouses. Sticky-card application is one of the popular methods, which can be time-consuming to obtain the data if the population density has high. The purpose of the results obtained by this method in this research was to detect the presence of a species in a particular habitat and to study the population fluctuation to determine the species density etc. Moreover, it could help the farmers that carried out control methods at the best time to gain maximum benefit. Therefore, farmers can save money and reduce pesticide use, which in the end will give benefits to the environment and human health.

40. U. Bernardo, G. Viggani and R. Sasso (2005)

Review : The greenhouse thrips, *Heliothrips haemorrhoidalis*, is a cosmopolitan pest of spontaneous and cultivated ornamental plants. High infestation causes severe defoliation and eventually the death of the plants. It is uniparental and reproduces by thelytokous parthenogenesis. Chemical control of this pest can be achieved by using systemic and translaminar insecticides, which are highly toxic and persistent. *T. semiluteus*, the solitary endoparasitoid is uniparental and, in preference, parasitizes the first or early second larvae of thrips. In this research study, some aspects of the biology of *T. semiluteus* were determined. Developmental time (egg to adult), potential fecundity, realized fecundity, progeny, daily rate of deposition of eggs and several demographic parameters were also evaluated. Result of this study confirm that the potential of *T. semiluteus* as a biological control agent of *H. haemorrhoidalis* and render it the best known parasitoid.

41.   K. Kavitha and K. Dharma Reddy (2014)

Review : A broad understanding of the various levels of interaction take place between plants – herbivores and their natural enemies. It is very much important for the development of biological control methods. In the natural ecosystems, plants and their arthropod pests have evolved a set of interactions with each other and also their natural enemies such as parasitoids, predators and pathogens. To enhance the effectiveness of pest management strategies this will also help in developing augmentative and alternative components in Insect Pest Management and will make it more sustainable. Therefore, tritrophy studies involving host plant, herbivore and natural enemy interactions are important and can give rise to significant advances in future biological control programme and integrated pest management strategies.

42.   J.S. Bale, J.C. Van Lenteren and F. Bigler (2008)

Review : Biological control is defined as the use of an organism to reduce the population density of another organism and thus includes the control of animals, weeds and diseases. The first major success in biological control occurred with exotic pests controlled by natural-enemy species from the country or area of origin of the pest (Classical control). Augmentative control has been successfully applied against a range of open-field and greenhouse pests, and conservation biological control schemes have been developed with indigenous predators and parasitoids. According to this research study the cost-benefit ratio for classical biological control is highly favourable (1:2-5:0) and for augmentative control is similar to that of insecticides (1:2-1:5), with much lower development costs. Results favoured; Biological control is a key component of a 'system approach' to integrated pest management, to counteract insecticide resistant pests, for withdrawal of chemicals and to minimize the usage of pesticides.

43. A. Bonet (2009)

Review : Parasitoid wasps are important insects. They are consumers (third & fourth trophic level) in the food web and play a magical role, in a multi trophic interaction context, in natural communities. Their high diversity (240,000 species) and major radiation in Hymenoptera relies on successful parasitism mode of life. This study was addressed to recent development in parasitoid evolutionary history, developmental strategies, endosymbionts, behavioural ecology and their role in natural and modified communities was done.

44. T. Sankaran (1986)

Review : Biological control of pests aims at suppression of insect pests of crops or other harmful organisms by using their natural enemies (parasites, predators or pathogens) which form part of the biotic environment. The bio-control programmes of the first half of this century resulted in a few notable successes in many areas. The use of parasites, predators and pathogens on an adequate scale in the field to maximize the benefits from biological control, could not be extended to other crop pests in various parts of the country for want of sufficient monetary, manpower and material resources. The aim of the present paper was to highlight the major developments of the last one-decade. It also highlights the current progress and comments on the future possibilities.

45. Alexandre Pires Aguiar (2012)

Review : A technique to dry Mount Hymenoptera (Hexapoda) from alcohol in a few seconds,and its application to other insect orders. It is the need to experiments that the insects are kept in alcohol storage and are later on mounted for experiments. Certain chemicals are used in this mounting process. But a new technique has been developed in which the insects, whether wet or soaked can be mounted directly from the alcohol storage. This technique is useful for many groups of Hymenoptera, Coleoptera, Mantodea, Neuroptera, Diptera and Orthoptera are such varieties of insects which have been successfully mounted with the help of this technique. A small electric air pump has been used for the purpose of drying them externally. With the help of 600 L/Min air current specimen's wings and pilosity are quickly and efficiently spread out and the position of antennae, legs and abdomen are correctly placed. This technique is useful for insects having the length greater than 3mm. An efficient mounting aid for such insects pressured in alcohol was lacking. Now, this new technique will help experiments in the insects of such length. Specimens should be pinned with minutents. This technique reduces the time needed for processing and needed for, processing and dry mounting of such insects.

46. D. Galazzi D, S. Maini and A.J.M. Loamans (1991-92)

Review : In this article possibilities of biological control of thrips, through the collection and rearing of an Italian strain of *C. menes* have been discovered. A survey of different genus's of thrips has been made and selected studies of parasitoids having control over the variety of thrips have been examined. In several Eurpean countries, a larval parasitoid of thrips called by the name, *Ceranisus menes* was found around 1990. It was seen on *Thrips tabaci* Lindeman in the southern part of Italy. A regular data analysis of the parasitoids attacking thrips revealed that *C. menes* can be used as a biological control agent for effective control on thrips. Studies related to the thrips of *F. occidentalis* character showed in Italy in 1987 that this is harmful in the protected crops as also in open fields. In order to control this harmful thrips two groups of predators amblyseids and anthocorids were used as biological control agents. But it was found that these predators were capable of keeping the infestation of the thrips sufficiently low so as to control the damage level. Therefore, it was further examined as to how the damage level of the thrips on the crop can be lowered. In this study, it has been ascertained that the Italian strain of *C. menes* may be planned as a new biological control strategy against phytophagous thrips.

47.   Yoshismi Hirose, Hiroshi Kajita, Masami Takagi, Shuji Okajima, Barpot Napompeth and Swami Buranapanichpan (1993)

Review : In this article the researcher has studied the effectiveness of the selected natural enemies in the native place of Thailand. The damaging effect of *T. plami* is observed and its control by the natural enemies has been watched and examined carefully. The researcher conducted a survey of thrips *palmi* Karny and tried to observe its natural enemies in their natural habitats. In Japan Classical Biological Control Programme was organized in 1987 and 1988. Certain parasitoids in egg form and larval stage were marked as natural enemies. The eulophid larval parasitoid *Ceranisus menes* was making more impact on the thrips. Even in the fields of Thailand, it was found most effective. Another effective natural enemy was the anthocorid larval predator *Billia* sp. but there were certain problems in the effective implementation of these two promising biological control agents.

48.   Xanxuan Zhang, Zi-Qianz Zhang, Jianzhen Lin and Zie Ji (2000)

Review : *Amblyseius cucumeris*, is a predatory mite of many insect pests and mites too. Its potential was studied in this review article under provided laboratory conditions, scientists evaluated it as a natural biocontrol agent against spidermites; *Schizotetranchyus nanjingensis*. This spider mite, was the pest of moso bamboo, in China. Results revealed that when predatory mite fed upon spider mites (females and eggs), then its life cycle was as long as its life cycle on its normal diet. The developmental period of *Amblyseius cucumeris* from egg to egg is 7.7 days and 7.8 days for 1st and 2nd generation simultaneously. In this manner, the number of preys consumed by the predatory mites increased with the density of preys. The number of eggs produced by predatory mites was directly correlated with consumed number of preys. Finally, the females of *Amblyseius cucumeris* were unable to invade into webnets of spider mites. On the other hand, the females were able to invade and liked to lay their egges in the broken nests of spider mites with existing holes.

49. Yuonne M. Van Houten, Mai Linn Ostile, Hans Hoogerbrugge and Karel Bolckmans (2005)

Review : Predatory mites are better known biological agent against thrips population. In this research study, researchers compared some predatory mites namely, *Amblyseius cucumeris, Iphiseius degenerans, A. andersoni* and *A. swirskii* with respect to their performance as a biocontrol agent. These biological agents were released in the separate greenhouses and at the same rates on the plants of sweet pepper. After successful releasing, experiments concluded that the *A. swirskii* and *I. degenerans* showed better establishment on sweet pepper. Sometimes high temperatures and low humidity conditions occurred and were adverse to the effect. Therefore, thrips control on Dutch sweet pepper greenhouse was less effective in Southern Europe. However, from the experiments it was clear that high temperature and humid conditions were favourable for the infestation of thrips and extra precautions have to be taken to control them.

50. David G. Riley, Shimat V. Joseph, Rajagopal Babu Srinivasan and Stanley Diffie (2011)

Review : Thrips belongs to the Phylum Thysanoptera of the family Thripidae. Thrips are known to transmit tospoviruses to their host plants. These tospoviruses belong to the genus *Tospovirus* under the family *Bunyaviridae* of the known species, there are only 14 species of thrips which are reported to transmit tospoviruses and may induce a suit of disease symptoms viz; leaf speackling, molting and necrotic lesions and finally give the result of wilting to infected host plant. Sometimes, infected plants may cause severe yield loss. According to this research, these tospoviruses created global trade problem in the U.S. and caused an estimated loss of $ 1.4 billion over 10 years. Earlier some serological and molecular techniques had led for the identification of such tospoviruses and presently, these modified techniques are used to know how vector spreads into newer areas. The initiation of new vector pathogen and its interaction between introduced and native thrips species has also been found in this research scenario. The aim of this manuscript research was to provide a comprehensive and modified list of the species of thrips, which were the vectors of tospoviruses. Along with this research, researchers have introduced the information related to common names of thrips species, distinguished characters, thrips distribution, their important host crops which were economically affected and tospovirus induced symptoms. Finally, the purpose of this research was to provide some basic biological information related to thrips species.

51. Affandi and Medina (2013)

Review : A completely randomized design, analysis of variance and least significant difference were used to design knew the age structure and sex ratio of *Scirtothrips dorsalis* Hood. Thirteen mango trees set in cross section were sampled and observed for the presence of *Scirtothrips dorsalis* including weeds under the mango canopy and four cardinal directions of border. Based on total population numbers, there was not significantly difference of age structure and sex ratio of *Scirtothrips dorsalis* associate with weeds inside the orchard including mango leaves and borders.

52. Kaur and Singh (2013)

Review: Field efficacy of different insecticidal and botanical formulations was evaluated for the management of sucking pests on *Capsicum* under net house conditions at Vegetable Research Farm. Sprays of different treatments such as Acephate @ 0.05%, Decis 2.8 EC @ 0.025%, confidor 17.8%SL @ 0.025% and 0.05% and neem soap @ 1.0% for the control of aphid and thrips.

53. Buckman *et al.* (2013)

Review: The monophyly of the suborders, include families and the recognized sub families, and investigated their relationships. Phylogenies were reconstructed based upon 5299 bp from five genetic loci: 18S ribosomal DNA, *Histone 3, Tubulin- alpha I* and *cytochrome oxidase c subunit* I.

54. **Varghese et al. (2013)**

Review: Bioefficacy of newer insecticides such as; Spiromesifen and prpargite used against the sucking pests of chilli and safety of these insecticides to natural enemy population in chilli ecosystem were evaluated at college of agriculture, Vellayani, Kerala.

55. **Toda et al. (2013)**

Review: In Japan a novel strain of *Scirtothrips dorsalis* attacking *Capsicum* crops. To differentiate the two strains, developed a multiplex- PCR method using the ribosomal ITS2 region. (**Toda et. al., 2013**).

56. **Packiam and Ignacimuthu (2013)**

Review: Formulation and evaluation of botanicals pesticidal formulations and neem oils used to control of *S. dorsalis* in peanut agricultural ecosystem.

57. **Ssemwogerere et al. (2013)**

Review: Information on species composition and occurrence of thrips on tomato and pepper as influenced by farmers' management practices by biological monitoring survey in central Uganda.

58. **Dogramaci et al. (2013)**

Review: Si solutions at 100, 300 or 500mg L-1 which made from potassium silicate were applied as foliar sprays or soil drenches to pepper (*Cpasicum annum* L.) plants, and their effects on chilli thrips (*Scirtothrips dorsalis*) populations.

59. **Aliakbarpour et al. (2012)**

Review: To check the seasonal abundance of most prevelant species *Thripshawaiiensis*and *Scirtothrips dorsalis* on commercial and control mango orchard during the flowering season of December 2008 - march 2009.

60. **Kumar et al. (2012)**

Review: Twelve different crops were found in Miami-Dade County, Florida, to be economically affected by *S. dorsalis* during scouting and sampling of various plant species. An open free choice host susceptibility test was conducted on 6 fruit hosts from the nursery.

61. **Mandal (2012)**

Review: Sprays of difentheuron with acetamiprid at 10 days interval and 30 days after transplanting of the crop, was used to control the population of thrips count.

62. **Purnima and Jagdish (2011)**

Review: At different stages of rose such as; bud, half opened and full opened flower to find out the efficacy of botanical and entomopathogens against *Scirtothrips dorsalis* Hood. Results revealed that among different stages of rose, half opened flower was found superior to control *Scirtothrips dorsalis*.

63. **Tatagar *et al.* (2011)**

Review: To find out the effect of border crop for the management of chilli leaf curl caused due to thrips and mites in Fields experiments were carried out for two years during *kharif* 2006 and 2007 at agricultural research station, haveri, Karnataka.

64. **Seal *et al.* (2010)**

The discovery of insecticides with different modes of action with addition of entomopathogenic fungus *Beauveriabassiana*, for rotational use against the resistance in *S. dorsalis* on 'Jalapeno' pepper, *Capsicum annum* L.

65. **Farris and Ciomperlik (2010)**

Review: DNA sequence data and polymerase chain reaction (PCR) were utilized to develop a molecular diagonistic marker for *Scirtothrips dorsalis*. The DNA sequence variation from the internal transcribed spacer 2 (ITS2) region of nuclear ribosomal DNA (r DNA) was analyzed from various thrips species, including *S.dorsalis* ITS2 r DNA. This diagonastic PCR assay provides a quick, simple, and reliable molecular technique to be used in the identification of *S. dorsalis*.

66. **Chandra *et al.* (2010)**

Review: The effect of Malathion toxicity on adult thrips *Scirtothrips dorsalis* and *Rhiphiphorothripscreentatus* (Thysanoptera: Thripidae).

Malathion has long life synthetic pesticide, which will cause maximum mortality in shortest time and remain effective for longer period.

67. Verma et al. (2010)

Review: The effect of Pyrethrin toxicity for adult T. tabaci and S.dorsalis (Thysanoptera: Thripidae) communities. Pyrethrin was recommended as biopesticide, because it has no damage to other insects of pollination and gain highest productivity without any hazardous conditions.

68. Venkanna et al. (2010)

Review: Bioefficacy of systemic insecticides of neonicotinoid group as foliar sprays was evaluated on groundnut sucking pests with reference to thrips (Scirtothrips dorsalis Hood), and leaf hoppers (EmpoascakerriPruthi) against conventional pesticide, monocrotophos and unsprayed control at 25 and 40 days of the crop stage.

69. Senapati et al. (2009)

Review: The effectiveness of different insecticides which were synthetic and biological in nature to reduce the population of thrips (Scirtothrips dorsalis Hood) infesting chilli (Capsicum frutescens) by ecofriendly manner. The use of four insecticides such as; acetamiprid 0.004%, thiamethoxam 0.005%, neem pesticide 0.4% and Bacillus thuringiensis. The result was that experiment in favour of highest marketable yield and higher cost benefit ratio.

70. Duraimurugan and Jagdish (2009)

Review: The chilli thrips, Scirtothrips dorsalis Hood mostly a leaf thrips has recently become a serious pest on rose flowers in India. Under laboratory conditions, the egg, first, second instar larvae, pre-pupal and pupal periods ranged from 3-5, 1-2.25 to 3.75, 0.75-1.50 and 3.25-4.75 days. Pupation occurred on the curled portion of the flower petals.

71. Gopal et al. (2009)

Peanut yellow spot virus (PYSV) was efficiently transmitted byScirtothripsdorsalis Hood in groundnut. Larvae could acquire the virus in 30 minutes and the maximum percentage transmission of 43.8% by individual insects. In this study PYSV persistently transmitted more than 75% of their life span.

72. **Jha et al. (2009)**

Review: Diel periodicity of the intra-plant dispersion of *Scirtothrips dorsalis* was observed on cotton, peanut, and pepper at 2- hour intervals every day, to find an association among their activity, dispersal, and micro climate of the habitat.

73. **Nietschke et al. (2008)**

Review: A weather- based mapping tool, NAPPFAST was used to predict potential establishment of *S. dorsalis* in North America. The analysis was based on degree-day model and cold temperature survival of *S. dorsalis*.

74. **Shinichi et al. (2007)**

Review: The reproduction period of overwintered adults of yellow tea thrips, *Scirtothrips dorsalis* Hood was investigated in tea groves and adjacent bigleaf podocarps trees. First generation adults moved to bigleaf podocarp trees in mid-May, and second-generation larvae occurred from late May to early June.

75. **Ananthakrishnan and Annadurai (2007)**

Review: The occurrence of thrips vectors in considerable numbers enables their functioning in a dual role as vectors and as direct crop pests. The resistance of thrips to pesticides has enabled quick transmission of viruses, the transient nature of their populations being essentially responsible for the infection.

76. **Matthew et al. (2006)**

Review: The use of three different CC trap base colours (blue, yellow, and white) with or without dichlorvos as a killing agent to the effective management of *Scirtothrips dorsalis* Hood. The Blue-D trap was studied in Taiwan and St. Vincent for attraction and capture of this pest.

77. **Satpathy et al. (2006)**

Review: Efficacy of methomyl against chilli thrips *Scirtothrips dorsalis* (Hood) was studied through a field experiment conducted in farmer's field during 2001-02 and 2002-2003 cropping season.

78. **Millawithanachchi et al. (2004)**

Review: Using improved chilli varieties, MI 1, MI 2, KA 2, Arunalu, MI Hot, IR, Thiwari and Hot Pepper twenty-eight hybrids of chilli were produced through the half Diallel genetic design.

79. **Umar et al. (2003)**

Review: Thripidae with three genera including *Scirtothrips, Megaleurothrips, Taeniothrips* with their species reported from Azad Jammu and Kashmir. The keys are provided for separation of genera. The collected speciemens have been identified and described in detail with keys and characters for identification along with illustration.

80. **Santharam et al. (2003)**

Review: For the management of thrips on chillies imidacloprid insecticide was applied which belong to the group of chloroniotiny1 as seed treatment, root dip and foliar spray.

81. **Anandam and Sabitha (2002)**

Review: Chilli (*Capsicum annum* L.) is affected by a large number of virusescausing mosaic disease. A field trial was conducted to know the role of barrier crops such as; maize, sorghum and sunflower in reducing its spread and increased the yield over control.

82. **Maris et al. (2002)**

Review: Different levels of thrips resistance were found in seven *Capsicum* accessions, based on the level of feeding damage, host preference and host suitability as tomato spotted wilt virus (TSWV).

83. **Fugro (2000)**

Review: The efficacy of several organic manures and organic pesticides with combination of inorganic fertilizers and chemical pesticides for the control of leaf curl (pepper leaf curl virus) and die-back caused by (*Collectotrichumcapsici*) diseases of chilli (*Cpsicum annum*). The organic and inorganic fertilizers and pesticides were needed to maximize the crop yield and manage chilli diseases to a satisfactory level.

84. Shibo (1997)

Review: The effects of insecticide application on the population density of *S. dorsalis* and on damage to grape clusters.

85. Nandakumar *et al.* (1996)

Review: Embryogenic suspension cultures of chilli were developed with an objective to induce somatic embryogenesis.

86. Tatara *et al.* (1992)

Review: To determine the relationship between the density of *Scirtothrips dorsalis* on Satsuma mandarin fruit and damage at harvest by regression analysis during 1984-88.

87. Gnanachandran and Sivayoganathan (1990)

Review: The non-chemical pest control methods are adopted in chilli and brinjal by jaffana farmers. Data were collected by a survey of a random sample of 180 farmers. To bring about better adaptation of non-chemical pest control, research on cropping systems and pest control methods adopted by traditional farmers should also take into consideration socio - economic factors.

88. Ghoneim K. (2014)

Review : *Tuta absoluta*, a tomato leaf miner is an invasive pest of South America and its native regions. *Tuta* sp. belongs to the order *Lepideoptera* of the family *Gelechiidae*. Moreover, it has expanded quickly year after year and hence, it becames economically important pest in the tomato producing countries. The infestation of this pest causes high-level of damag by attacking leaves and flowers of the tomato crop. It was seen that its infestaion usually starts with the developmental stage and might be seen upto the mature stages of plants. Mostly, some insecticides were used as a control method in North-Europe and North-Africa. But, according to present study some different predator species which belong to several insect orders and families as well as some arachnid species were used as biocontrol agents. By this study it has been ascertained that the use of biocontrol agents is a safe alternative to the chemical insecticides.

89. Saumya George and R.S. Giraddi (2007)

Review : Chilli is a useful versatile spice crop grown all over India. It is popularly known as 'Red pepper'. It is important cash crop in India as it can play an important role in earning valuable foreign exchange. India is the second largest exporter of chilli. In India, its production is about 10.70 lakh tonnes of the total production of chilli in India about 90% is consumed within the country. About 10% of the total chilli production is exported in the form of dry chilli, chilli powder and Oleoresins. Oleoresin is the important pigment which is found in chilli. Chilli is also liked by Indians for its aromatic and acidic flavour. Chilli crop is attacked by a multitude of insect pests and mites. These insect pests attack chilli at different growing stages from seedling to mature stages. Earlier many surveys were conducted by AV & DC in Asia to find out the impact of such insect pests. Surveys revealed some major pests that attack chilli namely, Aphids, Mites and Thrips. These insect pests affect their host crop by sucking on leaves and flowers. Farmers are using spray of chemical insecticides to get rid of the probable damage caused by these insects. These chemical sprays have increased over the years and have created severe diseases. These chemicals pollute our environment too. In this research article scientists considered non-chemical strategies in the form of organic amendments, botanical pesticides and some bioagents for controlling the insect pests with a view to avoiding damage to the crop. In this manner, the researchers have made an attempt to save the crop of chilli in eco-friendly manner. Therefore, this method will be completely safe to the consumers, who eat chilli on regular basis. In this research field experiments were conducted to study the effect of organic amendments on the activity of thrips and mites on chilli crop and optimistic results were gained.

90. Karel Bokkmans, Yuonne Van Houton and H. Hoogerbrugge (2005)

Review : This study related to the control of thrips through the release of predatory mites and parasitoids was conducted in a green house in Netherlands. Cage trials were conducted in the commercial greenhouse crops. The predatory mite, *Amblyseius swirskii* was used against *Frankliniella occidentalis* and *Bemisia tabaci* in sweet peppers. Experiments conducted in the green house showed that *Amblyseius swirskii* was effective on chilli thrips when predatory mites were released on the flowering sweet pepper plants. In all the trials it was realized and marked with satisfaction tht *Amblyseius swirskii* showed a very high numerical response to the presence the damage causing thrips in the chilli crop. In the study, it was realized that certain predatory mites which were generally recommended for the control of the thripss were economically expensive and therefore, hard to be brought in practical use. Mirid bugs (Miridae) such as macro-laphus Caliginosus Wagner are expensive. In this manner, their uses in the related crops are limited. So, it is the urgent need of time to explore less expensive biological control agents so that the crop may be safeguarded before the whiteflies enter to damage it. In this study as carried through several trials, it has been established that biological control of whiteflies with phytoseiid predatory mites can be a major step towards the benefit in the crop production specially in the areas where whiteflies and thrips are found in abundance. It can be economically reared in the large fields and can yield practical results.

91. **Jan M. Mari (2012)**

Review : This article is related to the experiments in the field study conducted at Kunsi, Hasul Rind Farm during 2011-12. The effect of Chrysoperia carnea Stephens and *Trichogramme chilonis* was seen on the population of insects pests in the chilli crop. In the experiments the pest population was left untreated at the first stage. Thereafter 1500 eggs of C. carnea were released. At the third stage, 1500 eggs of *T. chilonis* were released. In the results, it was found that the pest population was largely sucked and there was significant reduction in the damage effect. The natural enemies of thrips had their effect on the damage causing thrips in the chilli crop. Plot B and C were marked by releasing the eggs of *C. carnea* and *T. chilonis* respectively, whereas plot A was left untreated. When the results were compared, it was found that there was significant reduction in thrips, aphids, mites and whiteflies numbers due to release of *C. carnae*. It was measured at 57.31% in thrips, 70.86% in aphids and 65.12% in mites and 80% in whiteflies. On Tobacco caterpillar population, its impact was 53.18% and it was also observed that in the untreated plot there was abundance of the Tobacco caterpillar population along with gram pod borer. In the plot C the effect of the natural enemy of thrips, *T. chilonis* was observed and marked. It was reduced 74.14% and 89.38% respectively for gram pod borer and Tobacco caterpillar population respectively in comparison to what it was in the untreated plot. The results confirmed that *C. carnae* is effective a the early stage of the crop, whereas *T. chilonis* works better when the crop reaches budding and flowering stage. However, there are two natural enemies of thrips and other damage causing insects are very helpful in keeping the crop on the safer ground .

92.   G. Mikunthan (2008)

Review : A study of the associations of Mycopathogens with the pests of the crop of chilli was made in the Karnataka state of India. In order to complete this study, soil samples were taken from the fields of chilli crop. In all ten agro-ecological zones were selected. With the purpose of isolating fungal pathogens found in the soil, Larvae of *Tribolium castaneuma* and pupae of *Scirtothrips dorsalis* were used as soil baits. The diseased cadavers of pests were isolated with foilage sampling. With the help of different culture media, the fungi associated with diseased cadavers of insects was also isolated. In the experiments carried in this regard nineteen fungal species were discovered. Out of them thirteen were isolated from soil with the help of *T. castaneum* and *G. mellonella* as soil baits. Six fungal species viz. *Fusarium semitectum, Fusarium sp.* isolatd GM 15, *Neozygites floridana* from *S. dorsalis* and *Polyphagoratarsonemus latus, Nomuraea rilevi* from larvae of *S. litira* were reserved for foilage sampling. *Fusarium sp.* was recovered from the soil. It was also seen in the disease cadavers obtained from the foliage. In the laboratory conditions, a study of its pathogenicity was carried and the whole matter was studied against *S. dorsalis* and *P. latus*. Study concluded that on larvae of *S. dorsalis* the LC50 of *E. semitectum* and *Fusarium sp.* isolate G.M. 15 were 2.7 x 10^7 spores ml and 7.6 x 10^7 spores ml. It was on greater side on active stages. In this manner, it was concluded that fungus *Fusarium sp.* can serve as an effective biuocontrol agent in order to control thrips population in the chilli crop. It can be used as an effective tool in the pet management programme.

93. **C.M. Mannion, Andrew I. Derksam, Dakshina R. Seal, Lance S. Osborne and Cliff G. Martin (2013)**

Review : This study has been conducted with regard to Roses (*Rosa sp.*) which are known as the important ornamental hosts of chilli thrips *Scirtothrips dorsalis*. These thrips were found affecting cultivers of landscape roses which are popular in Florida. The names of these popular cultivers are "Angel Face", "Pink Summer Snow", "Radsunny", "Don Juan", "Radrazz", "Sun Flare" and "St. Patrick". It was thought that these would be helpful in developing techniques for integrated pest management of *S. dorsalis*. The researcher tried to evaluate the effects of 3 rates of fertilizer and cultivers on population densities of chilli thrips and host plant damage. It was seen that the population density of thrips was directly affected by fertilization rate, plant organ and cultivor. Different fertilizer rates were applied to mark differentiates in the total number of *S. dorsalis* and in numbers of flowers and buds produced by the plants. Difference in the damage rating were also noticed. Higher rate of floral growth showed more damage to *S. dorsalis*. In the rose plants, the highest density of *S. dorsalis* was found around buds. Thereafter, it was found in flowers and leaves. It was also observed that larger flowers were more infested with *S. dorsalis* than small flowers. Different cultivers of rose were almost equally susceptible to *S. dorsalis*. However, "Radcon", "Don Juan" and "Sun Flare" were comparatively more under the grip of *S. dorsalis*. In this manner, proneness of *S. dorsalis* was keenly observed in the rose flowers and buds of different varieites. These findings may help in reducing techniques of successful pest management through biological control.

94. D.J. Greathead (1986)

Review : Biological control is akin technology for the controlling of insect pests. The term biological control is typically used to describe the application of natural enemies i.e. Predators, Parasitoids and Pathogens. These natural enemies have enormous value in crop management. Behaviour modifying chemicals, some genetic manipulations and sterile insect techniques can create increasing tendencies in natural enemies. In this article *Greathead* defines the biological control as a process of using nature against nature. A method which have been made into Africa since 1928. He said that the different ways of applying natural enemies come under biological control. These ways could be outlined as Introduction, Augmentation, Inoculation, Inundation and Conservation. He also told some criteria for selecting such apporpriate controls. He discussed such apropriate strategies in relation to ecological theory. He concluded that pest life history strategy, crop type, stability of the agroecosystem provide a suitable guideline on the choice of biological control technique. He emphasized this biological technique by using some examples and preferred that the classical biological control should be best technique, when it would be used in an appropriate manner. Therefore, classical control programmes should be thoroughly researched and made to enhance the probability of success in favour of insect pests.

95. OEPP / EPPO – European and Mediterranean Plant Protection Organization (2005)

Review : *Scirtothrips* genus includes over 100 species. These species have been found in tropics and sub-tropics. According to Hoodle and Mound, some *Scirtothrips* species have thir economic importance. *Scirtothrips* sp. have rapid life cycle. They pass through 5 devlopmental stages. These stages follow an egg, 2 larval instars, 2 relatively inactive pupal stages that are called instars of thrips, and the last adult stage. The larval stages feed on plants and they are active. The inactive stages of thrips are winged and they feed on the calyxes of fruits. Thrips insert their eggs into young leaf tissues, stem and fruits. The species *Scirtothrips aurantii*, has recently been reported. This species is native to Africa and Yeman. This species is known as the pest of citrus plant specially (sweet orange), but somtimes they feed on the mango. So, it is referred as polyphagous species. Its wide range belongs to different plant families namely *Arachis, Asparagus, Gossypium, Musa, Ricinus* and *Vitis* genera. It was noted that *aurantii* species is the primary cause of banana fruit spotting disease. In this research article, one more species has been taken into consideration i.e. *Scirtothrips citri*. This species is known as the pest of citrus plant. It is one of the most important species from the genus *Scirtothrips* for the international agriculture. It is found that it feeds on more than 50 plant species in the genus viz. *Carya, Gossypium, Magnolia, Medicago, Phoenix, Rose* and *Vitis*. Due to this nature they are also referred as polyphagous species. From the research article EPPO revealed that the species *Scirtothrips dorsalis* is a general plant feeder of many cultivated plants namely, *Actinidia* sp., *Allium cepa, Arachis hypogaea, Capsicum, Citrus, Fragaria* etc. It was noted that its principal range was found as a serious pest of vegetables and rose plants in tropical region.

96. N. Nandini (2012)

Review : In this research article, studies showed that assessment of pest spectrum under chilli cultivation during Kharif season in the year 2009-2010. Chilli is a renumerative vegetable crop. It is popularly known as 'Bell Pepper'. In cultivated areas, Karnataka, Maharashtra, Tamil Nadu, Himachal Pradesh and in some hilly areas of Uttar Pradesh, the cultivation of chilli has gained popularity in the form of colour hybrids. On the other hand, the researchers encountered problems against chilli cultivation in the form of various biotic and abiotic factors. Due to such problem, farmers of various regions are not getting lucrative chilli yeild through its higher productivity. In this manner, some other problems as insect pests, mites and thrips create pecularity to green house condition. These insect pests created serious problems to their host crop and finally led to significant yield loss. Results concluded from the above facts were like this pest spectrum in relation to the extent of damage to the crop was assessed from seedling to fruiting stages.

For that assessment, the vermicompsot and neem cake was mixed thoroughly and were applied in the transplanted plantings as a fertilizer. Sampled assessed data were analyzed with the help of statistical analysis.

97.   Pavel Kindlmann and Katernia Houdkova (2006)

Review : Intraguild predation has become a major research aspect applicable biological control. It is also helpful for the conservation ecology. It occurs when the competition begins between two predators species against their prey. In intraguild predation one competitor also feeds upon another competitor in the absence of their prey. From the view of this research, it as assumed that in this predation process widespread interaction was done with in many, but not all communities of biological control agents. For the intraguild predation, the qualification of multipredator interactions and their consequencs on the target prey were needed. This intraguild – predation process is mainly based on Lotka – Volterra equations for the highlights and importance of population dynamics models. In this article, scientists used a simple model which developed for the stimulation of population dynamics of insects and prey. Results concluded the estimation of real strength in the proportion of predatory individuals that face a conflict with a heterospecific competitor at least once during their life. By the following deals, scientists were found the predictions on the population dynamics of both predatory species.

98.   Manika Gupta (2016)

Review : In our country, the role of agriculture has an extensive background. India is densely populated and its population has been increasing every year at the rate of 1.8%. By 2020, the population of India is expected to number around 130 crores. In India, Agriculture is a vital activity. It will be better to explain it as the backbone of our country. It also provides great opportunities in the field of self employment and labour employment. The word agriculture has its origin in Latin language. Its adaptation in Middle English gave its modern meaning. At present it plays a tremendous role in the human efforts to use natural resources for the betterment of human life. India is prominent country based on agriculture. It deserves the second rank worldwide on the basis of the farm input and other life forms for food, bio-fuel, medicinal products etc. Our area of research is in Western Uttar Pradesh, which is famous for its fertile soil, abundant water, and varied climate, cold wet and hot. In Western Uttar Pradesh agriculture is known as vital source of 'State Wealth' as far as the production of chilli is concerned. In Western Uttar Pradesh, the crop of chilli provides a lot of profit to the ordinary farmers. Yet insect pest problems are most important constraints for chilli and its production. The chilli pest *Scirothrips dorsalis* (Chilli thrips) commonly called yellow tea thrips, castor thrips, assam thrips can create serious problems before the farmers in getting good results of healthy crop of chilli. *S. dorsalis* in chilli causes losses upto 60%. Mostly thrips prefer young leaves and gather there in heavy infestation. They damage leaves, buds, fruits of chilli, and finally the production is affected heavily. Thrips excessive feeding on the leaves turns their colour from bronze to black. When the thrips attack leaves they curl upward and the entire growth of the plant is affected leaving it dwarf. Due to heavy infestation of thrips leaves with petioles detach from the stem, causing defoliation in infected plants and measurable result is yield loss. In order to meet the need of food grains for the growing population, we require accelerating agricultural production. These efforts of human beings have direct bearing on environment and the biological world. The urgent need in all philosophical conceptions is to use the age old experiences with latest scientific explorations. So, a systematic approach developed on the basis of the previous experiences and innovative techniques can boost agricultural productivity. In this manner, human needs may be supplied with economic viability and the quality of life may be maintained in the long run.

99. **Manika Gupta and Virendra Kumar (2016)**

Review : Chilli is widely cultivated throughout the warm temperate, tropical and sub-tropical countries. Archaeologists believe that *Capsicum* was used as a food as long as 9,000 years ago.. Chilli is known as a 'favourite item' in the use of cooking, curries, breads and appetizers. Although India is one of the biggest producers of chilli but the productivity rates and quality are not yet satisfactory. In order to compete in the international market effective measures are to be taken to increase its productivity and quality level. In India, the production of chilli is popular in all its corners, but there are several states which do not produce as much chilli as they consume. The production of chilli is found at large scale in those states, where climate is generally hot and dry. But chilli crops are heavily infected by pests during certain circumstances. Chilli is a much simpler and sensitive crop to cultivate. Cosmetic damage largely affects the chilli fruits, which is directly proportional to yield loss. On the other hand, pests cause indirect damage in the form of several deformities on twigs, leaves, buds, flowers and affect the growth of the plant. Minute insects i.e. Thrips fall in the order Thysanoptera and Family Thripidae. They show exploiting behaviour intermittently. Thrips plays an active role as a pest of many ornamental, vegetable and agricultural crops. Thrips merit attention cause direct and indirect cosmetic infestation and damage to their host plants. They feed on plant tissues by their mouth parts and its result is seen in the form of tissue scarification and depletion of host plant. Thrips prefers young plant tissue. Thrips damage shows the result of direct feeding on leaves flowers or fruits and transmission of viruses, as well as the contamination of plant product. Regular studies of environmental concerns have revealed varieties of insects and their behavioural patterns. It is on the basis of such studies that certain methods are evolved to curb the growth of insects. Thrips known to be parasitized by many hymenopterans which belongs to the Family; Eulophidae: Trichogramatidae and Myrnaridae. Parasitoids are mostly host specific and appeared more effective to prevent the population of thrips at low densities. In this study, we examined biological control of chilli thrips with a eulophid wasp; *Thripobius semiluiteus* and to determine the efficacy of selected parasitoid under experimental net house condition. Insect parasitoids have immature life stages and they develop on or within a single host, ultimately killing the host. Adult's parasidoids are free-living and may be predaceous. They are sometimes called parasites. Insect parasitoids only attack a particular life stage of one or several related species. Parasitism may occur at any trophic level of a food chain. These are organisms which lay eggs in or on the bodies of their hosts and complete their life cycle on host bodies as a result, of which hosts die. Results showed that *Thripobius semiluteus* was an effective parasitoid against thrips population. They parasitize only in their larval (immature) stages and lead free lives as adults. They usually consume all or most of the host's body and then pupate, either within the host or externally on the host. This parasitoid species maintained ≤ 4.56 thrips per plant and suggest that it could be used to lower the insect pests infestation in chilli field.

Chapter-3
Research Methodology

3.1 Research Methodology for Survey and Sampling
3.2 Collection of Natural Enemies from Different Outfields
3.3 Rearing of Sampled Insects
3.4 Establishment of Research Laboratory
3.5 Laboratory Bioassays
3.6 Methods of Stock Collection
3.7 Taxonomic Research Study
3.8 Laboratory Processes
3.9 Research Methodology for Cultivation of Chilli Plants
3.10 Biological Controlled Experiments of Thrips Under Net House Conditions

Chapter-3
Research Methodology

> **Biological Control - is a Term & Approach**
>
> It is a rapidly growing approach which brings together ecologists, entomologists, weed scientists, plant pathologists and microbiologists. Biological control is a manipulation through nature and natural agents.
>
> These are beneficial organisms that reduce pests and diseases. They are called **Farmers' Friends**. They can be conserved by taking care with farming practices.

Keeping in view the several existing hypothesis and research efforts of researchers published in various journals, experiments have been made on the host plant Chilli, *Capsicum annuum*. These experiments have been carried out on Chilli thrips, *S. dorsalis*, and their host plant chilii, *Capsicum annuum*. These experiments were carried out with reference to biological control with their natural enemies; predators and parasitoids.

Various aspects and legislative approaches have been considered to accomplish our research work. Points in favour of ecology and human health have been kept in mind in the following manner:

- Checked pest; Thrips population and level of fluctuation on Chilli crop by the exploring survey of Chilli outfields.
- Evaluated the natural occurrence of natural enemies from selected outfields.
- Maintained the research aspects of our objectives around the Tritrophic Interaction between Host plant; Chilli, Pest - Thrips and Natural Enemies: Predators and Parasitoids.
- Used Random Sampling and Tapping Methods to check the fluctuation and collection of thrips and their natural enemies ; Predators andParasitoids.
- Maintained Experiemntal Net house near research laboratory to controlled the infestation caused by thrips by augmentative biological control release.

- Performed Taxonomical and Biological studies of thrips and natural enemies.
- Explored and evaluated our methodological aspects on Chilli crop with the help of selected predators and parasitoids.

3.1 Research Methodology for Survey and Sampling

To accomplish our research objectives regarding checking of the pest population on Chilli crop and for the presence of their natural enemies, we selected those localities, where farmers used less pesticides. In this connection, the five selected sites in District Aligarh of Western Uttar Pradesh namely; Talib Nagar, Jalali, Tappal, Sumera, Kayamganj were taken.

The conditions of the decided outfields were in favour of sampling the chilli plants. Therefore, total five outfields covered our whole research. The surveys were made during the morning hours and in the favourable conditions of chilli crop. Survey was started in the month of August, 2014 and lasted on August, 2015. During survey, the growth in the population of chilli thrips on chilli crops was observed from seedling stage till harvesting stage. Survey and observation interval was calculated "fortnightly". In this manner, sufficient time was maintained from one time growth to another time growth.

For increasing demand in domestic and foreign markets, it deserves our serious consideration for higher production of agricultural crops. They form a part of daily diet in almost all households. These beneficial crops are prone to heavy infestations by thrips pests. In this regard we are concerned with the sampling of thrips. From the view of sampling we saw the population fluctuation of pests and its natural enemies.

Population fluctuation of Thrips Pests

It is important to check the presence of Thrips pests in the farmers' fields during our research. For this sampling method was applied at regular time intervals. In the regulatory survey, we chose some chilli plants for measuring the intensity of thrips population.

Figure 3.1: Chosen Some Chilli Plants

By Random Sampling Method : It is a method of sampling in which each individual plant is chosen randomly and entirely by chance. So, each individual has the same probability of being chosen at any stage.

By using this method, the sampling has been done against the presence of targeted insects. Sampling plants were chosen randomly as, four in every direction, North, South, East, and West and fifth in Central Area of the selected chilli outfield. In this regard, we were also concerned with the sampling of natural enemies. So, in the same manner, we sampled the natural enemies; predators and parasitoids of Thrips. From farmers' fields we selected five plants from one outfield. In order to complete the process, an initial experimental unit was established.

By Tapping Method – For Thrips: In order to get better and more remarkable results we applied tapping method. By this method, samples of insects were collected. We gently tapped the plant leaves, twigs and flowers on white sheet of paper. In this manner, the thrips concentrated on the leaves were dislodged and could be seen on the white paper.

Seasonal Population Fluctuations of *Scirtothrips dorsalis*: A survey was made to check the population fluctuations of *S. dorsalis* on chilli crop at changing environment. For this purpose a regular survey for collection of thrips was completed during second half of August 2014 to the first half of August 2015. Collected data was tabulated according to collection sites. Simple tabular analysis was made to work out seasonal population fluctuation of thrips species.

Figure 3.2: Tapping for Thrips

By Sticky Card Method: Thrips were collected from their host plant at the time of infestation. The Thrips adults were picked up with the aid of camel hair brush. It was transferred to glass vial containing 70% ethanol and brought to the research laboratory in D.S. College, Aligarh.

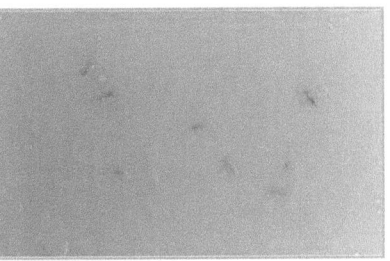

Figure 3.3: By sticky Card

3.2 Collection of Natural Enemies from Different Outfields

We collected a few natural enemies of thrips on the basis of our observations. Adults were directly handpicked from the chilli plants with the help of forceps.

Besides the collection of handpicked natural enemies, we maintained stock of insect fauna from other sources also with a view to finding out the method of biological control. A few species of parasitoids were collected for the purpose of exploration.

3.3 Rearing of Sampled Insects

The evaluation through the process of rearing and mass production of *S. dorsalis* was performed continuously under research laboratory of Zoology Department, D.S. College, Aligarh. In this manner, some numbers of thrips were also collected from different surveying sites of District Aligarh to maintaining our stock.

In the same manner the adults of selected predators and parasitoids were collected. On the other hand, a few species of selected predators and parasitoids were also collected during the rearing period of thrips under laboratory experiments when they were emerging from the infected eggs and larvae of the thrips. We also obtained some live species of natural enemies of thrips from the Department of Zoology, Aligarh Muslim University, Aligarh, Indian Agricultural Research Institute, and New Delhi and also from the National Bureau of Agriculturally important insects, Bangalore. In this manner, a stock of thrips (included all life stages) and their natural enemies were maintained for our experiments of biological control and its management against thrips population.

3.4 Establishment of Research Laboratory

The establishment of laboratory was a primary step to carry on our research work in the field of entomology with a view to discovering biological approaches with experiments on the host-plant pest i.e. thrips by their natural predators and parasitoids. The research laboratory was well equipped with appropriate scientific tools necessary for our purpose. The evaluated work was carried inside our research laboratory under suitable conditions, whichever plants were collected from the outfields at regular intervals and time-to-time transferred in research laboratory. Under laboratory conditions, we processed and assessed the biology, the preservation, and taxonomic identification of thrips species from the genus and their natural enemies.

Taxonomical Identification

After completed the gradual dehydration, the thrips species were identified and their morphological traits were compared using taxonomic characteristics. These specimen characteristics were identified with the help of dissecting microscope at a minimum of 10x magnification.

3.5 Laboratory Bioassays

The use of laboratory bioassays was also made for the mass production, culturing and rearing of selected samples. Some used bioassays were as followed:

(1) Petri Dishes

From the selected sites the infected or parasitized eggs and larvae of thrips were collected and put in petri dishes with the help of camel hair brush. After some time, these petri dishes were transferred in BOD incubator for emerging of adult thrips.

Figure 3.4: Use of Petri Dish

(2) Glass Vials

Glass vials with alcohol were used to preserve the different life stages; eggs, larvae, pupae and adults of thrips. It was also used to collect the life stages of selected natural enemies. These thrips and natural enemies were collected from the different selected sites of District Aligarh of Western Uttar Pradesh.

Figure 3.5: Use of Glass Vial

(3) Binocular Digital Microscope & Camera Lucida

For the study of biology and taxonomy of selected thrips specimens and their natural enemies we used the Binocular digital Microscope. We also applied the camera Lucida to draw the scratch diagrams of the insect specimens.

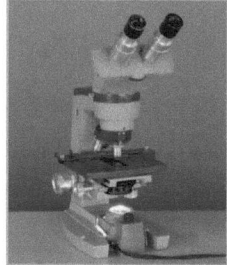 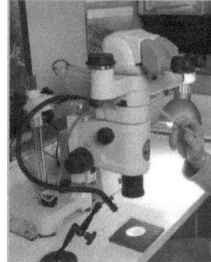

Figure 3.6 : Binocular Digital Microscope & Camera Lucida

(4) Rearing Cages & Plexi Glass Containers

These were made up of wooden cages. These wooden cages used for proper ventilation to thrips. Each cage opened by a sliding lid. A mesh in each cage was also fixed on lid to prevent the other insects.

Ventilated plexi glass containers with lids were purchased for rearing of thrips. Culture of thrips was also preserved in the containers for further scientific studies.

Figure 3.7: Rearing Cage

(5) BOD Incubator (Biological Oxygen Demand)

It is the method in which the amount of dissolve oxygen is needed by aerobic biological organisms against break down of organic material which present in a given sample at appropriate temperature over a specific period of time. BOD Incubator was used for rearing the selected specimens who were collected from selected outfields.

Figure 3.8 : BOD Incubator

(6) Insect collecting boxes

These were used for the preservation and to maintain our record of selected sample specimens which were collected from the different outfields.

3.6 Methods of Stock Collection

(1) Pinning of Predator Samples

This method was used for preserving and handling insects for the taxonomical studies. Pinned insects were stored quite safely. For this, various sizes of entomological pins were purchased and used.

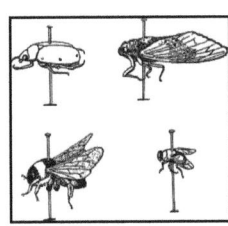

Figure 3.9 : Pinning of Hymenopterous insects

- Regarding this 2/3 portion of pin was used below the pinned insect and 1/3 above.
- In Hymenopterous insect predators were pinned through the centre of mesothorax.
- Some small collected insects were pinned by micropins; small sized pins.

(2) Spreading & Positioning by spreading board

The spreading board or setting board was used for some collected predatory insects. Predatory mites and minute pirate bugs were properly spread on the board under our laboratory conditions. In this process, the antennal was direct and frontal. Abdominal appendages and their ovipositor were directly backward. Wings were spread properly. Hind edge of forewing was at right angle to the body and hind wing appropriately matched with forewing. After this, the wing setting was done by pinning the paper stripes on the spreading board.

(3) Carding Method

Some collected small insects were placed on a white rectangular card; 5x8 mm or 5x12 mm. below this, card data labelled card was also placed in which we contained some information about the collection of the samples during the research survey. In this manner, information about identification of such insects, name of their host plant, date of collection, name of the collector etc was properly placed. Some small triangular cards were also used for pinning the small insects.

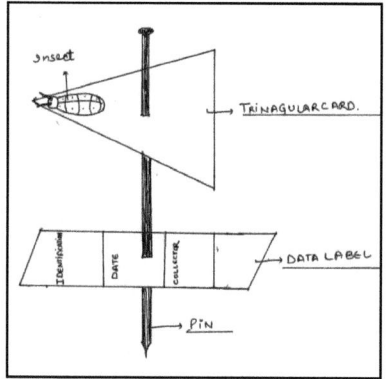 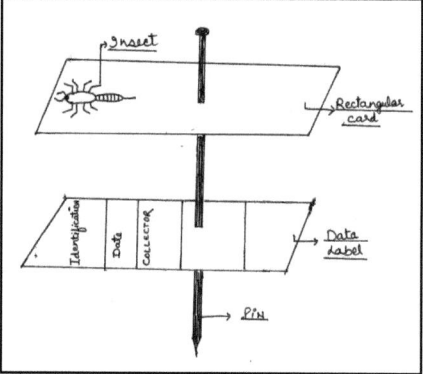

Figure 3.10 : Carding with Triangular and Rectangular Cards

3.7 Taxonomic Research Study

(1) Tools and Techniques

The evaluation and manipulation of specimens were arranged with the help of fine micropins, mounted with sealing wax on matchsticks. A pair of straight and with apex bent pins was used for taxonomic slide preparation. To accomplish our sample specimen's simple lifting tool was used and movement was successfully done. It was made of a small loop of fine wire. Alcohols of the dipped specimens were changed in laboratory dishes by using a fine glass pipette.

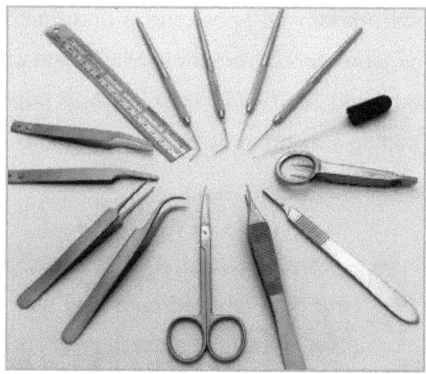 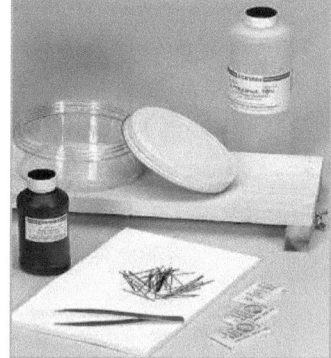

Figure 3.11 : Tools used for evaluation and manipuation of taxonomic study

(2) Process for Slide Preparation

For the taxonomical point of view, our objective was to prepare specimens on the slides with their natural shapes and colour. They were retained in a condition as close as possible to the natural position and living state. On the other hand, the body was clear and the surface details were visible.

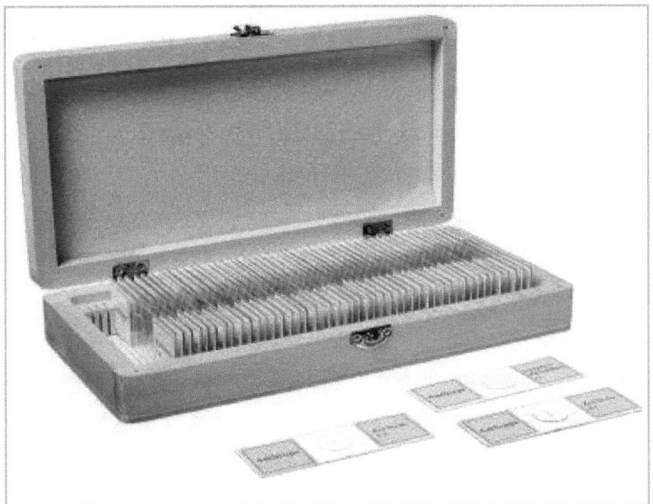

Figure 3.12 : Prepared Entomological slides for taxonomical study

For Routine Identification : The Routine Identification procedure was rapid and thus relatively inexpensive. The slides were prepared in such media solutions which could keep them intact for years, even though these slides were not permanent. It is a method which may be recommended for all routine identification work. In the present research work, this method was particularly used against small larvae and pupae and for small adults which were pale in colour. The Mounting and preserving steps were taken in the following manner:

1. Removal of the selected specimens from the collected fluid into 70% ethanol.
2. Specimens were reasonably flexible. So, they were attempted to open wings and straighten their antennae with the help of micropins.

3. After that we placed a drop of mountant solution over the specimen thrips and then they were covered with a cover slip of 13mm circle. Specimen thrips were placed in such a way that ventral slide was upper most.
4. When the processed mountant became dry, we labelled them in appropriate manner.

Thrips specimens were mounted on microscopic slides by following method.

From the preservative solution thrips were taken and placed in 10% solution of cold sodium hydroxide. The inter segmental ventral abdominal integument was punctured in order to aid the alkali solution quickly reach the body contents and to make the procedure more efficient and faster .Small Terebrantion insects remained there for 5-15 minutes. This depended on the sample colour of their integument.

After the removal of specimens from that caustic solution, they were placed in a solution of acetic alcohol (50% ethanol, 4 parts glacial acetic acid, 1 part) and there they were allowed to remain for at least 12 hours.

In this manner, specimens were transferred to 70% alcohol in order to remove the acetic acid contents from the specimens. The wings, legs and antennae were arranged in a systematic manner. Then, the processed specimens were placed in a "syracuse" watch glass in 95% alcohol and weighted down with a piece of cover glass. In this manner, specimens were allowed to harden at least over-night.

We arranged three small round bottomed porcelain dishes which were about 2.5 inches in diameters for the purpose of keeping the processed specimens intact.

- The first bottomed porcelain dish, contained 95% ethanol.
- The second dish contained half 95% ethanol and half amount of xylene.
- In the third porcelain dish we contained pure xylene.

These specimens were left in dish first for at least 10 minutes and then transferred to second dishe for 10 minutes.

In the process of mounting use the fine camel hair brush was used to avoid damage on the specimens when moved from one dish to another. After one minute, they were placed in the third dish and were allowed to remain for 1-1.5 minutes. It depended on the specimen and our observation.

In that series of mounting, a drop of thick balsam was placed on a slide and specimens were transferred to it. In order to hold the wings and legs in outspread position use of thick balsam was made. Efforts for it were made for proper arrangement of wings, legs, and antennae. This arrangement was necessary because we needed for extensive taxonomical study of thrips specimens during the research.

Suggestion: The cover slip which would be used should be small because large cover slips crush specimens and also need more mountant solution.

3.8 Laboratory Processes

(1) Maceration: The Process

The objective of this process in our research was to remove the body contents of sampled specimens. Maceration process was done by soaked specimens in a weak NaOH solution for a appropriate period. The preference of NaOH solution was in favour of less damage to the body surface compared to KOH solution. The treatment of Maceration process was maintained for the period which was necessary for our research programme. The process was charted according to the suggestions and advice of the entomological experts.

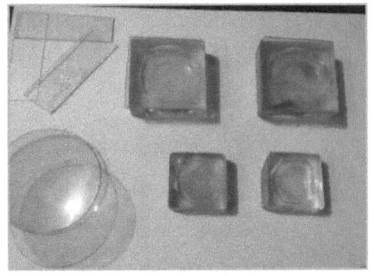

Figure 3.13 : Some Tools for Laboratory Processes

(2) Dehydration: A Process

The main purpose of this process is to remove water from specimens. The specimens clearing were improved by massaging each specimen gently with the help of the back of the bent needle. At this point we followed the following steps:

1. Replaced the 60% alcohol with 70% alcohol. It was left for 1 hour.

2. Unmacerated specimens were punctured for speeding up the entry of alcohols. It was the way to spread the insect legs, antennae and wings.
3. Then replaced with 80% alcohol. It was left out for 20 minutes.
4. Replaced with 90% alcohol. It was left for 10 minutes.
5. Replaced and performed with absolute alcohol. It was left for 5 minutes.
6. Some specimens showed the signs of collapsing. Then they were treated with the help of gentle massage and each one was stretched.

(3) Mounting of Thrips

Mounting was an essential procedure after the collection of insect's fauna. It was necessary, because, if the collected insects were not mounted immediately, they could become hard.

Before the procedure of Mounting, we relaxed such sampled insects. Relaxation was done in a wide mouth airtight jar with moist sand. Blotting paper was spread as a cover over the mouth of the jar.

Thrips were placed in a solution prepared with alcohol (10 parts); glycerine (1 part) and 1 part of acetic acid mixture. It helped distend the bodies of the most of the thrips. In this manner their body parts keep supplied.

Such prepared specimens were stored until they could be mounted on microscopic slides. Stored specimens were kept in the dark and preferably at temperature well below $0^0 C$ to prevent the loss of their integument colour.

3.9 Research Methodology for Cultivation of Chilli Plants

Before the cultivation of sampled chilli plants we prepared the soil with natural minerals and without any use of pesticides so that the growth of all chilli seedlings picked up from different areas could be seen in healthy environment. The Net House was naturally ventilated and climatically controlled. It was made free from weeds and grass at regular intervals so that the necessary minerals for the growth of chilli plants might not be wasted. In this manner, temperature, humidity, light and intensity of soil media with proper irrigation were kept suitable for the growth of the chilli plants.

In order to facilitate our research objectives, and to obtain the pure yield production of chilli, *Capsicum annuum* under established experimental net house conditions, chilli seedlings transplanted in sets of microplot. In total, seven experimental net houses were prepared for our controlled experiments.

We collected certain sample plants from different nurseries and transplanted then in the experimental microplots in order to find out results over damage on chilli plants with the help of our biological experiments.

Each nethouse was in size of 3x2 m with 4.5 ft height of the net. Chilli plants (*Capsicum annuum*) were transplanted on first week of January 2015 and first week of July in all micro-plots of the net house respectively. However, the cultivation of sampled plants arranged in a systematic linear fashion. Each plant was separated from the other at the distance of about 40-50 centimetres. The distance was maintained from row-to-row and plant-to-plant.

Note: Seedling plants were purchased from Krishna Nursery and Dev Nursery at G.T. Road, Aligarh.

Figure 3.14 : Cultivation of Chilli Plants under Net House

3.10 Biological Controlled Experiments of Thrips Under Net House Conditions

The evaluation of our research was done by the process of rearing. Biological controlled experiments were done under seven net house of 21 microplots. These microplots were covered by nylon net. Each biological treatment was replicated at three times with the help of selected predators and parasitoids. Each experiment was performed till the population of thrips reached at Economic Threshold Level (ETL). In this manner, we calculated the mortality in the experimental units.

In our experiments, the possibilities of use the predator; *Amblyseius cucumeris* and *Macrotracheliella nigra* and the parasitoids, *Thripobius semiluteus* and *Ceranisus menes* for the biological management of thrips on chill plants under net house condtions.

Figure 3.15 : Prepared Net House for Biological Controlled Experiments

Chapter 3: Research Methodology

Number of Figures: 15

Number of Charts: 0

Number of Tables: 0

Chapter-4
Experimental Analysis

4.1 Population Fluctuation
4.2 Sampling of Natural Enemies During Year 2014-15
4.3 Biological Controlled Experiments Under Net House Conditions
4.4 Results of Biological Controlled Experiments During the Year 2014-15 and 2015-16
4.5 Statistical Analysis of Biological Controlled Experiments
4.6 Discussion
4.7 Discussion of Biological Controlled Experiments

Chapter-4
Experimental Analysis

Thrips are globally distributed as key pests of Chilli; *Capsicum annuum* crop. Thrips may feed on chilli plants and may cause damage to chilli fruits. In order to fulfill our purpose regarding biological management of thrips on chilli crop, we checked the natural presence of thrips and their natural enemies in the outfields. In the survey, we used the method of sampling. In this connection five selected outfields namely; Talib Nagar, Jalali, Tappal, Sumera and Kayamganj were taken. The survey was done in the morning hours. It was done from August 2014 to August 2015. The population fluctuation count of thrips was calculated at fortnightly interval. In order to carry out research on the thrips and their natural enemies, we established experimental units on the basis of simple tabular analysis of the sampled plants. The presence of natural enemies was examined, at the time when they preyed and parasitized over the thrips population. We collected these natural enemies for research experiments and than used as consultants in the form of various predators and parasitoids at regular intervals.

In the present investigation, all the biological experiments were successfully carried out on *S. dorsalis* with their natural enemies, predators (*Amblyseius cucumeris* and *Macrotracheliella nigra*) and Parasitoids (*Thripobius semiluteus* and *Ceranisus menes*) on chilli crop.

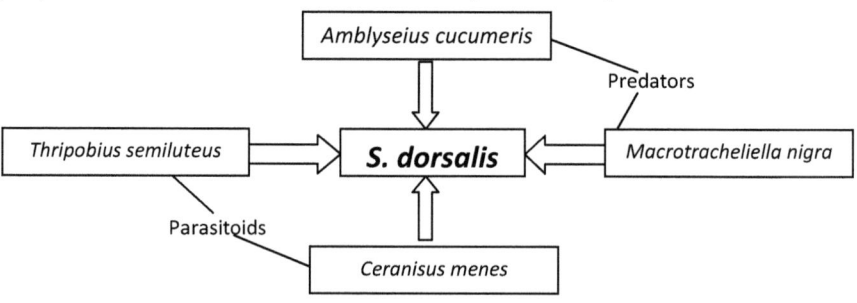

In order to find out the feasibility and viability of the biological control method to reduce the infestation of thrips, experiments were made in the net house conditions by releasing the natural enemies of thrips, of course, predators and parasitoids and the percentage of reduction in the population of thrips was observed and examined. Regular observations and examinations produced significant data which appeared proving our research hypothesis.

Figure 4.1 : Feasibility and Viability of Natural Enemies to reduce the infestation of thrips population on chilli plant

4.1 Population Fluctuation

Table 4.1
Population Fluctuation of Thrips on Chilli Fields in District Aligarh During the Year 2014-15

Season	Month		Population Count of *S. dorsalis*				
			Talib Nagar	Jalali	Tappal	Kayamganj	Sumera
Rainy	Aug. 14	II Half	96	106	115	120	135
	Sep. 14	I Half	50	43	60	63	56
		II Half	130	143	139	156	123
Autumn	Oct. 14	I Half	583	465	544	451	603
		II Half	390	473	339	416	480
	Nov. 14	I Half	30	22	21	18	16
		II Half	6	9	5	3	7
Winter	Dec. 14	I Half	3	2	4	2	3
		II Half	0	0	0	0	0
	Jan. 15	I Half	0	0	0	0	0
		II Half	0	0	0	0	0
	Feb. 15	I Half	3	2	0	0	5
		II Half	11	13	17	10	12
Summer	Mar. 15	I Half	32	40	43	29	23
		II Half	49	56	65	63	70
	Apr. 15	I Half	326	407	323	410	333
		II Half	460	465	389	408	445
	May 15	I Half	510	515	384	393	375
		II Half	289	229	245	303	309
	Jun. 15	I Half	160	170	153	139	146
		II Half	80	73	62	51	30
Rainy	Jul. 15	I Half	30	29	32	25	18
		II Half	13	8	6	9	8
	Aug. 15	I Half	27	58	60	83	58

Graph 4.1
Population Fluctuation of Thrips on Chilli Fields of Locality Talib Nagar in District Aligarh

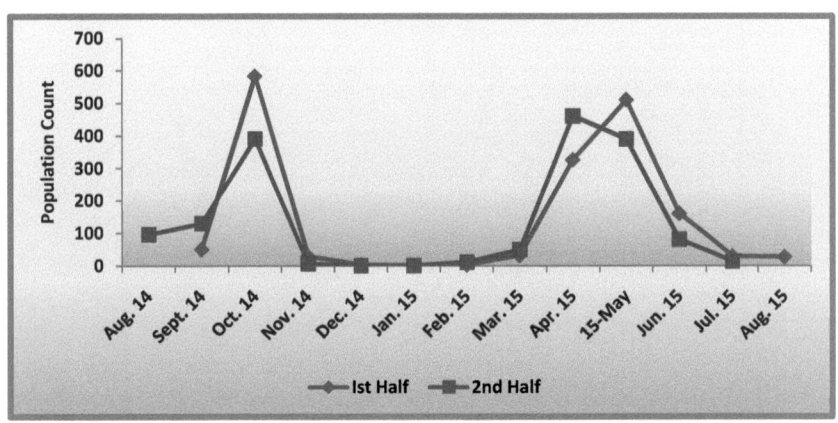

Graph 4.2
Population Fluctuation of Thrips on Chilli Fields of Locality Jalali in District Aligarh

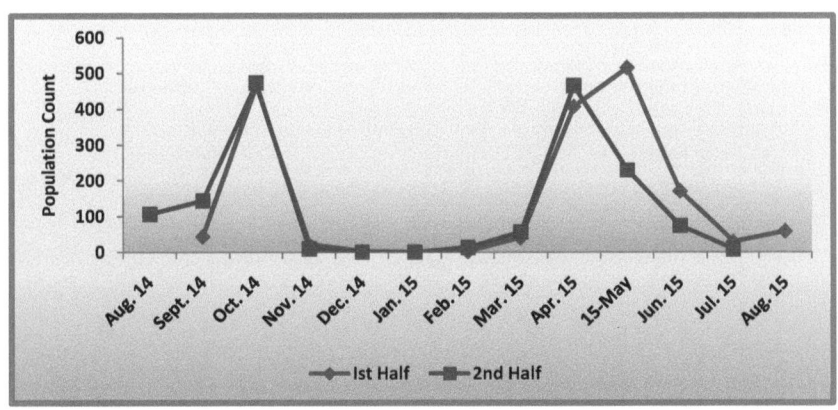

Graph 4.3
Population Fluctuation of Thrips on Chilli Fields of Locality Tappal in District Aligarh

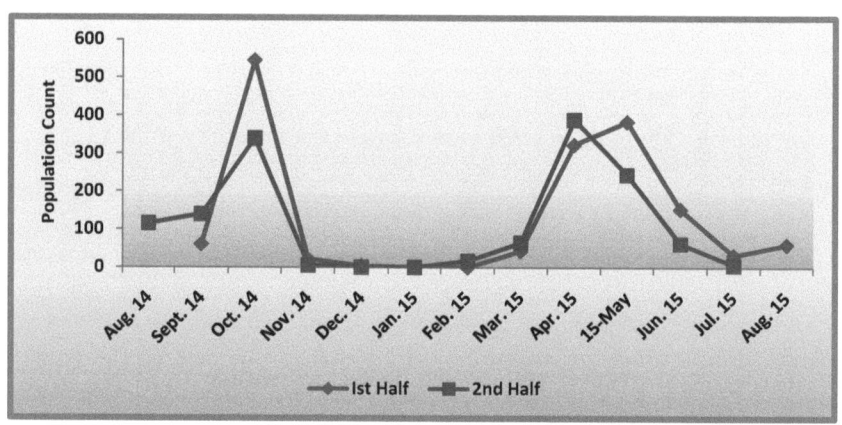

Graph 4.4
Population Fluctuation of Thrips on Chilli Fields of Locality Kayamganj in District Aligarh

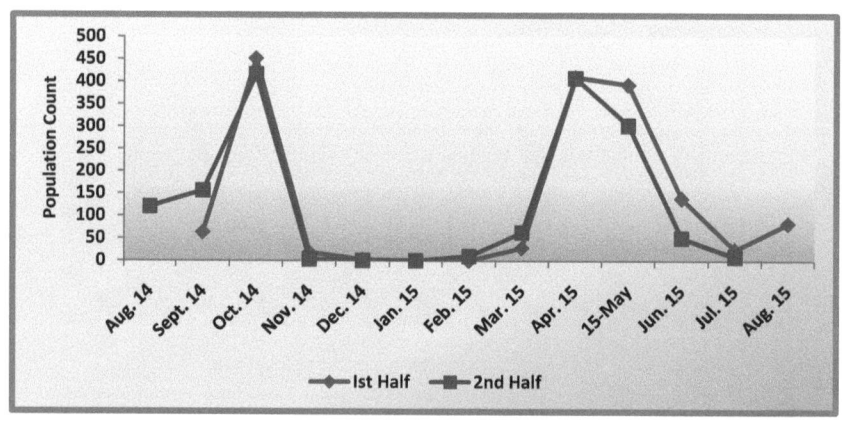

Graph 4.5
Population Fluctuation of Thrips on Chilli Fields of Locality Sumera in District Aligarh

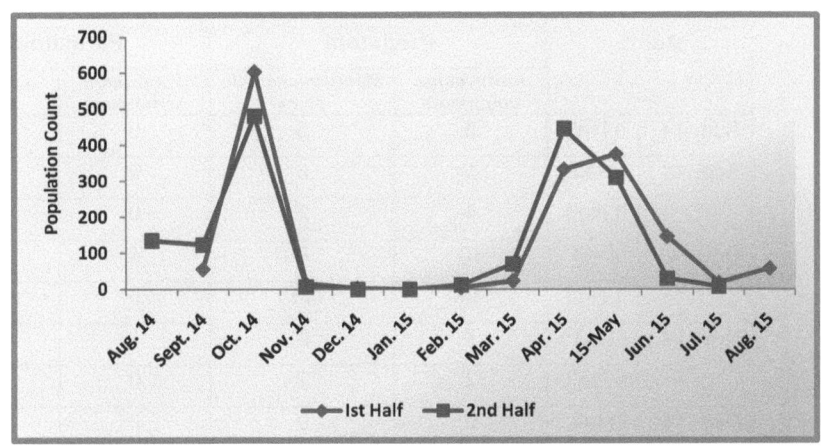

Figure 4.2 : Sample were taken during our research survey from different outfields for population fluctuation of thrips

4.2 Sampling of Natural Enemies During Year 2014-15

Table 4.2

Collection of Natural Enemies of Chilli Thrips : *Scirtothrips dorsalis* of Village Talib Nagar in District Aligarh during year 2014-15

Month		In Chilli Field (From Locality Talib Nagar)			
		Predators		Parasitoids	
		Amblyseius cucumeris	*Macrotracheliella nigra*	*Thripobius semiluteus*	*Ceranisus menes*
Aug. 14	II Half	6	5	0	0
Sep. 14	I Half	5	6	0	0
	II Half	4	2	0	1
Oct. 14	I Half	10	8	4	5
	II Half	14	11	5	7
Nov. 14	I Half	7	3	1	1
	II Half	2	2	0	0
Dec. 14	I Half	0	0	0	0
	II Half	0	0	0	0
Jan. 15	I Half	0	0	0	0
	II Half	0	0	0	0
Feb. 15	I Half	2	1	1	0
	II Half	4	5	2	1
Mar. 15	I Half	5	4	0	0
	II Half	7	6	0	0
April 15	I Half	15	13	7	6
	II Half	11	7	4	3
May 15	I Half	18	19	6	7
	II Half	20	16	7	5
June 15	I Half	9	10	1	2
	II Half	5	3	0	0
July 15	I Half	3	4	0	0
	II Half	1	2	0	0
Aug. 15	I Half	0	0	0	0

Figure 4.3 : Collection of natural enemies on the basis of sampling from different outfields during our research survey

Graph 4.6
Collection of Natural Enemies (Predator – *Amblyseius cucumeris*) of Chilli Thrips : *Scirtothrips dorsalis* of Village Talib Nagar in District Aligarh

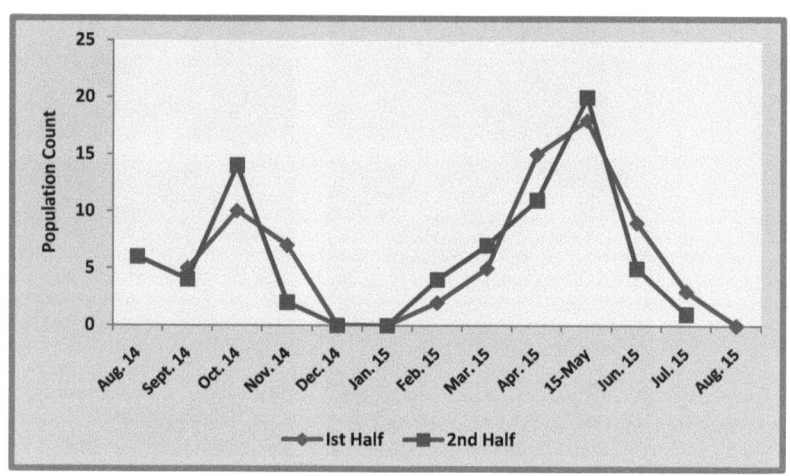

Graph 4.7
Collection of Natural Enemies (Predator – *Macrotracheliella nigra*) of Chilli Thrips : *Scirtothrips dorsalis* of Village Talib Nagar in District Aligarh

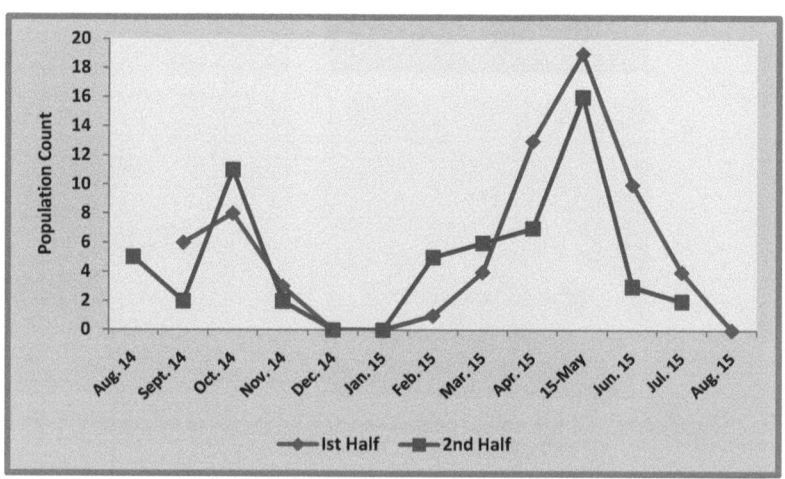

Graph 4.8
Collection of Natural Enemies (Parasitoids– *Thripobius semiluteus*) of Chilli Thrips : *Scirtothrips dorsalis* of Village Talib Nagar in District Aligarh

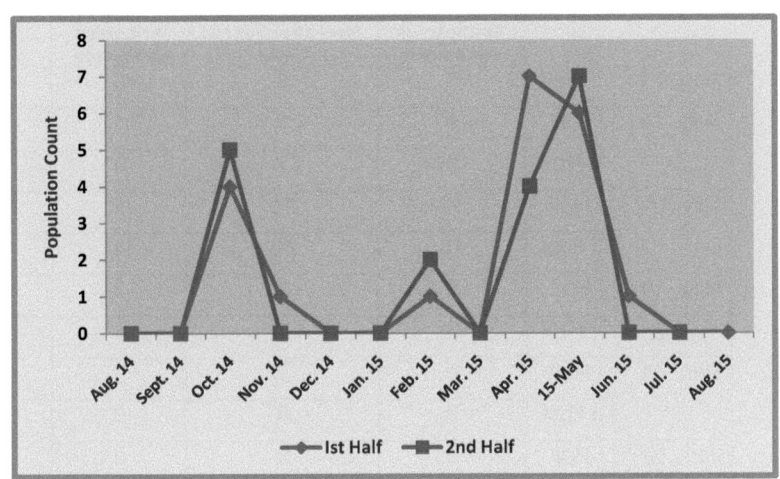

Graph 4.9
Collection of Natural Enemies (Parasitoids– *Ceranisus menes*) of Chilli Thrips : *Scirtothrips dorsalis* of Village Talib Nagar in District Aligarh

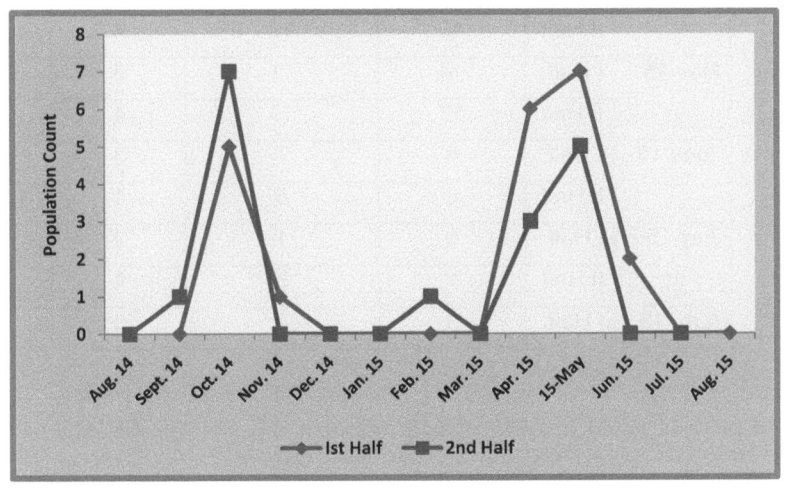

Table 4.3

Collection of Natural Enemies of Chilli Thrips : *Scirtothrips dorsalis* of Village Tappal in District Aligarh during year 2014-15

Month		In Chilli Field (From Locality Tappal)			
		Predators		Parasitoids	
		Amblyseius cucumeris	*Macrotracheliella nigra*	*Thripobius semiluteus*	*Ceranisus menes*
Aug. 14	II Half	5	8	0	0
Sep. 14	I Half	6	9	1	0
	II Half	3	2	3	1
Oct. 14	I Half	12	15	2	2
	II Half	16	13	4	3
Nov. 14	I Half	6	8	3	2
	II Half	3	6	1	0
Dec. 14	I Half	0	0	0	0
	II Half	0	0	0	0
Jan. 15	I Half	0	0	0	0
	II Half	0	0	0	0
Feb. 15	I Half	3	2	1	1
	II Half	5	1	3	3
Mar. 15	I Half	7	8	4	5
	II Half	10	12	2	0
April 15	I Half	17	16	6	5
	II Half	15	13	4	7
May 15	I Half	14	17	8	4
	II Half	12	19	4	2
June 15	I Half	8	7	3	0
	II Half	6	5	1	0
July 15	I Half	0	1	2	4
	II Half	1	3	4	2
Aug. 15	I Half	3	2	0	0

Graph 4.10
Collection of Natural Enemies (Predator – *Amblyseius cucumeris*) of Chilli Thrips : *Scirtothrips dorsalis* of Village Tappal in District Aligarh

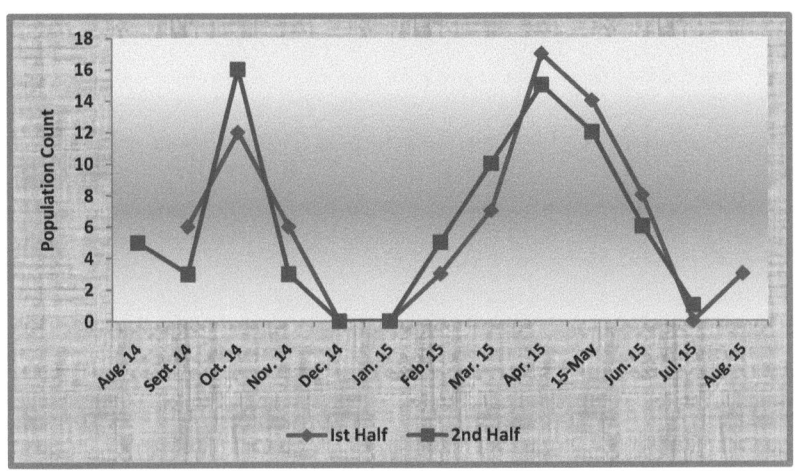

Graph 4.11
Collection of Natural Enemies (Predator – *Macrotracheliella nigra*) of Chilli Thrips : *Scirtothrips dorsalis* of Village Tappal in District Aligarh

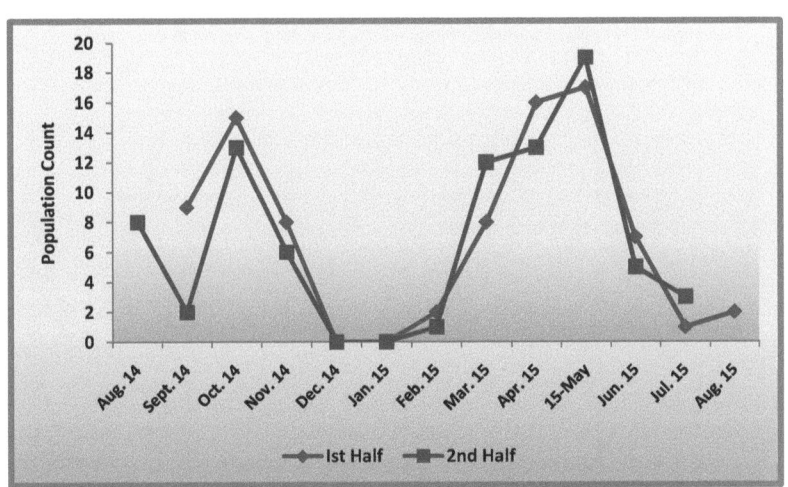

Graph 4.12
Collection of Natural Enemies (Parasitoids– *Thripobius semiluteus*) of Chilli Thrips : *Scirtothrips dorsalis* of Village Tappal in District Aligarh

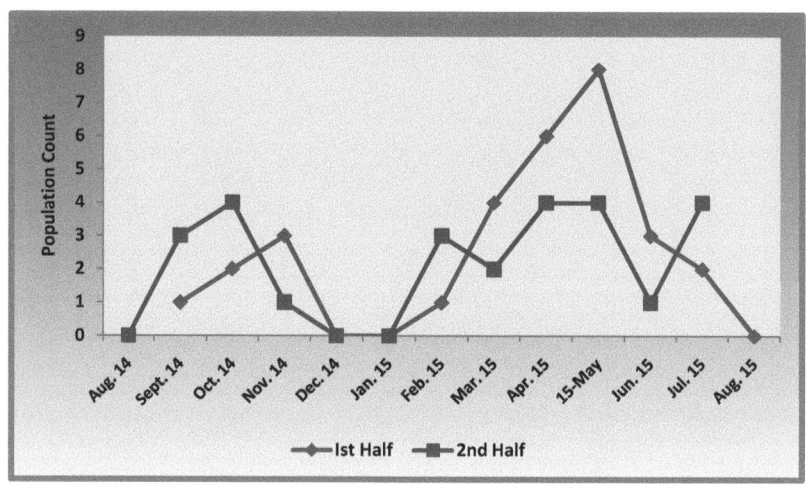

Graph 4.13
Collection of Natural Enemies (Parasitoids– *Ceranisus menes*) of Chilli Thrips : *Scirtothrips dorsalis* of Village Tappal in District Aligarh

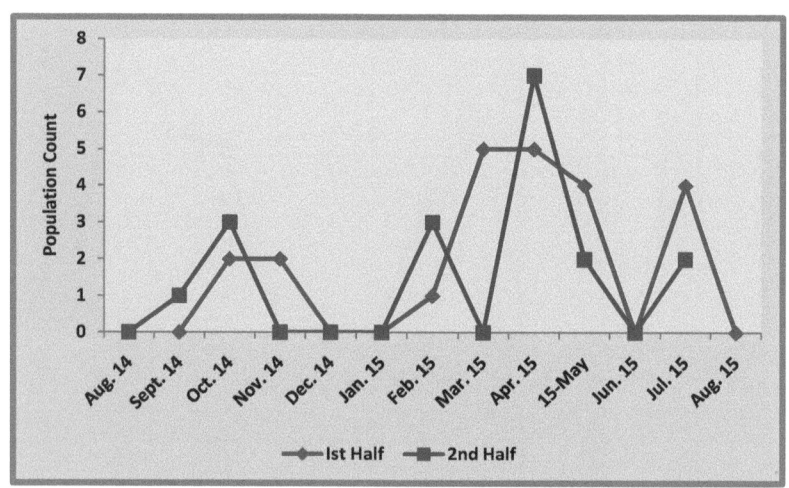

Table 4.4

Collection of Natural Enemies of Chilli Thrips : *Scirtothrips dorsalis* of Village Sumera in District Aligarh during year 2014-15

Month		In Chilli Field (From Locality Sumera)			
		Predators		Parasitoids	
		Amblyseius cucumeris	*Macrotracheliella nigra*	*Thripobius semiluteus*	*Ceranisus menes*
Aug. 14	II Half	8	6	1	1
Sep. 14	I Half	6	13	2	2
	II Half	11	9	1	3
Oct. 14	I Half	8	11	2	2
	II Half	14	13	3	4
Nov. 14	I Half	12	9	1	1
	II Half	6	5	0	0
Dec. 14	I Half	2	1	0	0
	II Half	0	0	0	0
Jan. 15	I Half	0	0	0	0
	II Half	0	0	0	0
Feb. 15	I Half	0	2	0	0
	II Half	3	1	0	0
Mar. 15	I Half	7	4	1	3
	II Half	6	3	2	4
April 15	I Half	13	9	2	4
	II Half	11	12	6	9
May 15	I Half	16	13	4	2
	II Half	14	19	5	4
June 15	I Half	6	4	0	0
	II Half	2	3	0	0
July 15	I Half	1	2	0	0
	II Half	0	1	0	0
Aug. 15	I Half	2	3	1	0

Graph 4.14
Collection of Natural Enemies (Predator – *Amblyseius cucumeris*) of Chilli Thrips : *Scirtothrips dorsalis* of Village Sumera in District Aligarh

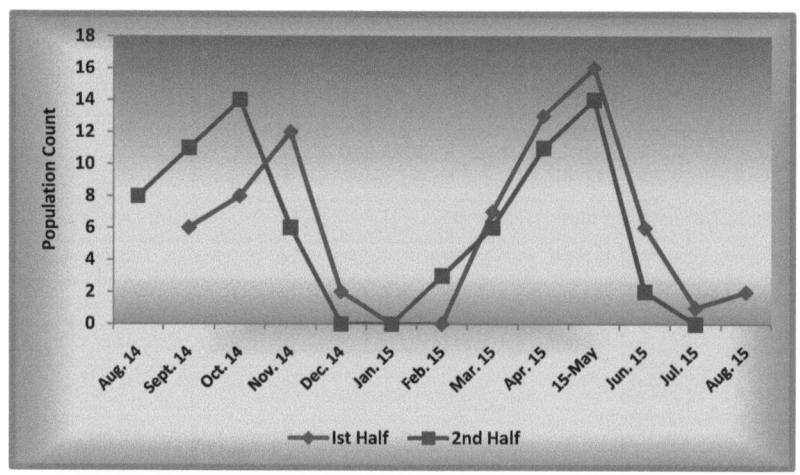

Graph 4.15
Collection of Natural Enemies (Predator – *Macrotracheliella nigra*) of Chilli Thrips : *Scirtothrips dorsalis* of Village Sumera in District Aligarh

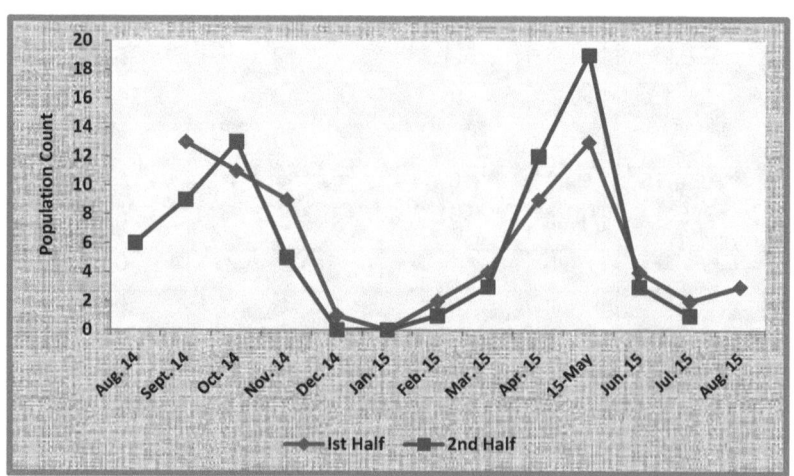

Graph 4.16
Collection of Natural Enemies (Parasitoids– *Thripobius semiluteus*) of Chilli Thrips : *Scirtothrips dorsalis* of Village Sumera in District Aligarh

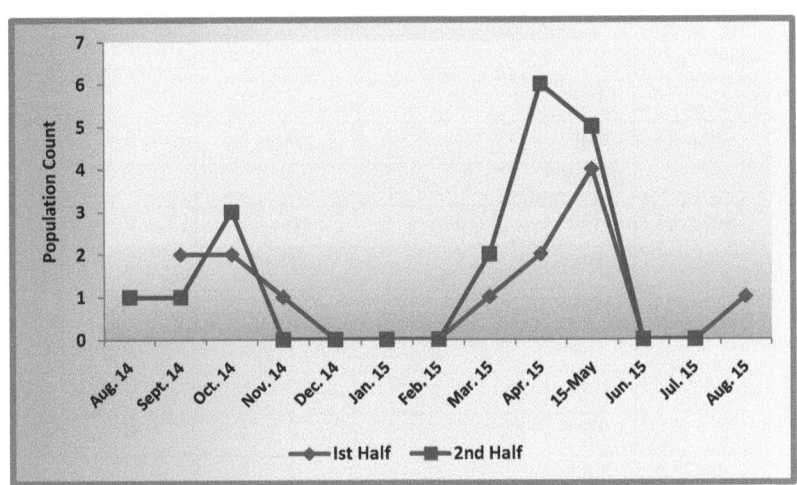

Graph 4.17
Collection of Natural Enemies (Parasitoids– *Ceranisus menes*) of Chilli Thrips : *Scirtothrips dorsalis* of Village Sumera in District Aligarh

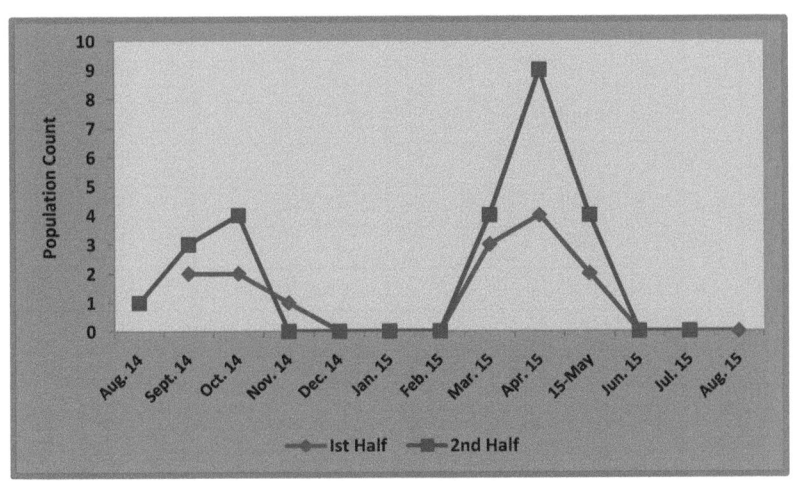

Table 4.5
Collection of Natural Enemies of Chilli Thrips : *Scirtothrips dorsalis* of Village Jalali in District Aligarh during year 2014-15

Month		In Chilli Field (From Locality Jalali)			
		Predators		Parasitoids	
		Amblyseius cucumeris	*Macrotracheliella nigra*	*Thripobius semiluteus*	*Ceranisus menes*
Aug. 14	II Half	4	3	0	0
Sep. 14	I Half	7	5	1	2
	II Half	9	3	0	3
Oct. 14	I Half	13	6	0	1
	II Half	15	7	2	0
Nov. 14	I Half	5	2	1	1
	II Half	0	1	3	2
Dec. 14	I Half	1	0	0	0
	II Half	0	0	0	0
Jan. 15	I Half	0	0	0	0
	II Half	0	0	0	0
Feb. 15	I Half	0	0	1	3
	II Half	2	0	2	1
Mar. 15	I Half	6	2	0	4
	II Half	5	3	3	2
April 15	I Half	18	10	5	4
	II Half	14	9	3	5
May 15	I Half	16	6	0	1
	II Half	13	3	2	0
June 15	I Half	3	3	3	2
	II Half	5	2	6	1
July 15	I Half	1	2	0	0
	II Half	0	0	0	0
Aug. 15	I Half	3	1	0	0

Graph 4.18
Collection of Natural Enemies (Predator – *Amblyseius cucumeris*) of Chilli Thrips : *Scirtothrips dorsalis* of Village Jalali in District Aligarh

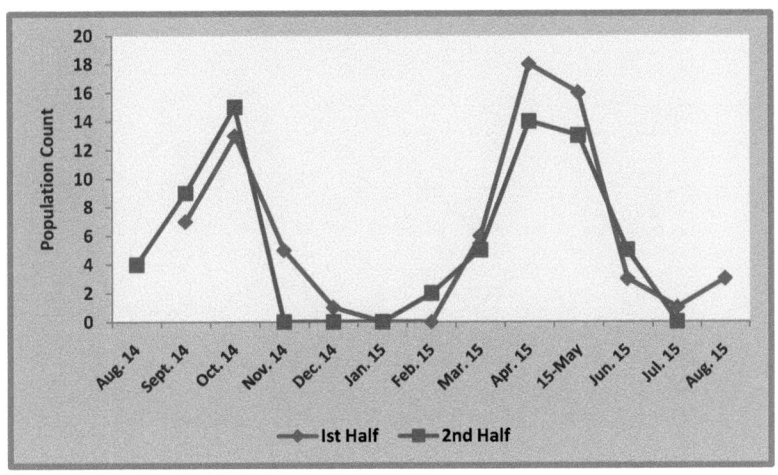

Graph 4.19
Collection of Natural Enemies (Predator – *Macrotracheliella nigra*) of Chilli Thrips : *Scirtothrips dorsalis* of Village Jalali in District Aligarh

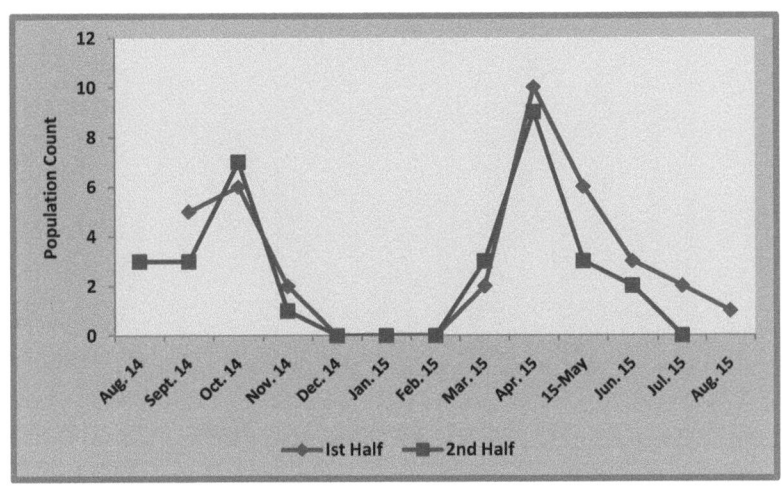

Graph 4.20
Collection of Natural Enemies (Parasitoids– *Thripobius semiluteus***) of Chilli Thrips :** *Scirtothrips dorsalis* **of Village Jalali in District Aligarh**

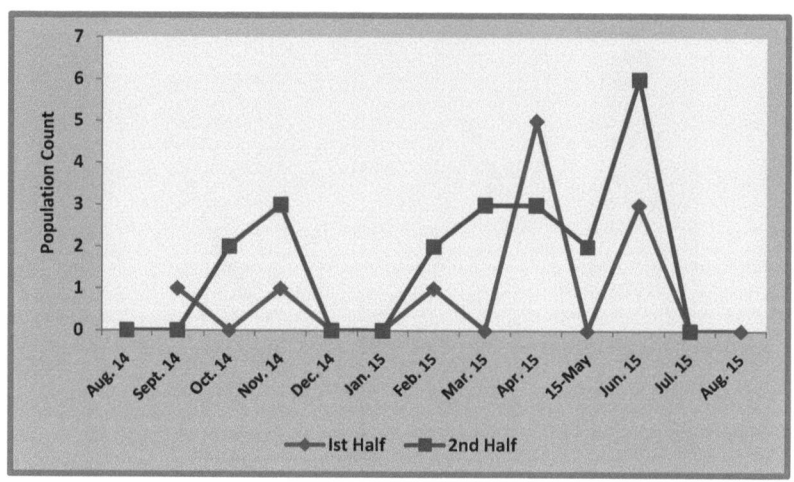

Graph 4.21
Collection of Natural Enemies (Parasitoids– *Ceranisus menes***) of Chilli Thrips :** *Scirtothrips dorsalis* **of Village Jalali in District Aligarh**

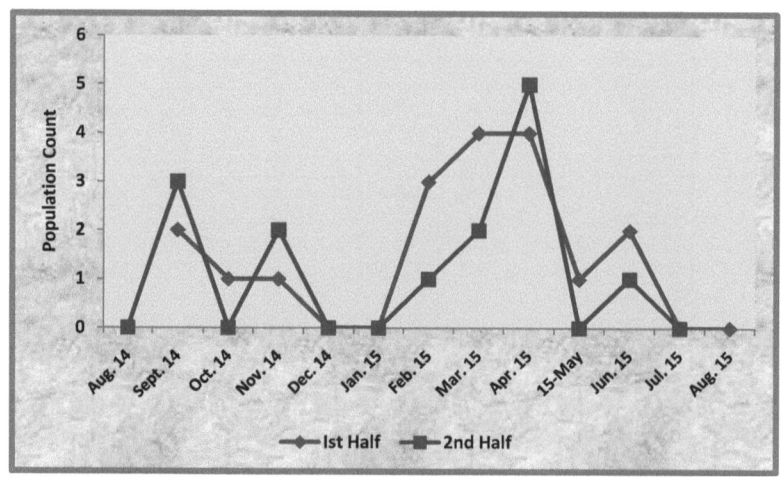

Table 4.6

Collection of Natural Enemies of Chilli Thrips : *Scirtothrips dorsalis* of Village Kayamganj in District Aligarh during year 2014-15

Month		In Chilli Field (From Locality Kayamganj)			
		Predators		Parasitoids	
		Amblyseius cucumeris	*Macrotracheliella nigra*	*Thripobius semiluteus*	*Ceranisus menes*
Aug. 14	II Half	3	0	0	0
Sep. 14	I Half	5	1	0	0
	II Half	7	2	1	1
Oct. 14	I Half	10	7	1	2
	II Half	17	8	3	1
Nov. 14	I Half	6	0	3	1
	II Half	0	0	1	0
Dec. 14	I Half	0	0	0	0
	II Half	0	0	0	0
Jan. 15	I Half	0	0	0	0
	II Half	0	0	0	0
Feb. 15	I Half	2	0	0	1
	II Half	3	1	3	0
Mar. 15	I Half	5	4	3	2
	II Half	7	2	5	2
April 15	I Half	15	11	3	2
	II Half	16	7	2	1
May 15	I Half	13	3	1	2
	II Half	16	1	2	0
June 15	I Half	5	1	0	0
	II Half	2	0	1	0
July 15	I Half	0	1	0	0
	II Half	0	1	1	0
Aug. 15	I Half	5	3	0	0

Graph 4.22
Collection of Natural Enemies (Predator – *Amblyseius cucumeris***) of Chilli Thrips :** *Scirtothrips dorsalis* **of Village Kayamganj in District Aligarh**

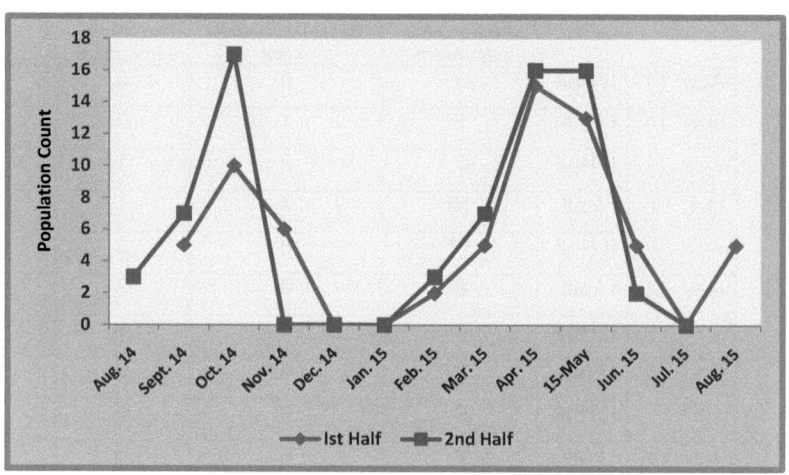

Graph 4.23
Collection of Natural Enemies (Predator – *Macrotracheliella nigra***) of Chilli Thrips :** *Scirtothrips dorsalis* **of Village Kayamganj in District Aligarh**

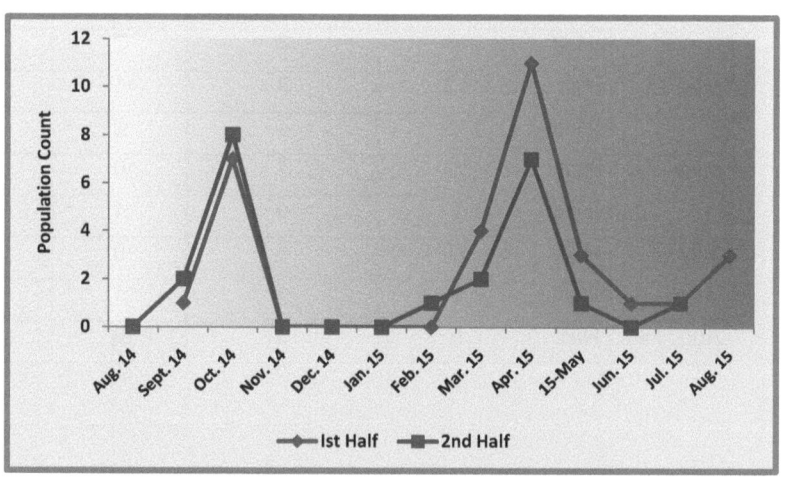

Graph 4.24
Collection of Natural Enemies (Parasitoids– *Thripobius semiluteus*) of Chilli Thrips : *Scirtothrips dorsalis* of Village Kayamganj in District Aligarh

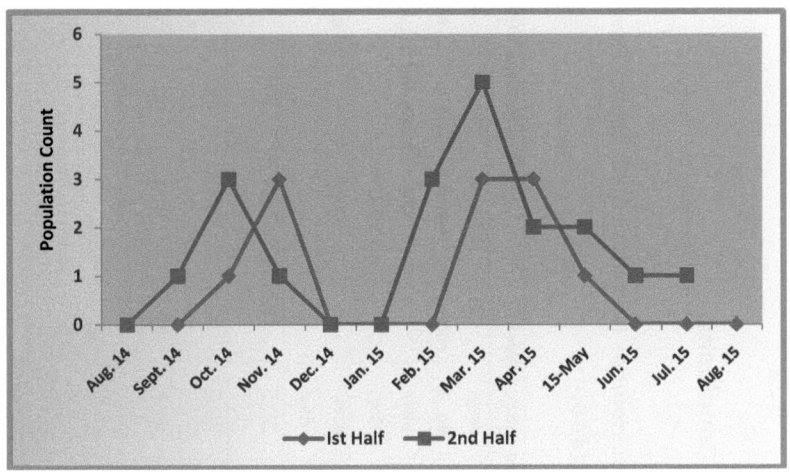

Graph 4.25
Collection of Natural Enemies (Parasitoids– *Ceranisus menes*) of Chilli Thrips : *Scirtothrips dorsalis* of Village Kayamganj in District Aligarh

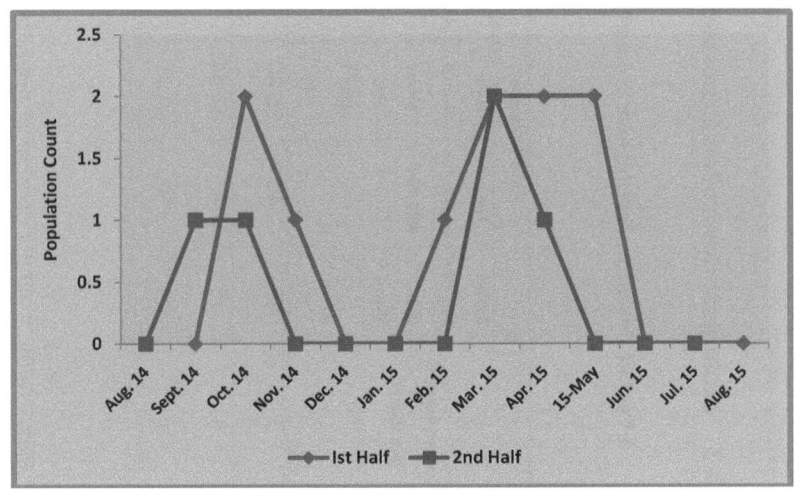

4.3 Biological Controlled Experiments Under Net House Conditions

Table 4.7

Effect of Predator; *Amblyseius cucumeris* on the population of *Scirtothrips dorsalis* in Chilli field of First Net House during the year 2014-15

Experimental Microplot	Initial No. of S.d.*/ Plant	First Release			Second Release (After 1 Week)				Third Release (After 1 Week)			
		No. of A.c. release**	Avg. No. of alive S.d. after 1 day	% reduction	No. of A.c. release	Avg. No. of alive S.d. after 1 day	No. of remaining S.d.*/ Plant	% reduction	No. of A.c. release	Avg. No. of alive S.d. after 1 day	No. of remaining S.d.*/ Plant	% reduction
I	45	40	35.60	20.8	40	36.5	70.4	48.5	40	11.60	36.5	68.2
II	45	45	32.54	27.6	45	30.2	63.5	52.4	45	8.70	33.50	74.02
III	45	50	29.71	33.9	50	23.6	57.6	59.02	50	5.41	30.6	82.32

* Mean Value, ** All the number of predators were releases in each micro-plot

S.d. = *Scirtothrips dorsalis*, A.c = *Amblyseius cucumeris*

Graph 4.26
Effect of Predator; *Amblyseius cucumeris* on the population of *Scirtothrips dorsalis* in Chilli field of First Net House during the year 2014-15 (I Microplot)

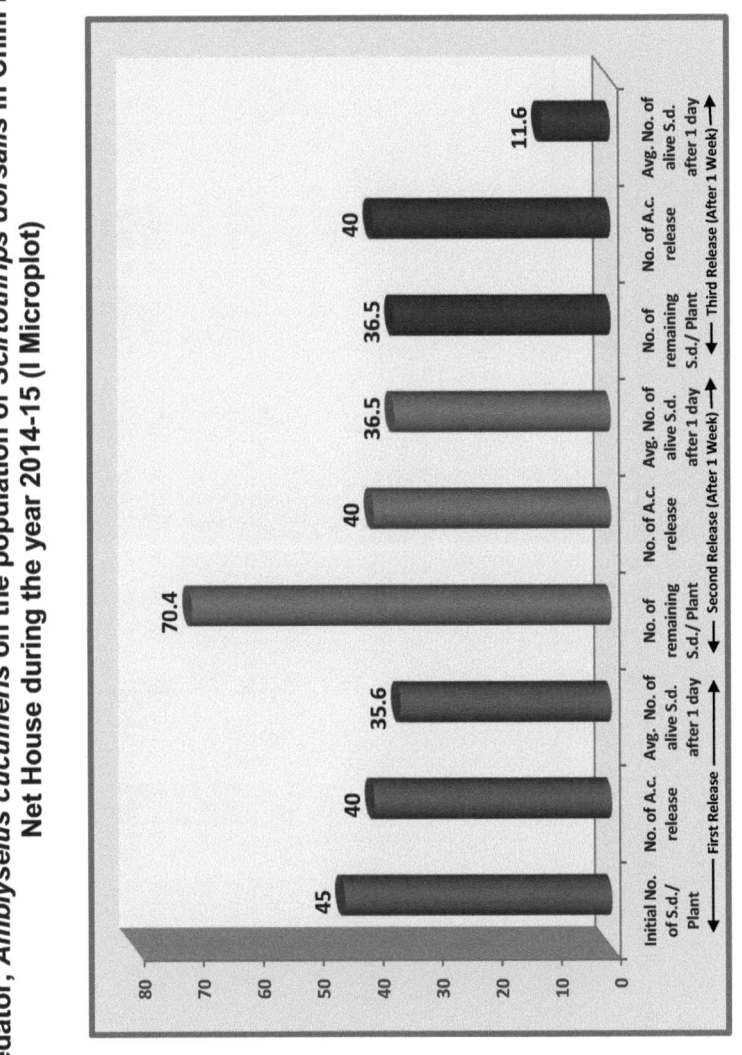

S.d. = *Scirtothrips dorsalis*, A.c = *Amblyseius cucumeris*

Graph 4.27
Effect of Predator; *Amblyseius cucumeris* on the population of *Scirtothrips dorsalis* in Chilli field of First Net House during the year 2014-15 (II Microplot)

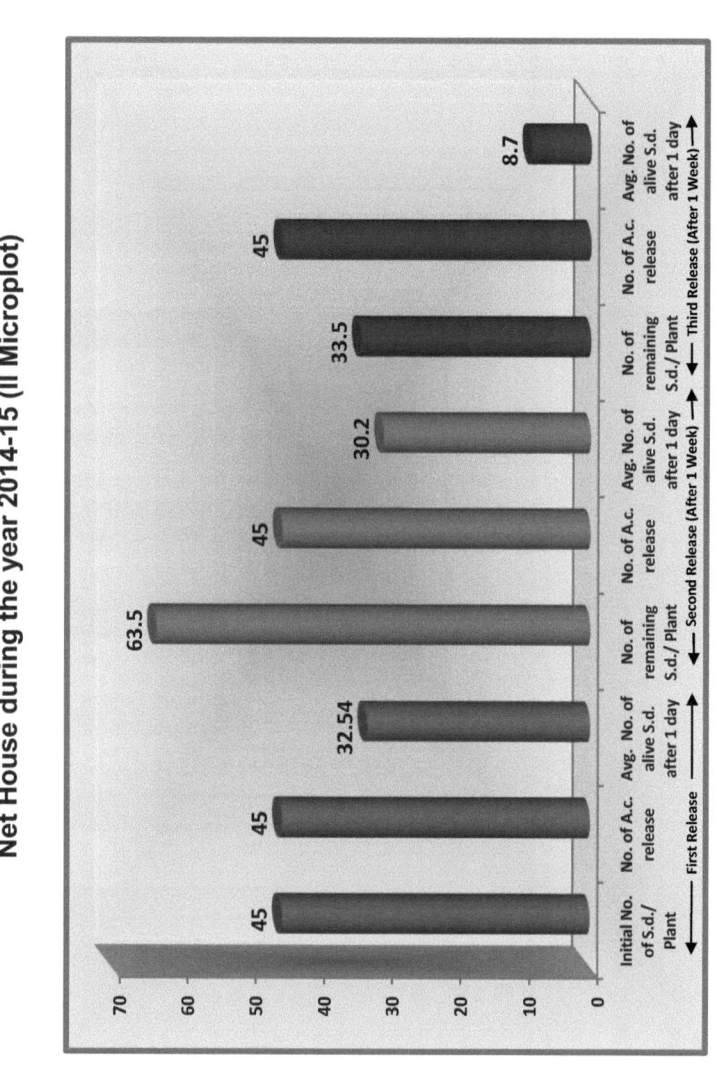

S.d. = *Scirtothrips dorsalis*, A.c = *Amblyseius cucumeris*

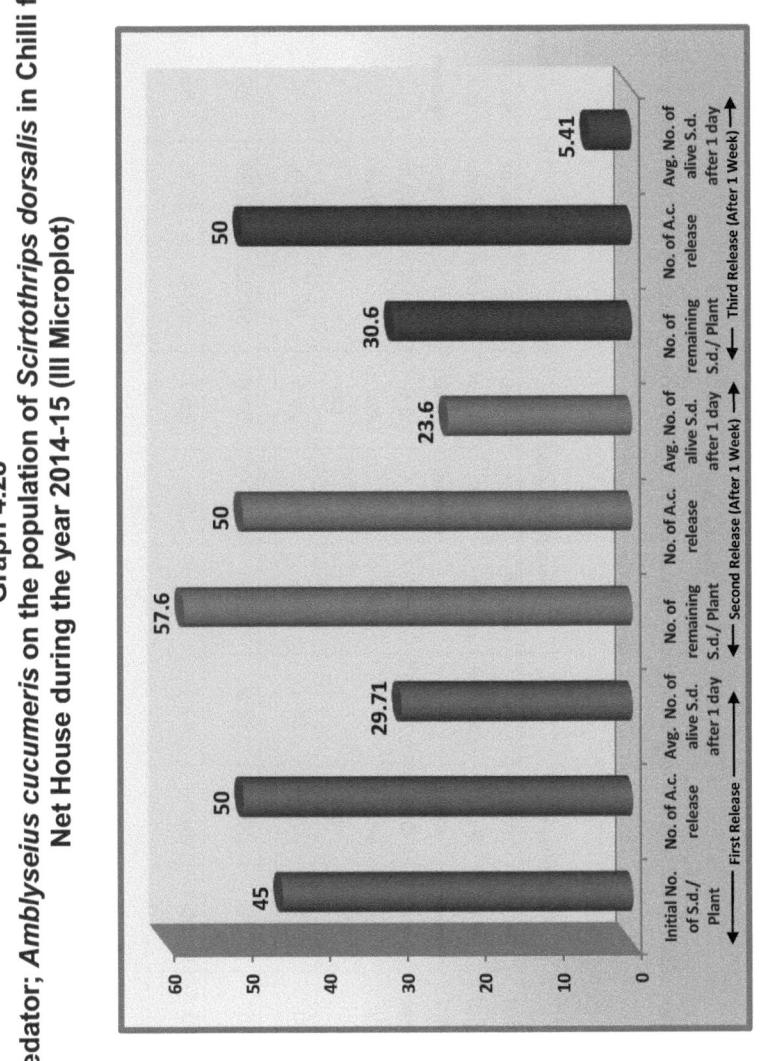

Graph 4.28

Effect of Predator; *Amblyseius cucumeris* on the population of *Scirtothrips dorsalis* in Chilli field of First Net House during the year 2014-15 (III Microplot)

S.d. = *Scirtothrips dorsalis*, A.c = *Amblyseius cucumeris*

Table 4.8

Effect of Predator; *Macrotracheliella nigra* on the population of *Scirtothrips dorsalis* in Chilli field of Second Net House during the year 2014-15

Experimental Microplot	Initial No. of S.d.*/ Plant	First Release				Second Release (After 1 Week)				Third Release (After 1 Week)			
		No. of M.n. release**	Avg. No. of alive S.d. after 1 day		% reduction	No. of remain-ing S.d.*/ Plant	No. of M.n. release	Avg. No. of alive S.d. after 1 day	% reduction	No. of remain-ing S.d.*/ Plant	No. of M.n. release	Avg. No. of alive S.d. after 1 day	% reduction
I	45	45	33.64		25.2	70.4	45	39.6	43.7	36.5	45	13.50	63.01
II	45	50	30.19		32.9	63.5	50	31.6	50.23	33.50	50	7.69	77.04
III	45	55	26.50		41.1	57.6	55	26.7	53.6	30.6	55	4.34	85.8

* Mean Value, ** All the number of predators were releases in each micro-plot

S.d. = *Scirtothrips dorsalis*, M.n = *Macrotracheliella nigra*

Graph 4.29
Effect of Predator; *Macrotracheliella nigra* on the population of *Scirtothrips dorsalis* in Chilli field of Second Net House during the year 2014-15 (I Microplot)

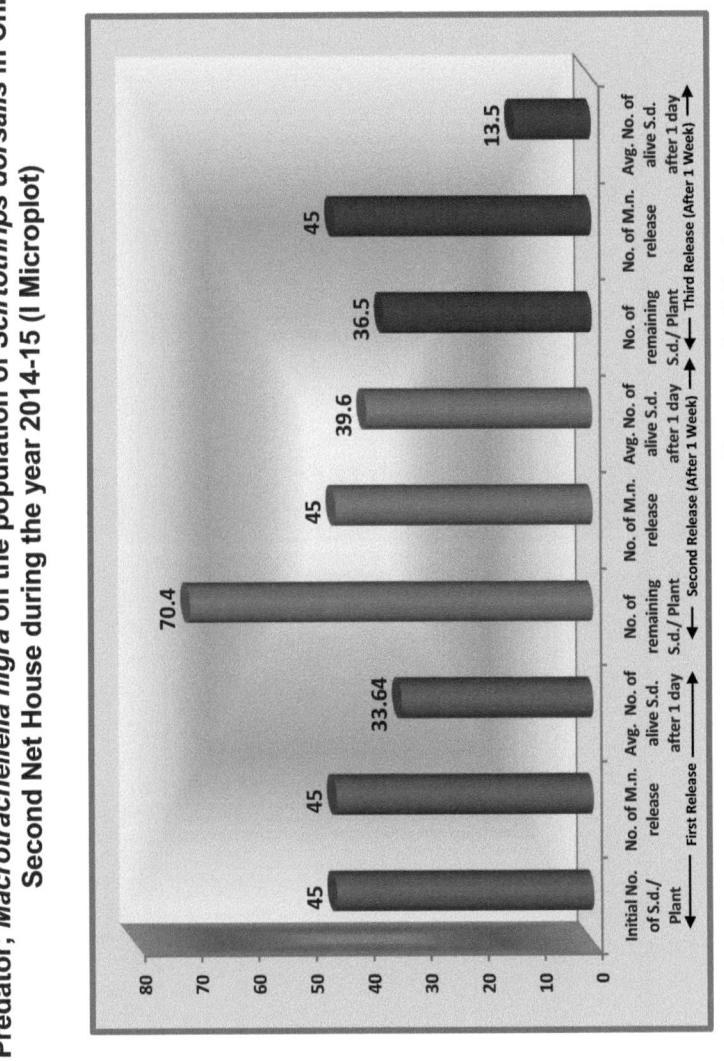

S.d. = *Scirtothrips dorsalis*, M.n = *Macrotracheliella nigra*

Graph 4.30
Effect of Predator; *Macrotracheliella nigra* on the population of *Scirtothrips dorsalis* in Chilli field of Second Net House during the year 2014-15 (II Microplot)

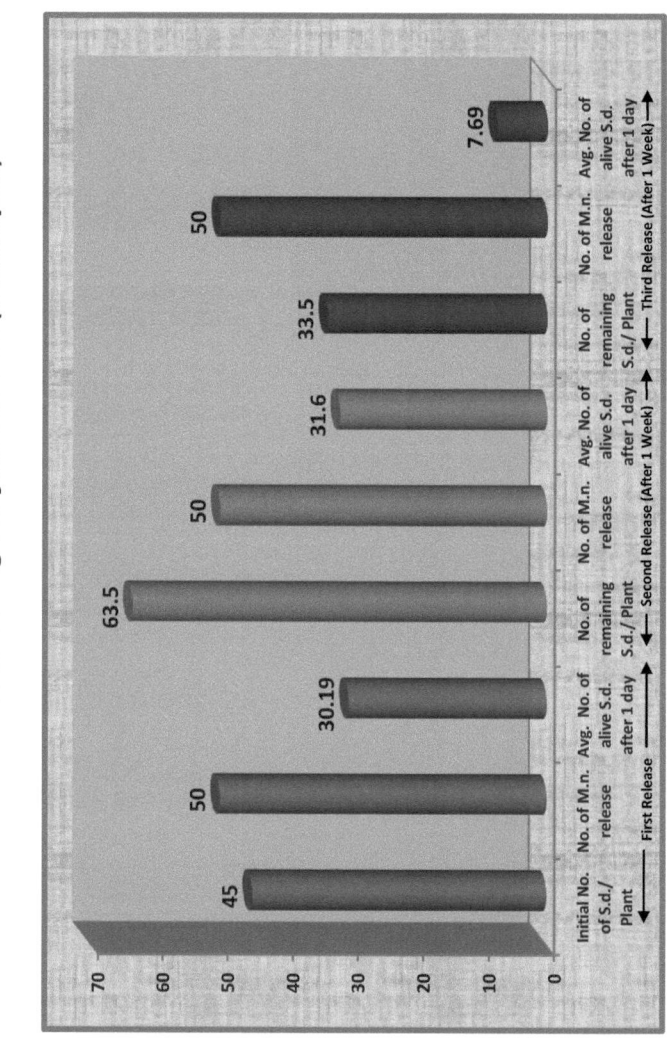

S.d. = *Scirtothrips dorsalis*, M.n = *Macrotracheliella nigra*

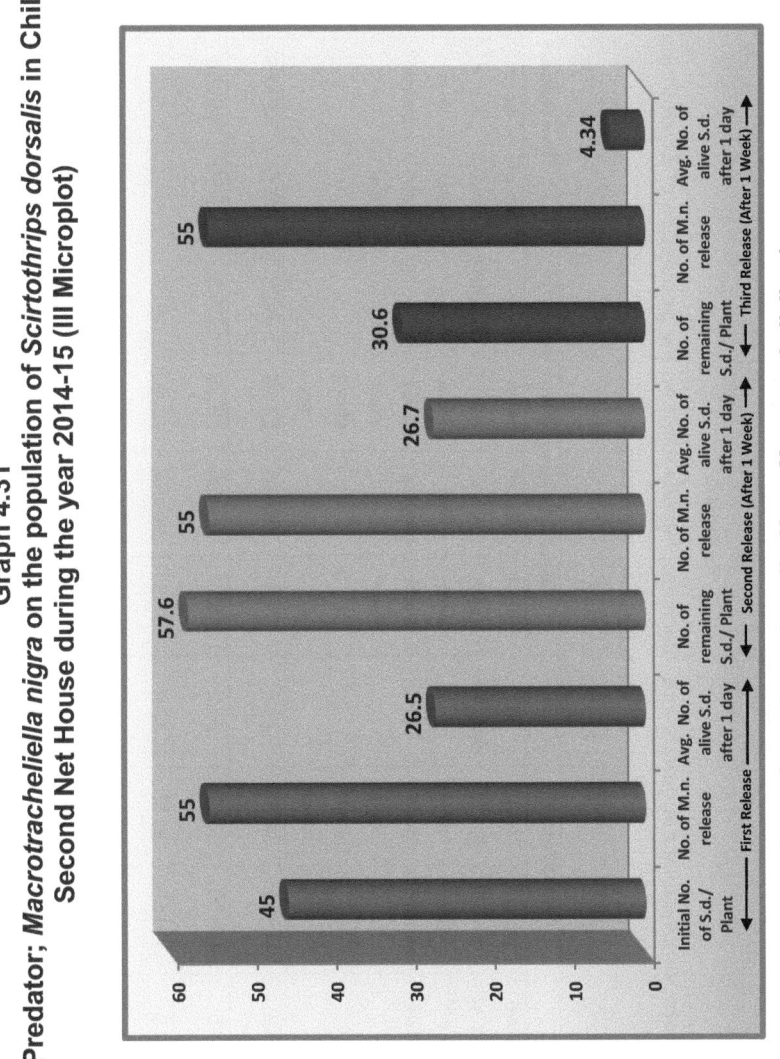

Graph 4.31
Effect of Predator; *Macrotracheliella nigra* on the population of *Scirtothrips dorsalis* in Chilli field of Second Net House during the year 2014-15 (III Microplot)

S.d. = *Scirtothrips dorsalis*, M.n = *Macrotracheliella nigra*

Table 4.9
Effect of Predators; *Amblyseius cucumeris* and *Macrotracheliella nigra* on the population of *Scirtothrips dorsalis* in Chilli Field of Third Net House during the year 2014-15

Experimental Microplot	Initial No. of S.d.*/ Plant	First Release			Second Release (After 1 Week)				Third Release (After 1 Week)			
		No. of A.c. + M.n. release**	Avg. No. of alive S.d./ plant after 1 day	% reduction	No. of remai-ning S.d.*/ Plant	No. of A.c. + M.n. release**	Avg. No. of alive S.d./ plant after 1 day	% reduction	No. of remai-ning S.d.*/ Plant	No. of A.c. + M.n. release**	Avg. No. of alive S.d./ plant after 1 day	% reduction
I	45	40+45	34.36	23.6	70.4	40+45	22.26	68.3	36.5	40+45	4.10	88.7
II	45	45+50	33.20	26.2	63.5	45+50	15.36	75.8	33.50	45+50	2.67	92.0
III	45	50+55	30.16	32.9	57.6	50+55	11.10	80.7	--	--	--	--

* Mean Value, ** All the number of predators were releases in each micro-plot

S.d. = *Scirtothrips dorsalis*, A.c. = *Amblyseius cucumeris*, M.n. = *Macrotracheliella nigra*

Graph 4.32
Effect of Predators; *Amblyseius cucumeris* and *Macrotracheliella nigra* on the population of *Scirtothrips dorsalis* in Chilli Field of Third Net House during the year 2014-15 (I Microplot)

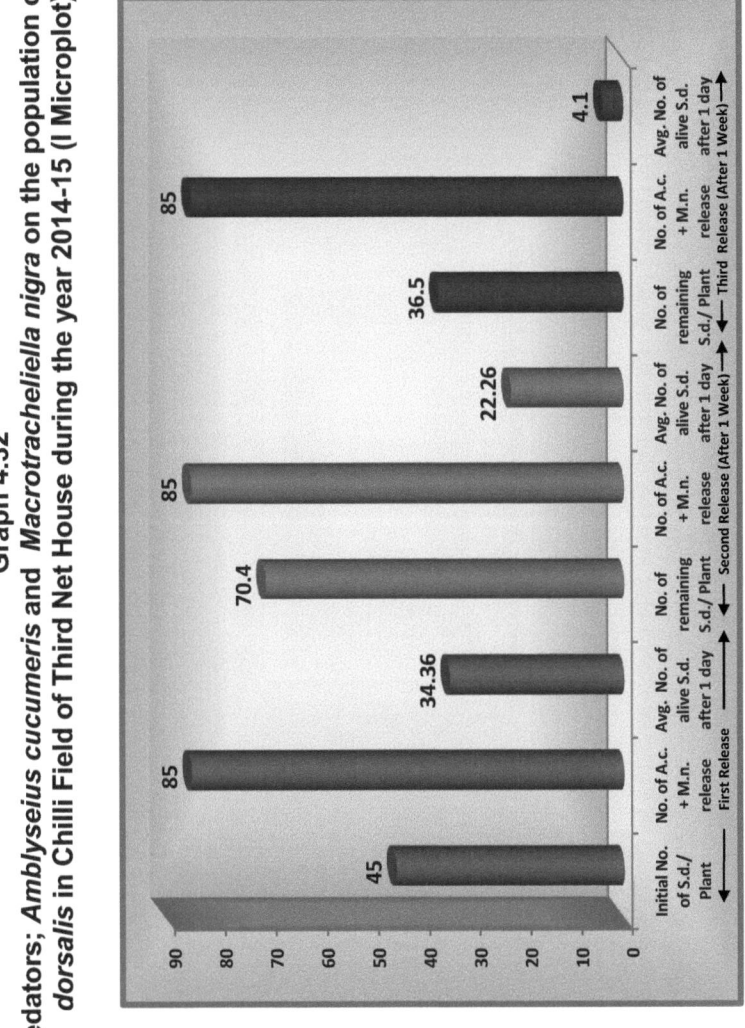

S.d. = *Scirtothrips dorsalis*, A.c. = *Amblyseius cucumeris*, M.n. = *Macrotracheliella nigra*

Graph 4.33

Effect of Predators; *Amblyseius cucumeris* and *Macrotracheliella nigra* on the population of *Scirtothrips dorsalis* in Chilli Field of Third Net House during the year 2014-15 (II Microplot)

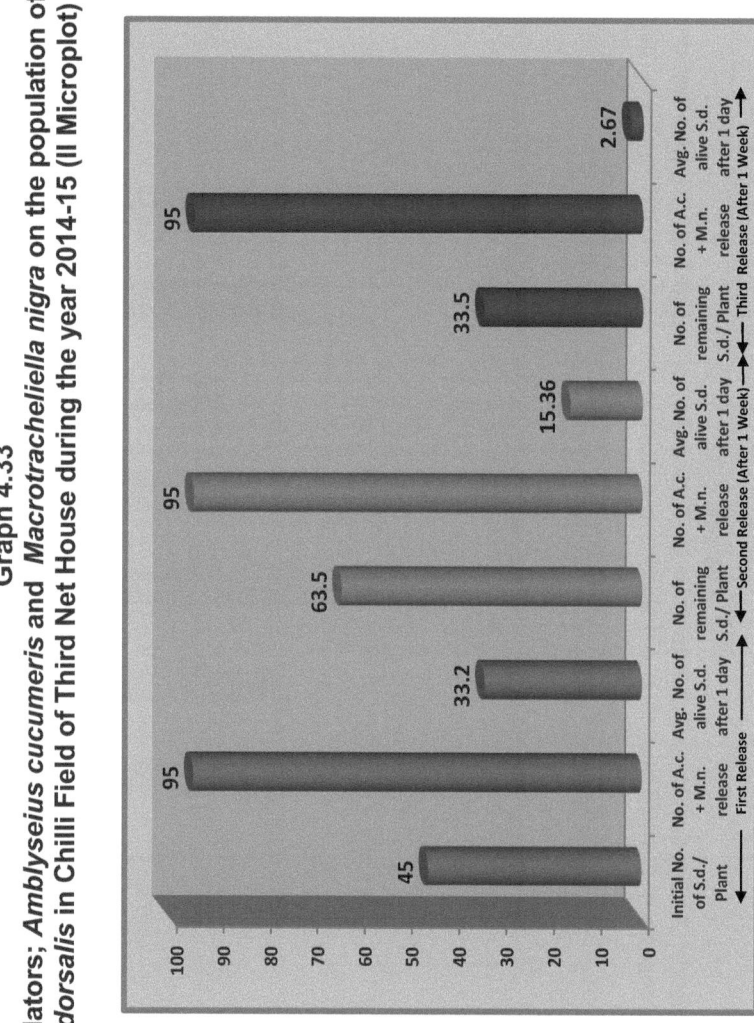

S.d. = *Scirtothrips dorsalis*, A.c. = *Amblyseius cucumeris*, M.n. = *Macrotracheliella nigra*

Graph 4.34
Effect of Predators; *Amblyseius cucumeris* and *Macrotracheliella nigra* on the population of *Scirtothrips dorsalis* in Chilli Field of Third Net House during the year 2014-15 (III Microplot)

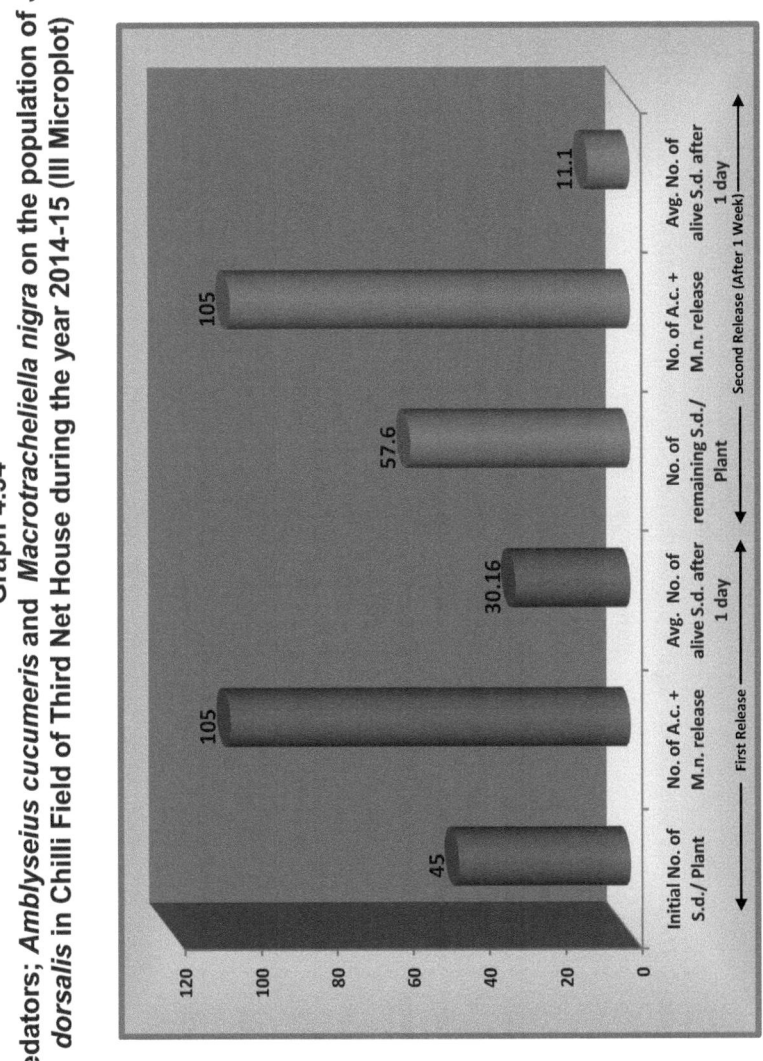

S.d. = *Scirtothrips dorsalis*, A.c. = *Amblyseius cucumeris*, M.n. = *Macrotracheliella nigra*

Table 4.10

Effect of Parasitoid; *Thripobius semiluteus* on the population of *Scirtothrips dorsalis* in Chilli field of Fourth Net House during the year 2014-15

Experimental Microplot	Initial No. of S.d.*/ Plant	First Release				Second Release (After 1 Week)				Third Release (After 1 Week)			
		No. of T.s. release**	Avg. No. of alive S.d. after 3 days	% reduction		No. of T.s. release	No. of remai-ning S.d.*/ Plant	Avg. No. of alive S.d. after 3 days	% reduction	No. of T.s. release	No. of remai-ning S.d.*/ Plant	Avg. No. of alive S.d. after 3 days	% reduction
I	45	15	40.19	10.6		15	70.4	39.96	43.23	15	36.5	10.41	71.4
II	45	20	36.68	18.4		20	63.5	33.80	46.7	20	33.50	6.45	80.7
III	45	25	32.16	28.5		25	57.6	25.56	55.6	25	30.6	4.56	85.09

* Mean Value, ** All the number of parasitoids were releases in each micro-plot

S.d. = *Scirtothrips dorsalis*, T.s. = *Thripobius semiluteus*

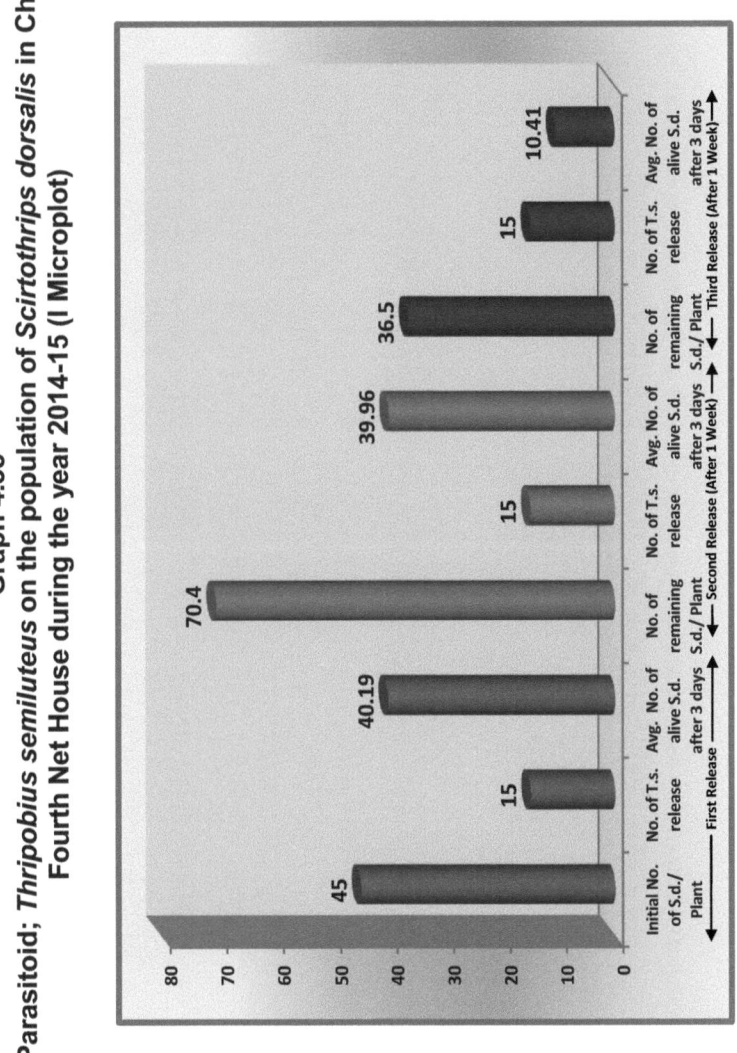

Graph 4.35
Effect of Parasitoid; *Thripobius semiluteus* on the population of *Scirtothrips dorsalis* in Chilli field of Fourth Net House during the year 2014-15 (I Microplot)

S.d. = *Scirtothrips dorsalis*, T.s. = *Thripobius semiluteus*

Graph 4.36
Effect of Parasitoid; *Thripobius semiluteus* on the population of *Scirtothrips dorsalis* in Chilli field of Fourth Net House during the year 2014-15 (II Microplot)

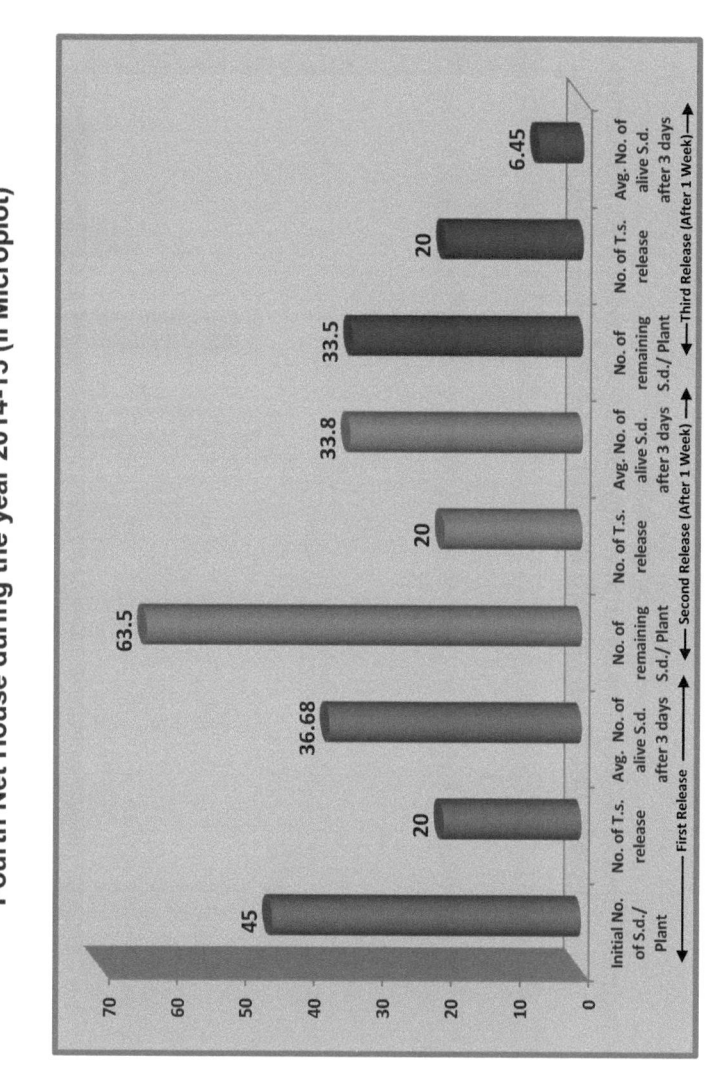

S.d. = *Scirtothrips dorsalis*, T.s. = *Thripobius semiluteus*

Graph 4.37
Effect of Parasitoid; *Thripobius semiluteus* **on the population of** *Scirtothrips dorsalis* **in Chilli field of Fourth Net House during the year 2014-15 (III Microplot)**

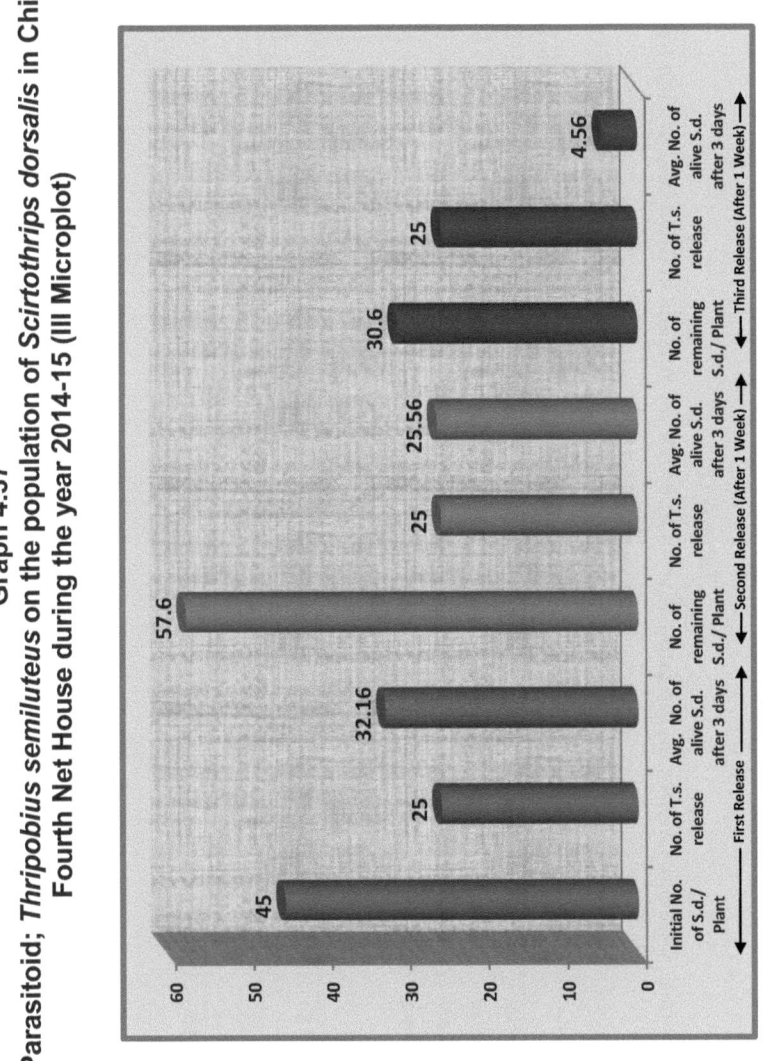

S.d. = *Scirtothrips dorsalis*, T.s. = *Thripobius semiluteus*

Table 4.11

Effect of Parasitoid; *Ceranisus menes* on the population of *Scirtothrips dorsalis* in Chilli field of Fifth Net House during the year 2014-15

Experimental Microplot	Initial No. of S.d.*/ Plant	First Release				Second Release (After 1 Week)				Third Release (After 1 Week)			
		No. of C.m. release**	Avg. No. of alive S.d. after 3 days	% reduction		No. of remaining S.d.*/ Plant	No. of C.m. release**	Avg. No. of alive S.d. after 3 days	% reduction	No. of remaining S.d.*/ Plant	No. of C.m. release**	Avg. No. of alive S.d. after 3 days	% reduction
I	45	20	39.61	11.9		70.4	20	33.7	52.1	36.5	20	11.5	68.4
II	45	25	35.54	21.02		63.5	25	28.34	55.3	33.50	25	8.37	75.01
III	45	30	32.43	27.93		57.6	30	20.26	64.8	30.6	30	6.50	78.75

* Mean Value, ** All the number of parasitoids were releases in each micro-plot

S.d. = *Scirtothrips dorsalis*, C.m. = *Ceranisus menes*

Graph 4.38
Effect of Parasitoid; *Ceranisus menes* on the population of *Scirtothrips dorsalis* in Chilli field of Fifth Net House during the year 2014-15 (I Microplot)

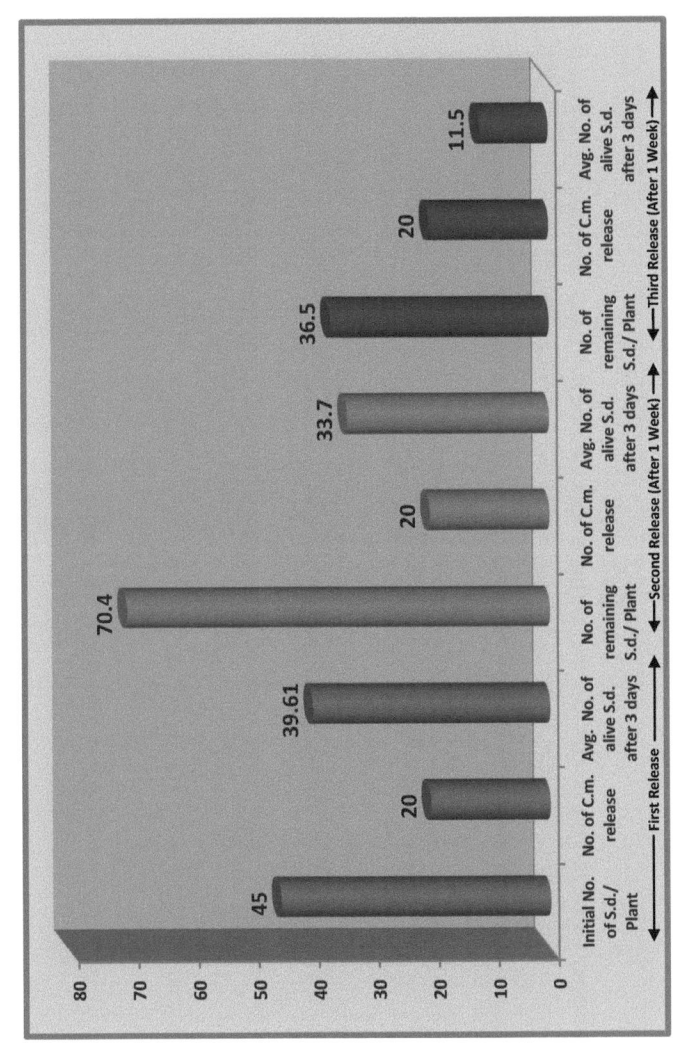

S.d. = *Scirtothrips dorsalis*, C.m. = *Ceranisus menes*

Graph 4.39
Effect of Parasitoid; *Ceranisus menes* on the population of *Scirtothrips dorsalis* in Chilli field of Fifth Net House during the year 2014-15 (II Microplot)

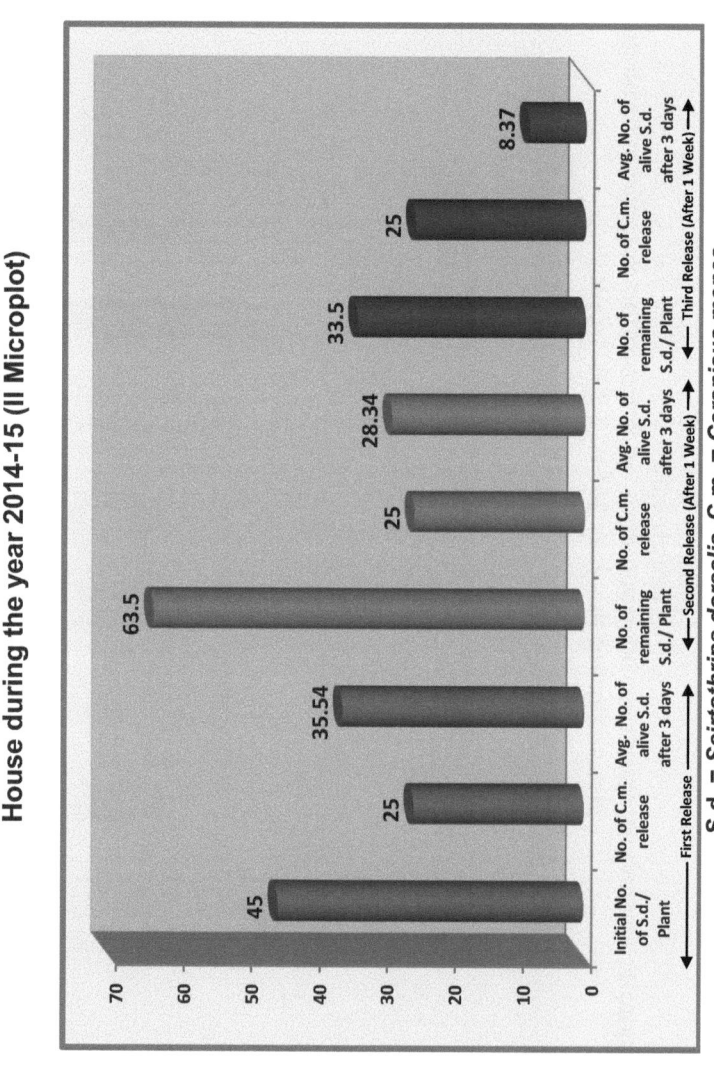

S.d. = *Scirtothrips dorsalis*, C.m. = *Ceranisus menes*

Graph 4.40
Effect of Parasitoid; *Ceranisus menes* on the population of *Scirtothrips dorsalis* in Chilli field of Fifth Net House during the year 2014-15 (III Microplot)

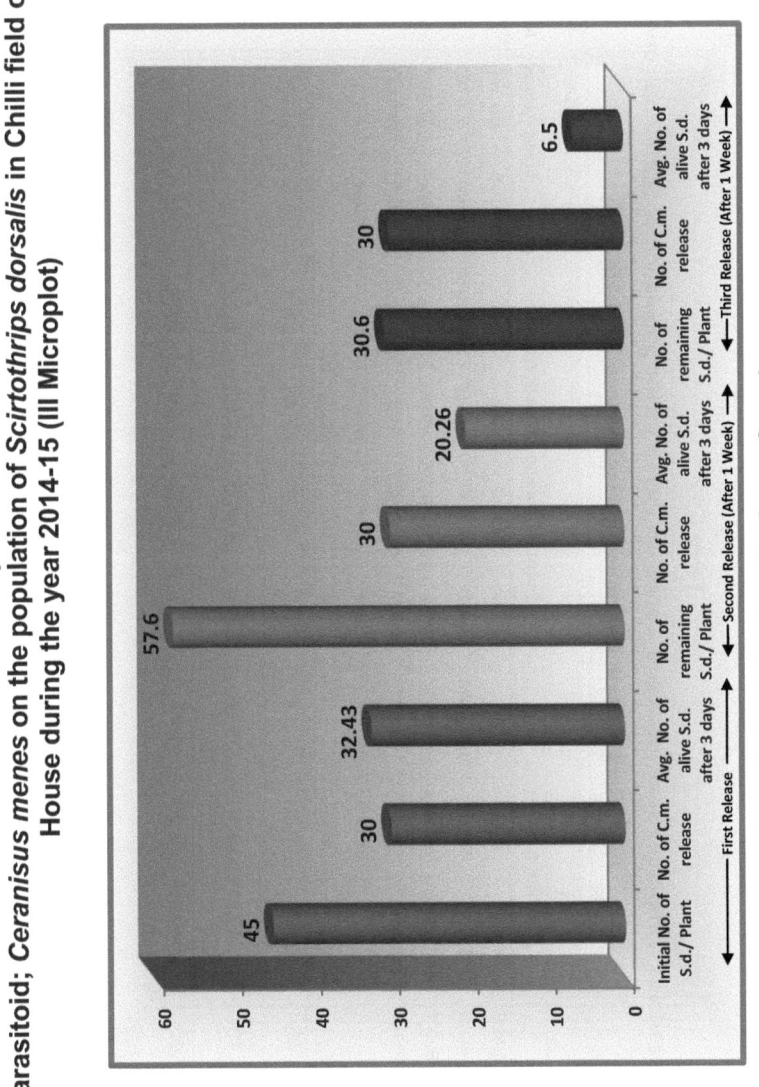

S.d. = *Scirtothrips dorsalis*, C.m. = *Ceranisus menes*

Table 4.12

Effect of Parasitoids; *Thripobius semiluteus* and *Ceranisus menes* on the population of *Scirtothrips dorsalis* in Chilli Field of Sixth Net House during the year 2014-15

Experimental Microplot	Initial No. of S.d.*/ Plant	First Release				Second Release (After 1 Week)				Third Release (After 1 Week)			
		No. of T.s. + C.m. release**	Avg. No. of alive S.d. after 3 days	% reduction		No. of remaining S.d.*/ Plant	No. of T.s. + C.m. release**	Avg. No. of alive S.d. after 3 days	% reduction	No. of remaining S.d.*/ Plant	No. of T.s. + C.m. release**	Avg. No. of alive S.d. after 3 days	% reduction
I	45	15+20	35.20	21.7		70.4	15+20	23.46	66.6	36.5	15+20	4.36	88.05
II	45	20+25	30.10	33.11		63.5	20+25	10.30	83.7	---	---	---	---
III	45	25+30	25.30	43.7		57.6	25+30	7.10	87.6	---	---	---	---

* Mean Value, ** All the number of parasitoids were releases in each micro-plot

S.d. = *Scirtothrips dorsalis*, T.s. = *Thripobius semiluteus*, C.m. = *Ceranisus menes*

Graph 4.41
Effect of Parasitoids; *Thripobius semiluteus* and *Ceranisus menes* on the population of *Scirtothrips dorsalis* in Chilli Field of Sixth Net House during the year 2014-15 (I Microplot)

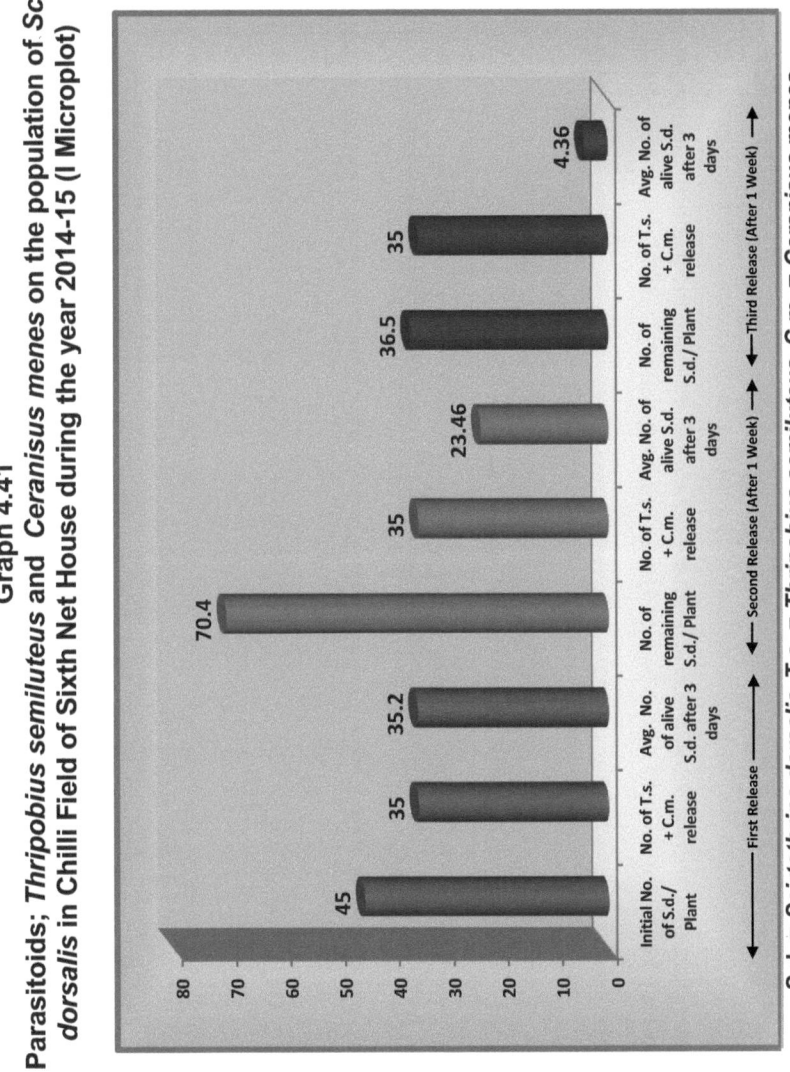

S.d. = *Scirtothrips dorsalis*, T.s. = *Thripobius semiluteus*, C.m. = *Ceranisus menes*

145 | Chapter 4 | Experimental Analysis

Graph 4.42

Effect of Parasitoids; *Thripobius semiluteus* and *Ceranisus menes* on the population of *Scirtothrips dorsalis* in Chilli Field of Sixth Net House during the year 2014-15 (II Microplot)

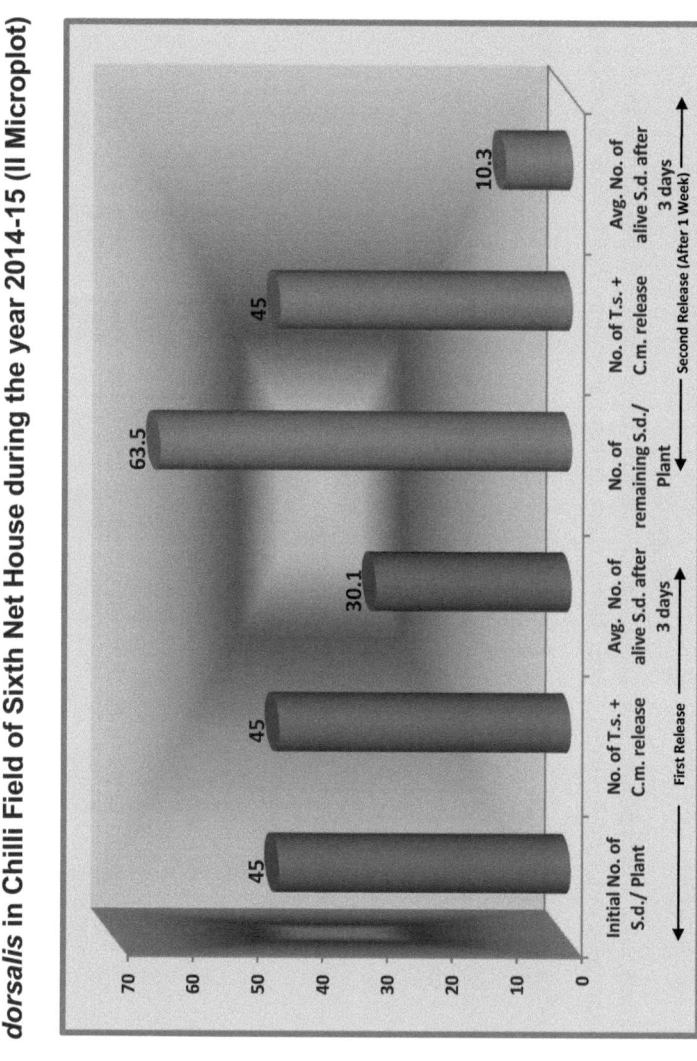

S.d. = *Scirtothrips dorsalis*, T.s. = *Thripobius semiluteus*, C.m. = *Ceranisus menes*

Graph 4.43

Effect of Parasitoids; *Thripobius semiluteus* and *Ceranisus menes* on the population of *Scirtothrips dorsalis* in Chilli Field of Sixth Net House during the year 2014-15 (III Microplot)

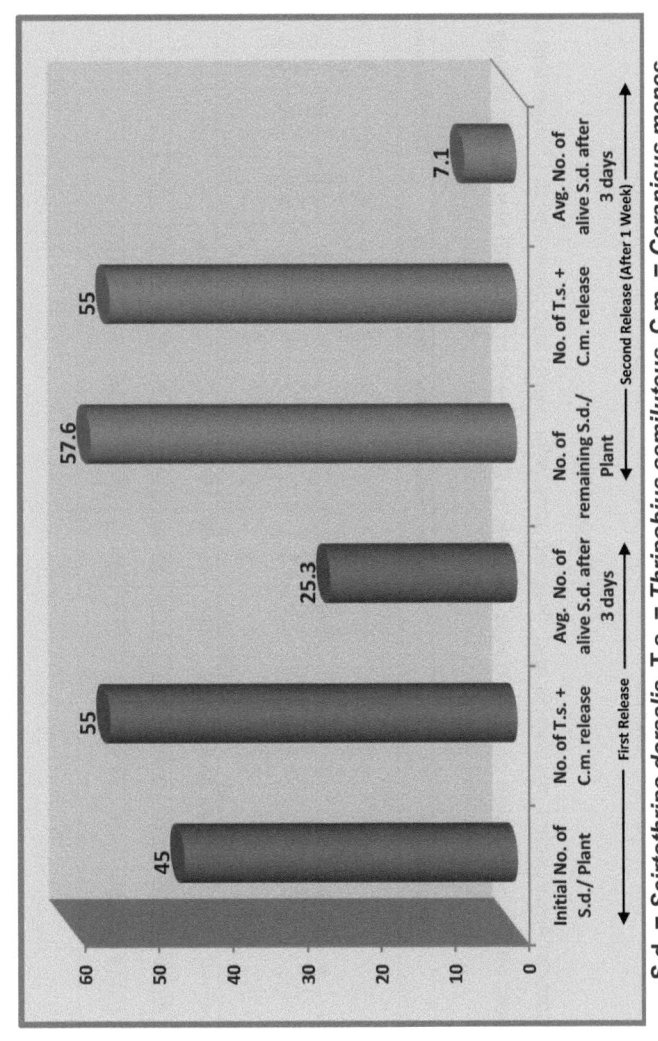

S.d. = *Scirtothrips dorsalis*, T.s. = *Thripobius semiluteus*, C.m. = *Ceranisus menes*

Table 4.13

Effect of Predators (*Amblyseius cucumeris* and *Macrotracheliella nigra*) and parasitoids (*Thripobius semiluteus* and *Ceranisus menes*) on the population of *Scirtothrips dorsalis* in Chilli Field of Seventh Net House during the year 2014-15

Experimental Micro plot	Initial No. of S.d.*/plant	First Release							Second Release (After 1 Week)							Third Release (After 1 Week)					
		No. of A.c. + M.n. release**	No. of T.s. + C.m. release**	Avg. No. of alive S.d. after 1 day	Avg. No. of alive S.d. after 3 days	% reduction			No. of remaining S.d.*/plant	No. of A.c. + M.n. release**	No. of T.s. + C.m. release**	Avg. No. of alive S.d. after 1 day	Avg. No. of alive S.d. after 3 days	% reduction		No. of remaining S.d.*/plant	No. of A.c. + M.n. release**	No. of T.s. + C.m. release**	Avg. No. of alive S.d. after 1 day	Avg. No. of alive S.d. after 3 days	% reduction
I	45	40+45	15+20	33.04	26.15	41.8			70.4	40+45	15+20	15.05	9.37	86.6		--	--	--	--	--	--
II	45	45+50	20+25	32.40	25.05	44.3			63.5	45+50	20+25	13.56	5.30	91.6		--	--	--	--	--	--
III	45	50+55	25+30	33.26	24.10	46.4			57.6	50+55	25+30	11.63	3.67	93.6		--	--	--	--	--	--

* Mean Value, ** All the number of predators and parasitoids were releases in each micro-plot

S.d. = *Scirtothrips dorsalis*, A.c. = *Amblyseius cucumeris*, M.n. = *Macrotracheliella nigra*

T.S. = *Thripobius semiluteus*, C.m. = *Ceranisus menes*

Graph 4.44
Effect of Predators (*Amblyseius cucumeris* and *Macrotracheliella nigra*) and parasitoids (*Thripobius semiluteus* and *Ceranisus menes*) on the population of *Scirtothrips dorsalis* in Chilli Field of Seventh Net House during the year 2014-15 (I Microplot)

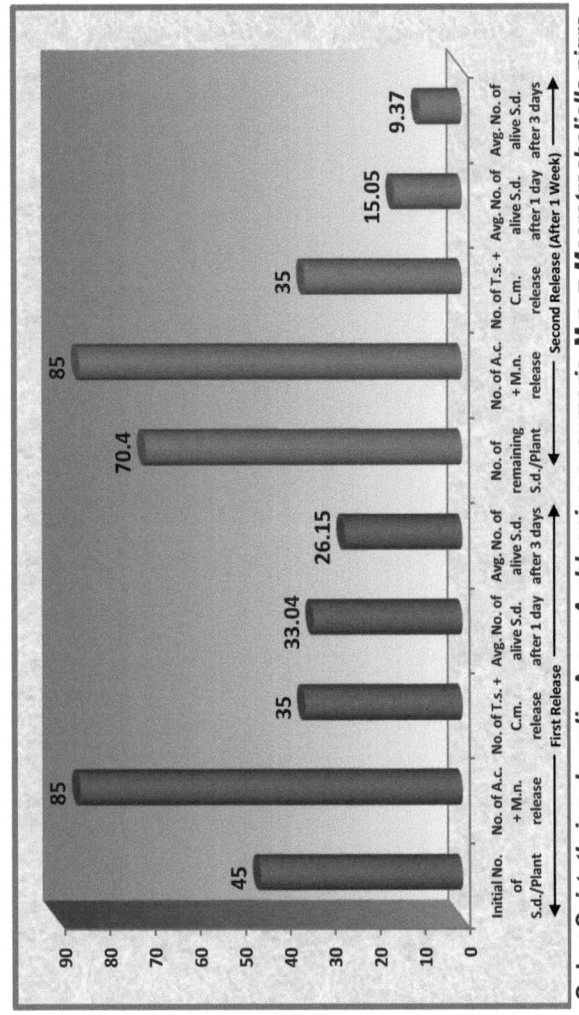

S.d. = *Scirtothrips dorsalis*, A.c. = *Amblyseius cucumeris*, M.n. = *Macrotracheliella nigra* T.S. = *Thripobius semiluteus*, C.m. = *Ceranisus menes*

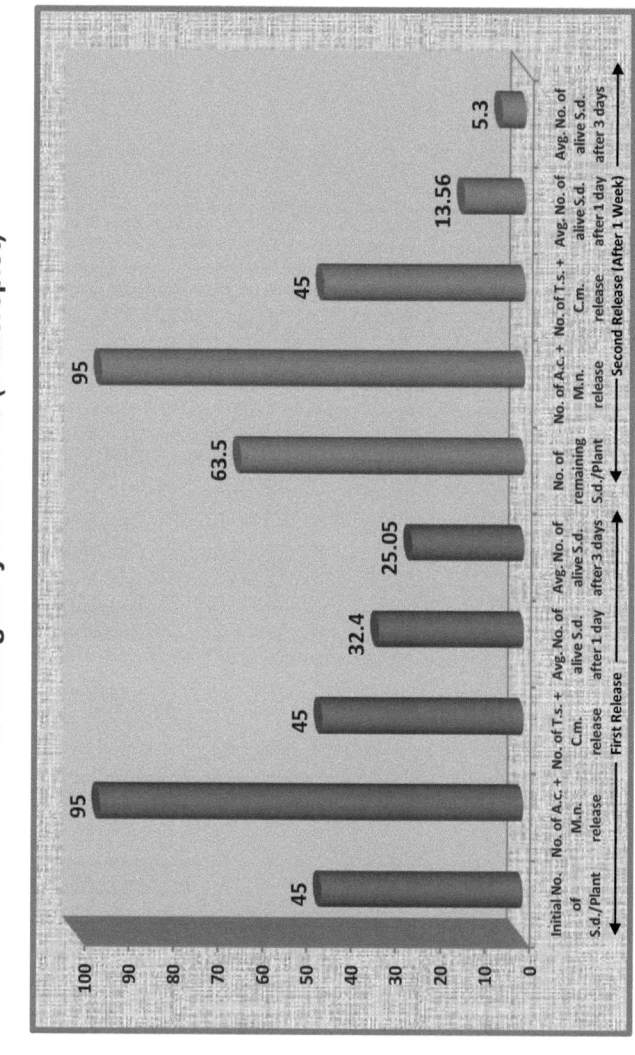

Graph 4.45 Effect of Predators (*Amblyseius cucumeris* and *Macrotracheliella nigra*) and parasitoids (*Thripobius semiluteus* and *Ceranisus menes*) on the population of *Scirtothrips dorsalis* in Chilli Field of Seventh Net House during the year 2014-15 (II Microplot)

S.d. = *Scirtothrips dorsalis*, A.c. = *Amblyseius cucumeris*, M.n. = *Macrotracheliella nigra* T.S. = *Thripobius semiluteus*, C.m. = *Ceranisus menes*

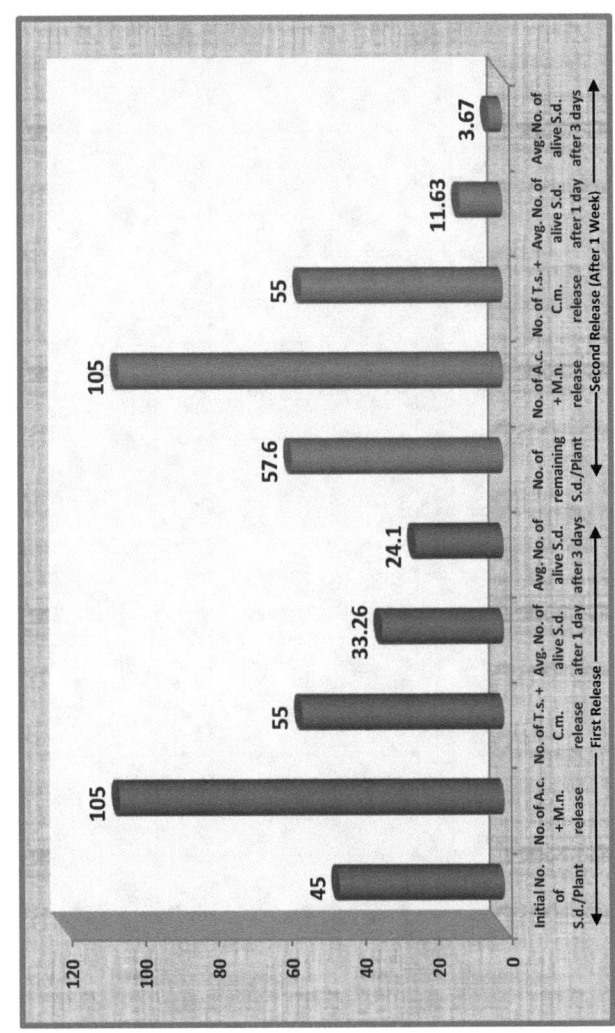

Graph 4.46
Effect of Predators (*Amblyseius cucumeris* and *Macrotracheliella nigra*) and parasitoids (*Thripobius semiluteus* and *Ceranisus menes*) on the population of *Scirtothrips dorsalis* in Chilli Field of Seventh Net House during the year 2014-15 (III Microplot)

S.d. = *Scirtothrips dorsalis*, A.c. = *Amblyseius cucumeris*, M.n. = *Macrotracheliella nigra*
T.S. = *Thripobius semiluteus*, C.m. = *Ceranisus menes*

Table 4.14

Effect of Predator; *Amblyseius cucumeris* on the population of *Scirtothrips dorsalis* in Chilli field of First Net House during the year 2015-16

Experimental Microplot	Initial No. of S.d.*/Plant	First Release			Second Release (After 1 Week)				Third Release (After 1 Week)			
		No. of A.c. release**	Avg. No. of alive S.d. after 1 day	% reduction	No. of remaining S.d.*/Plant	No. of A.c. release	Avg. No. of alive S.d. after 1 day	% reduction	No. of remaining S.d.*/Plant	No. of A.c. release	Avg. No. of alive S.d. after 1 day	% reduction
I	55	50	45.16	17.8	73.3	50	39.6	45.9	39.73	50	13.06	67.12
II	55	55	41.36	24.8	65.6	55	32.14	51.0	34.06	55	8.34	75.51
III	55	60	38.34	30.2	56.5	60	24.36	56.8	29.16	60	4.43	84.80

* Mean Value, ** All the number of predators were releases in each micro-plot

S.d. = *Scirtothrips dorsalis*, A.c = *Amblyseius cucumeris*

Graph 4.47
Effect of Predator; *Amblyseius cucumeris* on the population of *Scirtothrips dorsalis* in Chilli field of First Net House during the year 2015-16 (I Microplot)

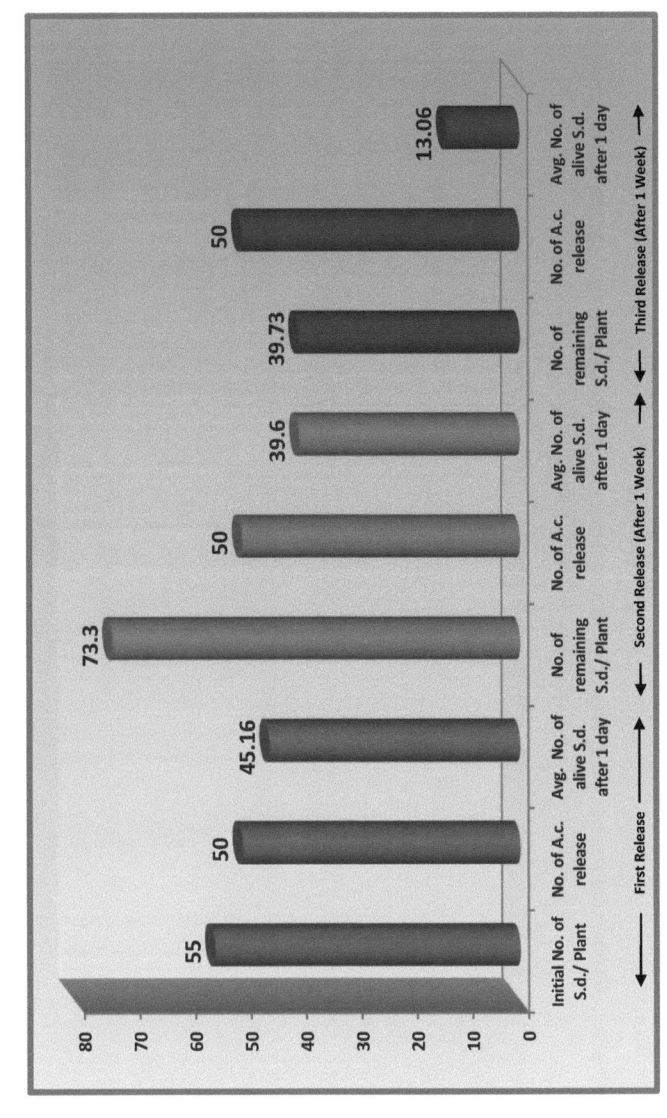

S.d. = *Scirtothrips dorsalis*, A.c = *Amblyseius cucumeris*

Graph 4.48
Effect of Predator; *Amblyseius cucumeris* on the population of *Scirtothrips dorsalis* in Chilli field of First Net House during the year 2015-16 (II Microplot)

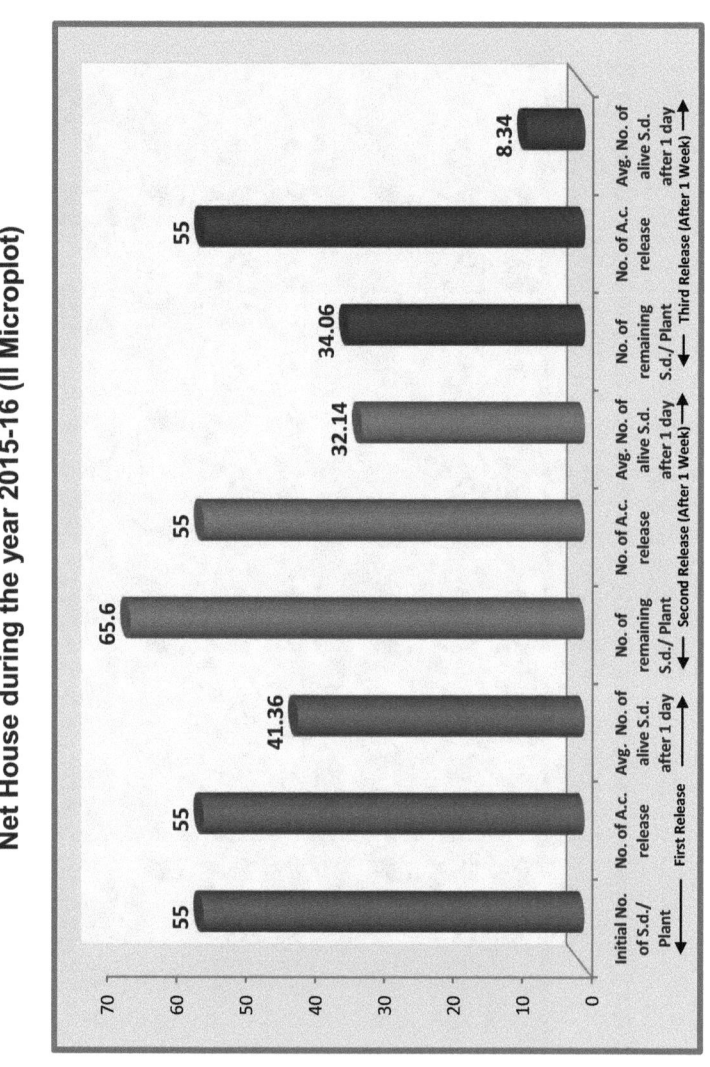

S.d. = *Scirtothrips dorsalis*, A.c = *Amblyseius cucumeris*

Graph 4.49
Effect of Predator; *Amblyseius cucumeris* **on the population of** *Scirtothrips dorsalis* **in Chilli field of First Net House during the year 2015-16 (III Microplot)**

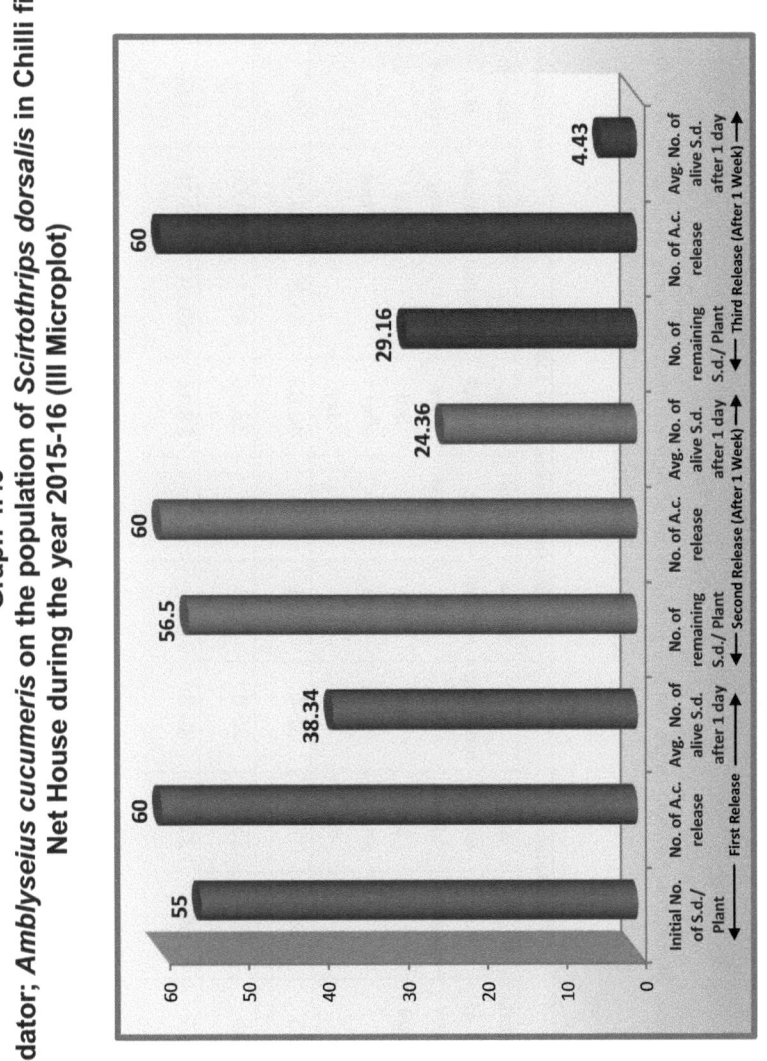

S.d. = *Scirtothrips dorsalis*, A.c = *Amblyseius cucumeris*

Table 4.15

Effect of Predator; *Macrotracheliella nigra* on the population of *Scirtothrips dorsalis* in Chilli field of Second Net House during the year 2015-16

Experimental Microplot	Initial No. of S.d.*/ Plant	First Release			Second Release (After 1 Week)				Third Release (After 1 Week)			
		No. of M.n. release**	Avg. No. of alive S.d. after 1 day	% reduction	No. of remaining S.d.*/ Plant	No. of M.n. release	Avg. No. of alive S.d. after 1 day	% reduction	No. of remaining S.d.*/ Plant	No. of M.n. release	Avg. No. of alive S.d. after 1 day	% reduction
I	55	55	37.17	32.4	73.3	55	42.63	41.8	39.73	55	15.30	61.49
II	55	60	35.40	35.6	65.6	60	33.16	49.4	34.06	60	7.46	78.09
III	55	65	33.04	39.9	56.5	65	23.14	59.04	29.16	65	4.03	86.1

* Mean Value, ** All the number of predators were releases in each micro-plot

S.d. = *Scirtothrips dorsalis*, M.n = *Macrotracheliella nigra*

Graph 4.50

Effect of Predator; *Macrotracheliella nigra* on the population of *Scirtothrips dorsalis* in Chilli field of Second Net House during the year 2015-16 (I Microplot)

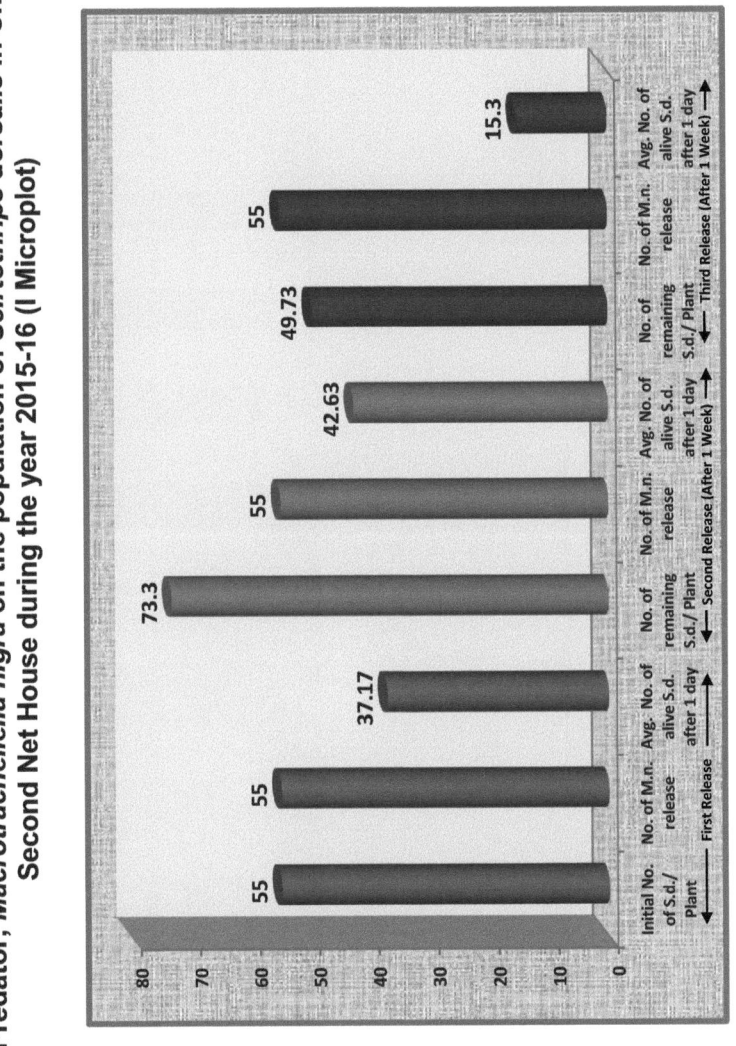

S.d. = *Scirtothrips dorsalis*, M.n = *Macrotracheliella nigra*

Graph 4.51
Effect of Predator; *Macrotracheliella nigra* on the population of *Scirtothrips dorsalis* in Chilli field of Second Net House during the year 2015-16 (II Microplot)

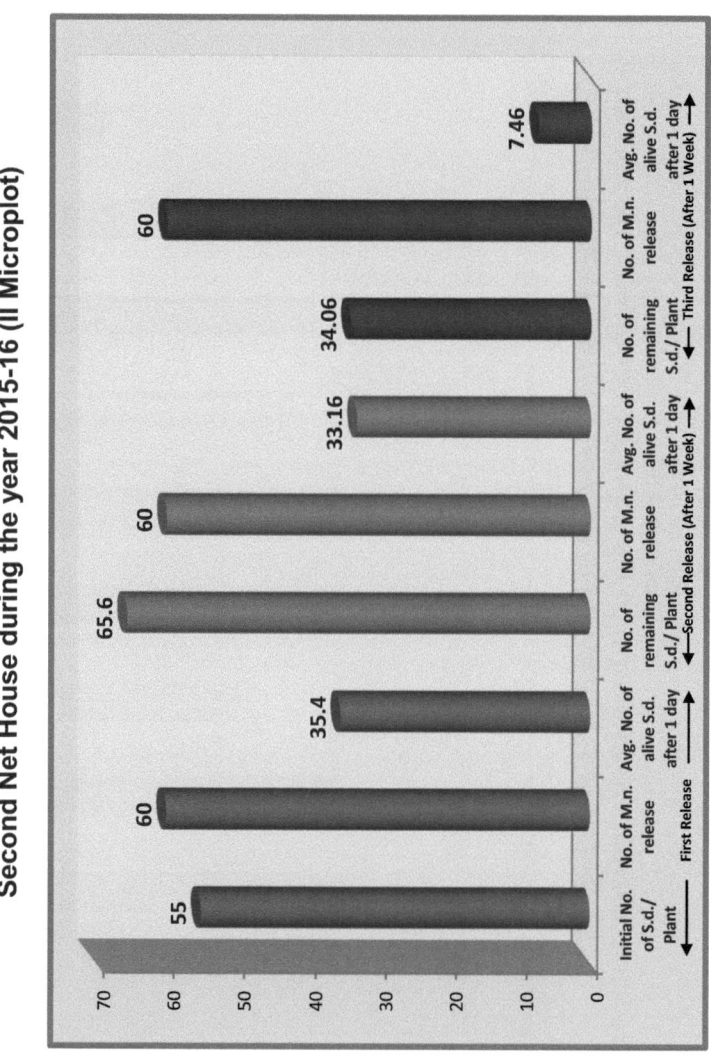

S.d. = *Scirtothrips dorsalis*, M.n = *Macrotracheliella nigra*

Graph 4.52

Effect of Predator; *Macrotracheliella nigra* on the population of *Scirtothrips dorsalis* in Chilli field of Second Net House during the year 2015-16 (III Microplot)

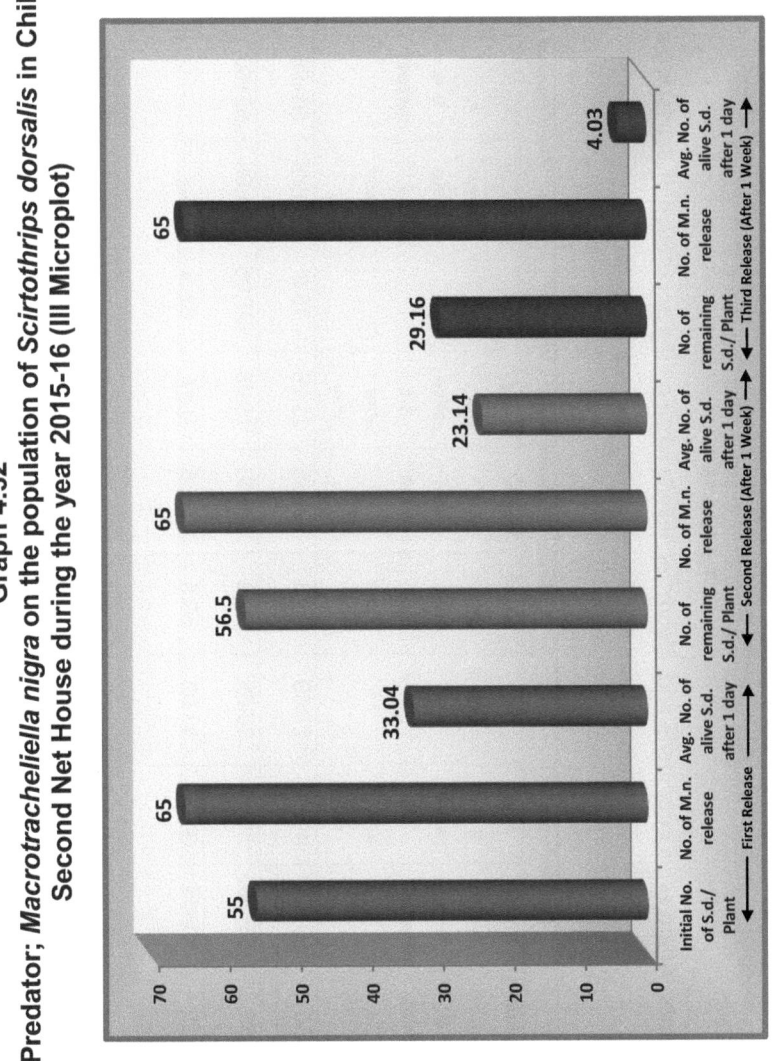

S.d. = *Scirtothrips dorsalis*, M.n = *Macrotracheliella nigra*

Table 4.16
Effect of Predators; Amblyseius cucumeris and Macrotracheliella nigra on the population of Scirtothrips dorsalis in Chilli Field of Third Net House during the year 2015-16

Experimental Microplot	Initial No. of S.d.*/ Plant	First Release			Second Release (After 1 Week)				Third Release (After 1 Week)			
		No. of A.c. + M.n. release**	Avg. No. of alive S.d. after 1 day	% reduction	No. of remaining S.d.*/ Plant	No. of A.c. + M.n. release**	Avg. No. of alive S.d. after 1 day	% reduction	No. of remaining S.d.*/ Plant	No. of A.c. + M.n. release**	Avg. No. of alive S.d. after 1 day	% reduction
I	55	50+55	40.10	27.09	73.3	50+55	21.26	70.9	39.73	50+55	5.25	86.78
II	55	55+60	38.14	30.65	65.6	55+60	15.05	77.05	34.06	55+60	3.13	90.8
III	55	60+65	36.34	33.92	56.5	60+65	9.87	82.5	--	--	--	--

* Mean Value, ** All the number of predators were releases in each micro-plot
S.d. = Scirtothrips dorsalis, A.c. = Amblyseius cucumeris, M.n. = Macrotracheliella nigra

Graph 4.53
Effect of Predators; *Amblyseius cucumeris* and *Macrotracheliella nigra* on the population of *Scirtothrips dorsalis* in Chilli Field of Third Net House during the year 2015-16 (I Microplot)

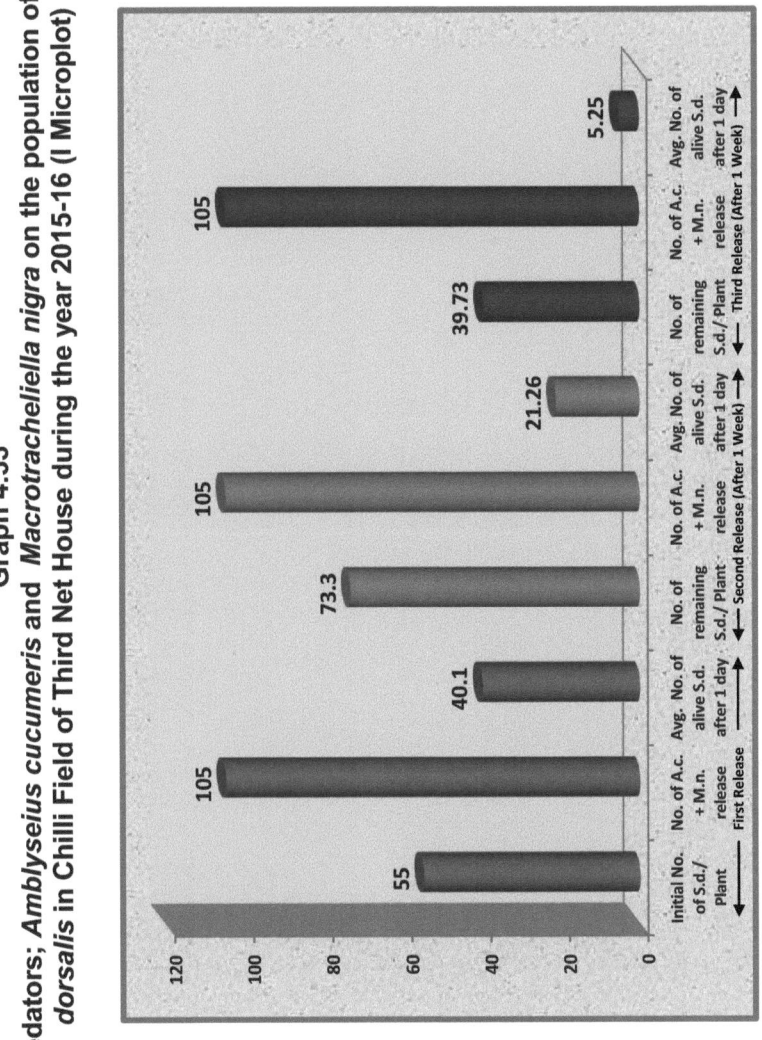

S.d. = *Scirtothrips dorsalis*, A.c. = *Amblyseius cucumeris*, M.n. = *Macrotracheliella nigra*

Graph 4.54

Effect of Predators; *Amblyseius cucumeris* and *Macrotracheliella nigra* on the population of *Scirtothrips dorsalis* in Chilli Field of Third Net House during the year 2015-16 (II Microplot)

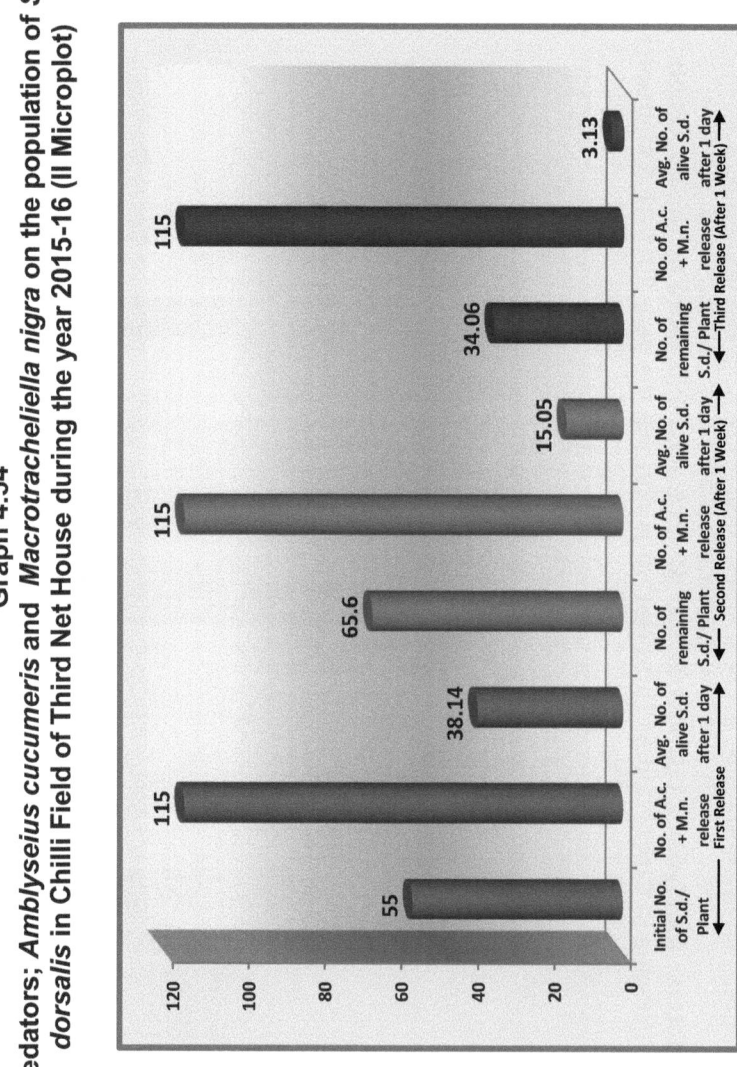

S.d. = *Scirtothrips dorsalis*, A.c. = *Amblyseius cucumeris*, M.n. = *Macrotracheliella nigra*

162 | Chapter 4 | Experimental Analysis

Graph 4.55
Effect of Predators; *Amblyseius cucumeris* and *Macrotracheliella nigra* on the population of *Scirtothrips dorsalis* in Chilli Field of Third Net House during the year 2015-16 (III Microplot)

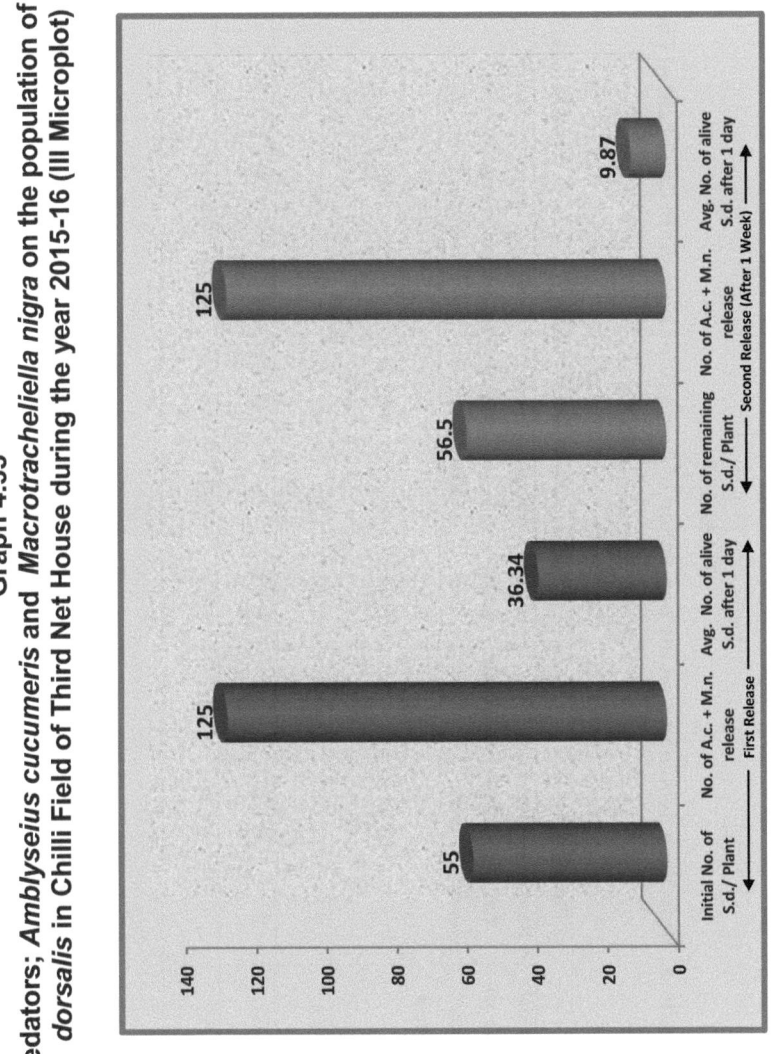

S.d. = *Scirtothrips dorsalis*, A.c. = *Amblyseius cucumeris*, M.n. = *Macrotracheliella nigra*

Table 4.17

Effect of Parasitoid; *Thripobius semiluteus* on the population of *Scirtothrips dorsalis* in Chilli field of Fourth Net House during the year 2015-16

Experimental Microplot	Initial No. of S.d.*/ Plant	First Release			Second Release (After 1 Week)				Third Release (After 1 Week)			
		No. of T.s. release**	Avg. No. of alive S.d. after 3 days	% reduction	No. of remaining S.d.*/ Plant	No. of T.s. release	Avg. No. of alive S.d. after 3 days	% reduction	No. of remaining S.d.*/ Plant	No. of T.s. release	Avg. No. of alive S.d. after 3 days	% reduction
I	55	20	49.19	10.56	73.3	20	43.16	41.1	39.73	20	12.30	69.04
II	55	25	45.30	17.6	65.6	25	32.70	50.15	34.06	25	7.19	78.8
III	55	30	39.36	28.4	56.5	30	26.30	53.4	29.16	30	3.56	87.79

* Mean Value, ** All the number of parasitoids were releases in each micro-plot

S.d. = *Scirtothrips dorsalis*, T.s. = *Thripobius semiluteus*

Graph 4.56
Effect of Parasitoid; *Thripobius semiluteus* on the population of *Scirtothrips dorsalis* in Chilli field of Fourth Net House during the year 2015-16 (I Microplot)

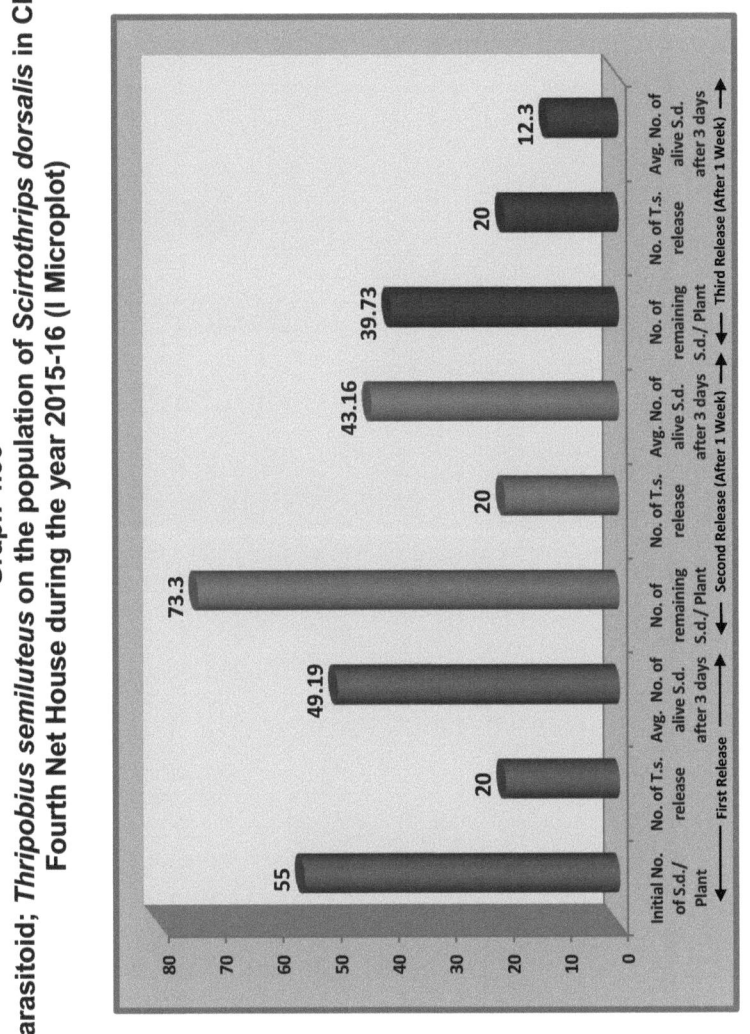

S.d. = *Scirtothrips dorsalis*, T.s. = *Thripobius semiluteus*

Graph 4.57
Effect of Parasitoid; *Thripobius semiluteus* on the population of *Scirtothrips dorsalis* in Chilli field of Fourth Net House during the year 2015-16 (II Microplot)

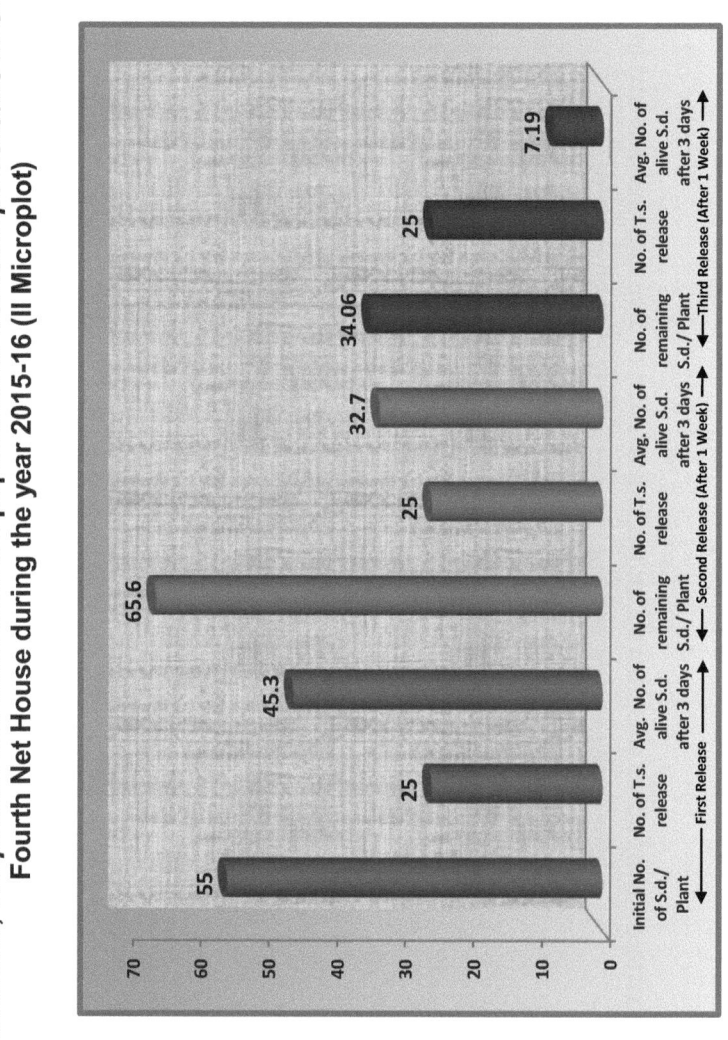

S.d. = *Scirtothrips dorsalis*, T.s. = *Thripobius semiluteus*

Graph 4.58
Effect of Parasitoid; *Thripobius semiluteus* on the population of *Scirtothrips dorsalis* in Chilli field of Fourth Net House during the year 2015-16 (III Microplot)

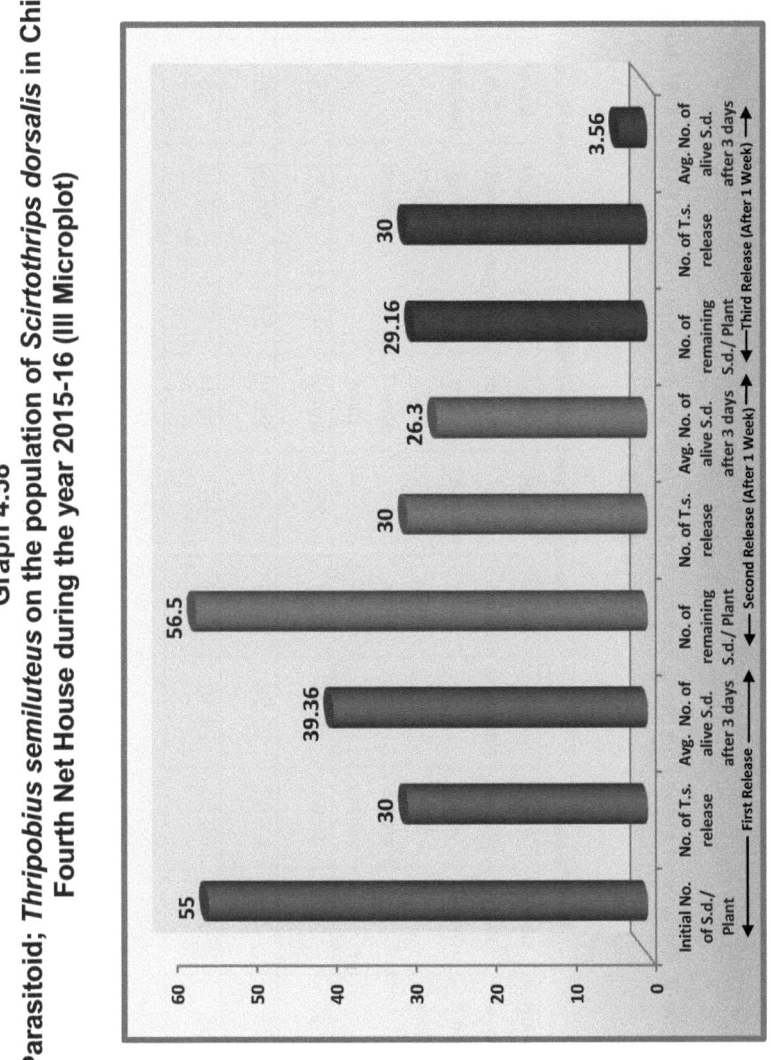

S.d. = *Scirtothrips dorsalis*, T.s. = *Thripobius semiluteus*

Table 4.18

Effect of Parasitoid; *Ceranisus menes* on the population of *Scirtothrips dorsalis* in Chilli field of Fifth Net House during the year 2015-16

Experimental Microplot	First Release				Second Release (After 1 Week)				Third Release (After 1 Week)			
	Initial No. of S.d.*/ Plant	No. of C.m. release**	Avg. No. of alive S.d. after 3 days	% reduction	No. of remaining S.d.*/ Plant	No. of C.m. release**	Avg. No. of alive S.d. after 3 days	% reduction	No. of remaining S.d.*/ Plant	No. of C.m. release**	Avg. No. of alive S.d. after 3 days	% reduction
I	55	25	46.61	15.25	73.3	25	40.16	45.21	39.73	25	13.65	65.64
II	55	30	42.03	23.58	65.6	30	30.10	54.11	34.06	30	7.70	77.3
III	55	35	39.50	28.18	56.5	35	18.34	67.5	29.16	35	4.90	83.1

* Mean Value, ** All the number of parasitoids were releases in each micro-plot

S.d. = *Scirtothrips dorsalis*, C.m. = *Ceranisus menes*

Graph 4.59
Effect of Parasitoid; *Ceranisus menes* on the population of *Scirtothrips dorsalis* in Chilli field of Fifth Net House during the year 2015-16 (I Microplot)

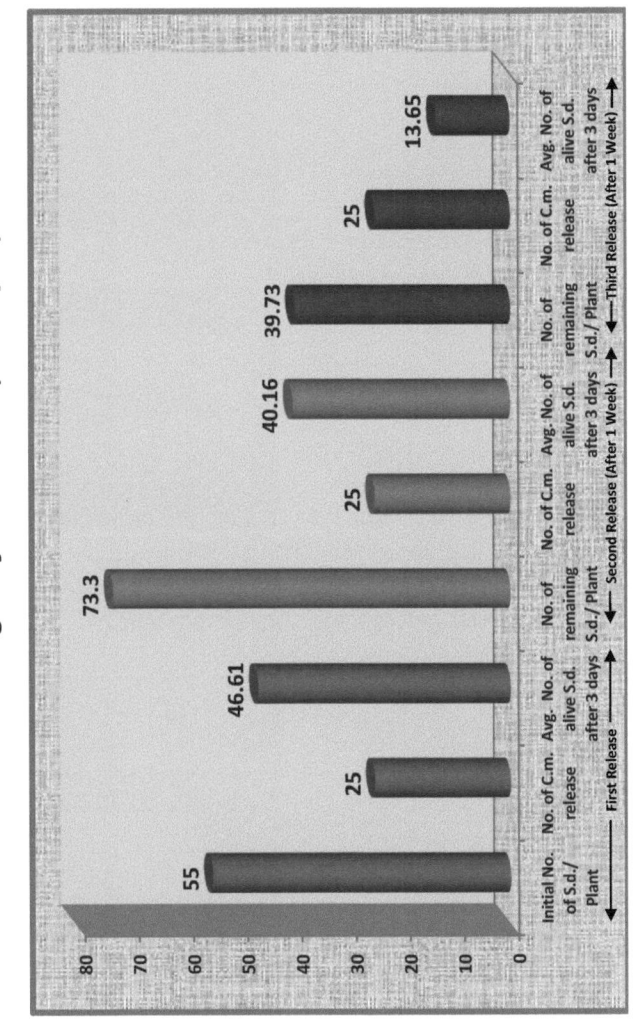

S.d. = *Scirtothrips dorsalis*, C.m. = *Ceranisus menes*

169 | Chapter 4 | Experimental Analysis

Graph 4.60
Effect of Parasitoid; *Ceranisus menes* on the population of *Scirtothrips dorsalis* in Chilli field of Fifth Net House during the year 2015-16 (II Microplot)

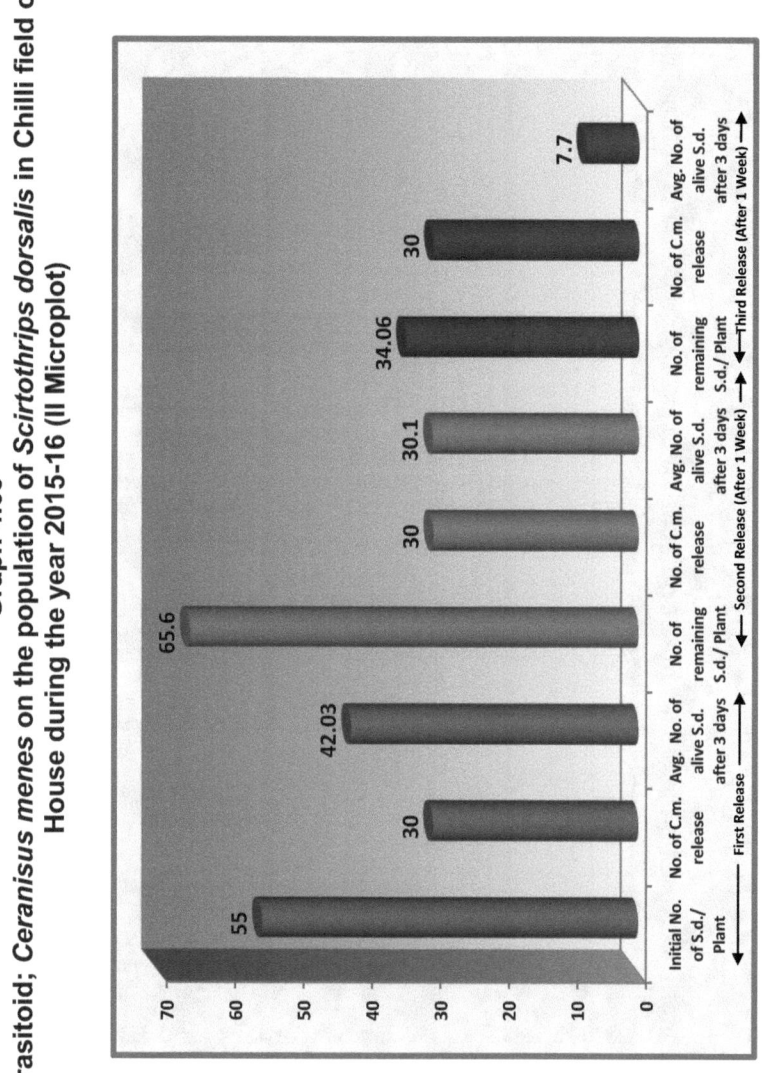

S.d. = *Scirtothrips dorsalis*, C.m. = *Ceranisus menes*

Graph 4.61
Effect of Parasitoid; *Ceranisus menes* on the population of *Scirtothrips dorsalis* in Chilli field of Fifth Net House during the year 2015-16 (III Microplot)

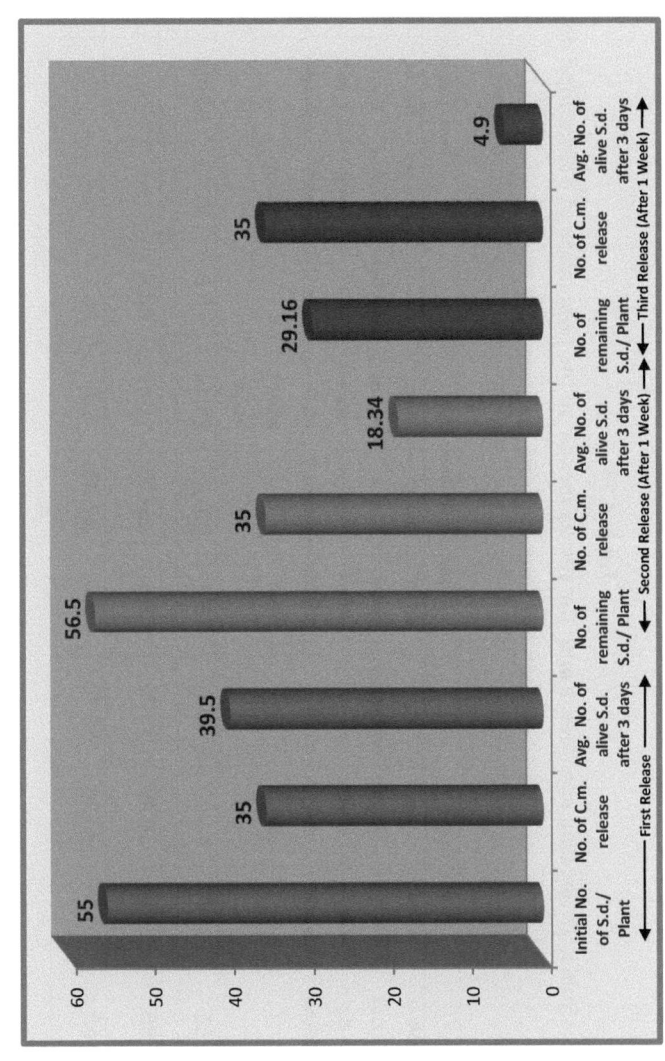

S.d. = *Scirtothrips dorsalis*, C.m. = *Ceranisus menes*

Table 4.19

Effect of Parasitoids; *Thripobius semiluteus* and *Ceranisus menes* on the population of *Scirtothrips dorsalis* in Chilli Field of Sixth Net House during the year 2015-16

Experimental Microplot	Initial No. of S.d.*/ Plant	First Release			Second Release (After 1 Week)				Third Release (After 1 Week)			
		No. of T.s. + C.m. release**	Avg. No. of alive S.d. after 3 day	% reduction	No. of remaining S.d.*/ Plant	No. of T.s. + C.m. release**	Avg. No. of alive S.d. after 3 days	% reduction	No. of remaining S.d.*/ Plant	No. of T.s. + C.m. release**	Avg. No. of alive S.d. after 3 days	% reduction
I	55	20+25	44.20	19.6	73.3	20+25	25.16	65.6	39.73	20+25	5.06	87.26
II	55	25+30	36.45	33.72	65.6	25+30	11.19	82.9	---	---	---	---
III	55	30+35	33.16	39.70	56.5	30+35	8.06	85.7	---	---	---	---

* Mean Value, ** All the number of parasitoids were releases in each micro-plot
S.d. = *Scirtothrips dorsalis*, T.s. = *Thripobius semiluteus*, C.m. = *Ceranisus menes*

Graph 4.62

Effect of Parasitoids; *Thripobius semiluteus* and *Ceranisus menes* on the population of *Scirtothrips dorsalis* in Chilli Field of Sixth Net House during the year 2015-16 (I Microplot)

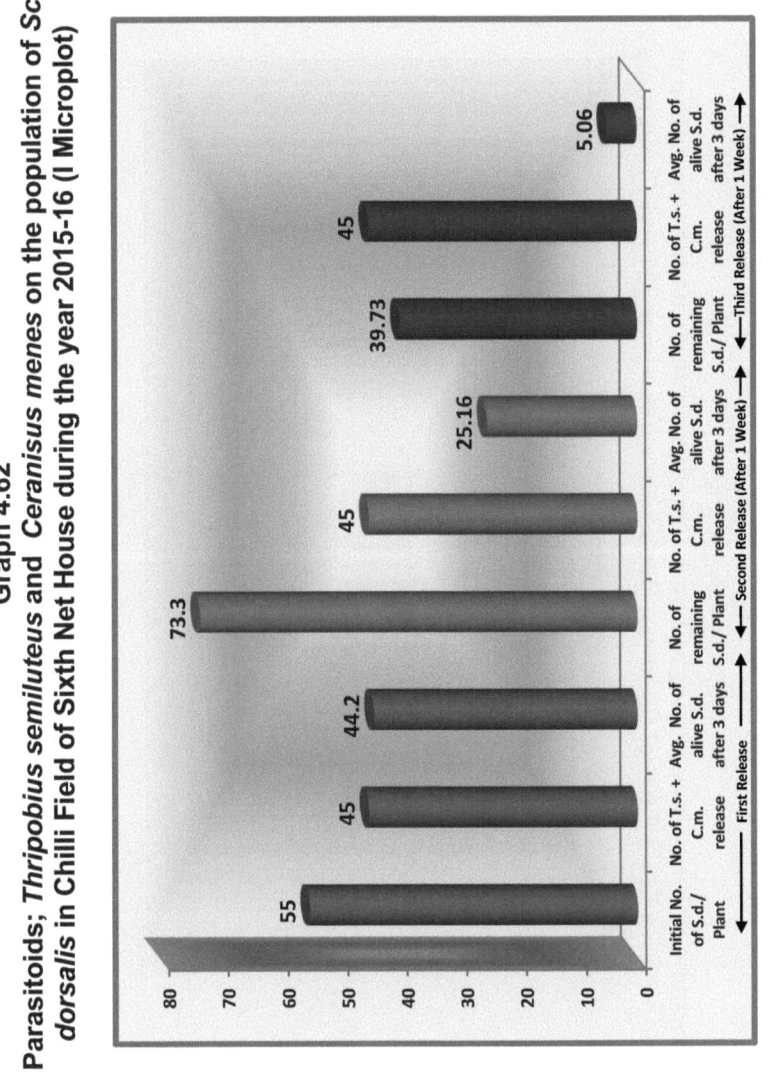

S.d. = *Scirtothrips dorsalis*, T.s. = *Thripobius semiluteus*, C.m. = *Ceranisus menes*

Graph 4.63
Effect of Parasitoids; *Thripobius semiluteus* and *Ceranisus menes* on the population of *Scirtothrips dorsalis* in Chilli Field of Sixth Net House during the year 2015-16 (II Microplot)

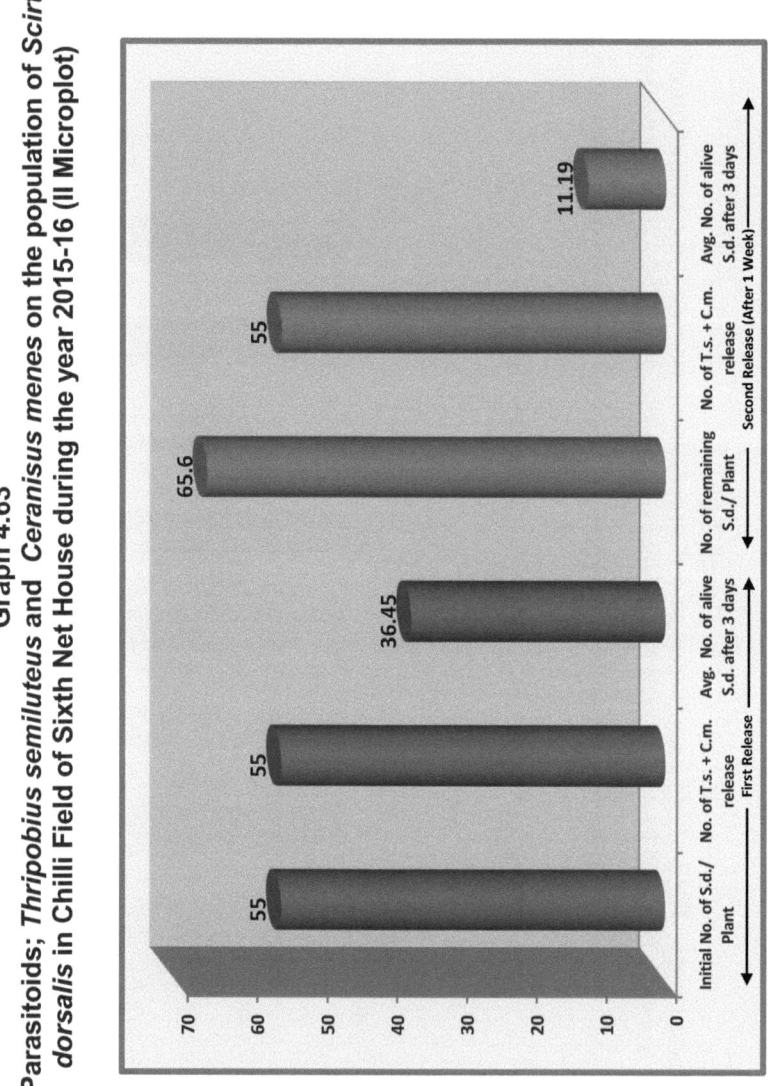

S.d. = *Scirtothrips dorsalis*, T.s. = *Thripobius semiluteus*, C.m. = *Ceranisus menes*

Graph 4.64
Effect of Parasitoids; *Thripobius semiluteus* and *Ceranisus menes* on the population of *Scirtothrips dorsalis* in Chilli Field of Sixth Net House during the year 2015-16 (III Microplot)

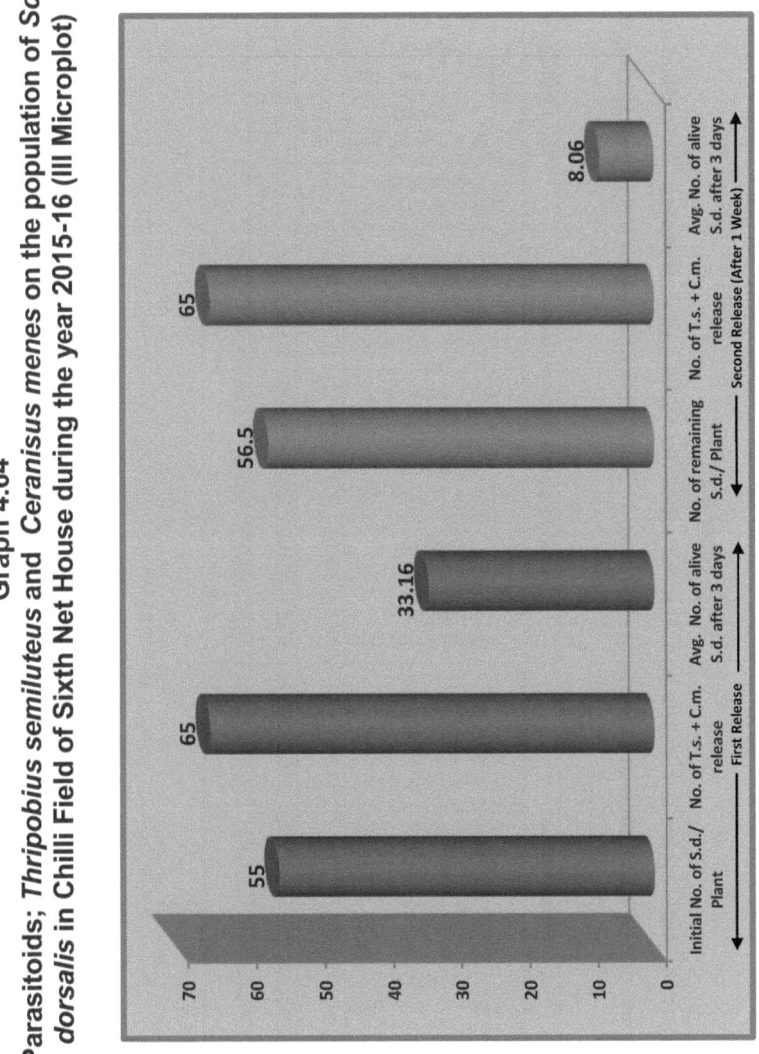

S.d. = *Scirtothrips dorsalis*, T.s. = *Thripobius semiluteus*, C.m. = *Ceranisus menes*

Table 4.20

Effect of Predators (*Amblyseius cucumeris* and *Macrotracheliella nigra*) and parasitoids (*Thripobius semiluteus* and *Ceranisus menes*) on the population of *Scirtothrips dorsalis* in Chilli Field of Seventh Net House during the year 2015-16

Experimental Micro plot	Initial No. of S.d.*/ plant	First Release							Second Release (After 1 Week)							Third Release (After 1 Week)					
		No. of A.c. + M.n. release**	No. of T.s. + C.m. release**	Avg. No. of alive S.d. after 1 day	Avg. No. of alive S.d. after 3 days	% reduction	No. of remaining S.d.*/plant		No. of A.c. + M.n. release**	No. of T.s. + C.m. release**	Avg. No. of alive S.d. after 1 day	Avg. No. of alive S.d. after 3 days	% reduction	No. of remaining S.d.*/plant		No. of A.c. + M.n. release**	No. of T.s. + C.m. release**	Avg. No. of alive S.d. after 1 day	Avg. No. of alive S.d. after 3 days	% reduction
I	55	50+55	20+25	36.10	30.39	44.7	73.3		50+55	20+25	14.10	8.30	88.6	-		-	-	-	-	-
II	55	55+60	25+30	30.36	27.30	50.3	65.6		55+60	25+30	13.06	6.75	89.7	-		-	-	-	-	-
III	55	60+65	30+35	31.32	22.03	59.9	56.5		60+65	30+35	11.13	3.35	94.07	-		-	-	-	-	-

* Mean Value, ** All the number of predators and parasitoids were releases in each micro-plot

S.d. = *Scirtothrips dorsalis*, A.m. = *Amblyseius cucumeris*, M.n. = *Macrotracheliella nigra*

T.S. = *Thripobius semiluteus*, C.m. = *Ceranisus menes*

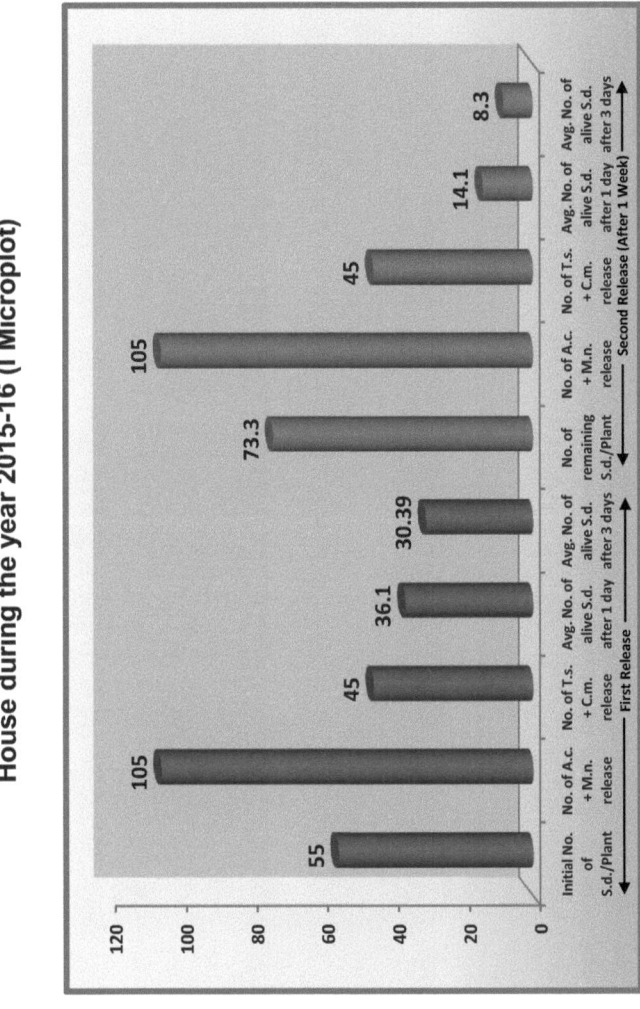

Graph 4.65
Effect of Predators (*Amblyseius cucumeris* and *Macrotracheliella nigra*) and parasitoids (*Thripobius semiluteus* and *Ceranisus menes*) on the population of *Scirtothrips dorsalis* in Chilli Field of Seventh Net House during the year 2015-16 (I Microplot)

S.d. = *Scirtothrips dorsalis*, A.m. = *Amblyseius cucumeris*, M.n. = *Macrotracheliella nigra*
T.S. = *Thripobius semiluteus*, C.m. = *Ceranisus menes*

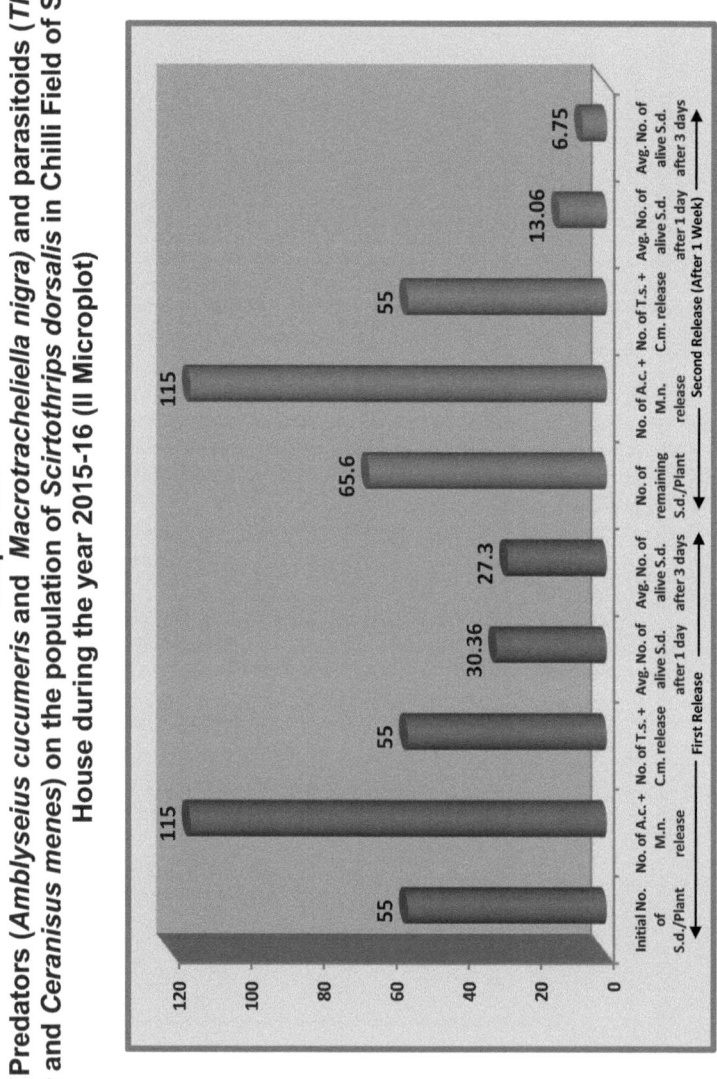

Graph 4.66
Effect of Predators (*Amblyseius cucumeris* and *Macrotracheliella nigra*) and parasitoids (*Thripobius semiluteus* and *Ceranisus menes*) on the population of *Scirtothrips dorsalis* in Chilli Field of Seventh Net House during the year 2015-16 (II Microplot)

S.d. = *Scirtothrips dorsalis*, A.m. = *Amblyseius cucumeris*, M.n. = *Macrotracheliella nigra*
T.S. = *Thripobius semiluteus*, C.m. = *Ceranisus menes*

Graph 4.67

Effect of Predators (*Amblyseius cucumeris* and *Macrotracheliella nigra*) and parasitoids (*Thripobius semiluteus* and *Ceranisus menes*) on the population of *Scirtothrips dorsalis* in Chilli Field of Seventh Net House during the year 2015-16 (III Microplot)

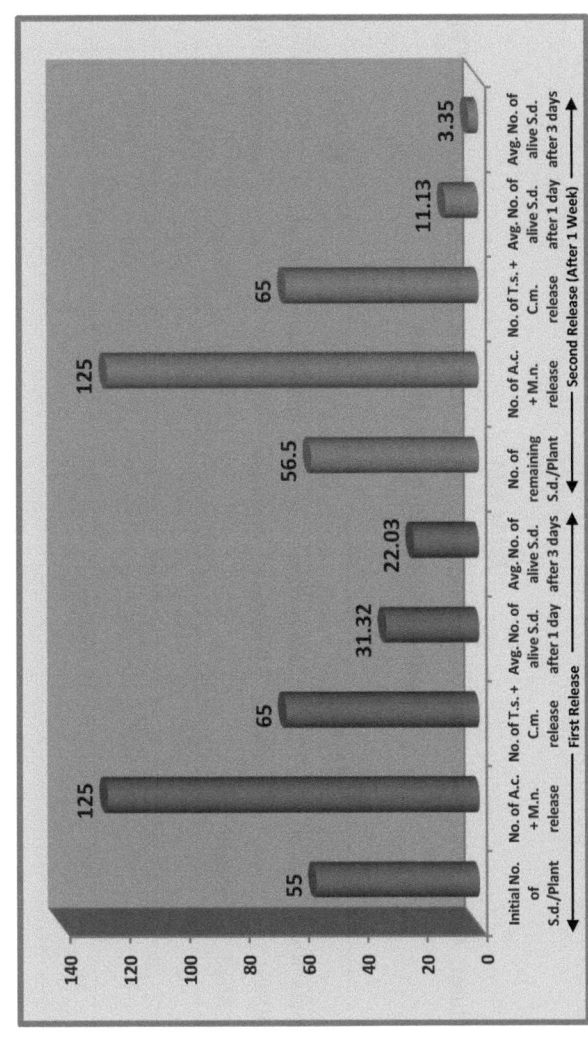

S.d. = *Scirtothrips dorsalis*, A.m. = *Amblyseius cucumeris*, M.n. = *Macrotracheliella nigra*
T.S. = *Thripobius semiluteus*, C.m. = *Ceranisus menes*

4.4 Results of Biological Controlled Experiments During the Year 2014-15 and 2015-16

Table 4.21

Effect of Predator; *Amblyseius cucumeris* on the population of *Scirtothrips dorsalis* in Chilli field of First Net House during the year 2014-15 & 2015-16

Experimental Microplot	2014-15						2015-16					
	Initial No. of S.d.*/ Plant	No. of A.c. release**	% reduction			Initial No. of S.d.*/ Plant	No. of A.c. release**	% reduction				
			I release	II release	III release			I release	II release	III release		
I	45	40	20.8	48.5	68.2	55	50	17.8	45.9	67.12		
II	45	45	27.6	52.4	74.02	55	55	24.8	51.0	75.51		
III	45	50	33.9	59.02	82.32	55	60	30.2	56.8	84.80		

* Mean Value, ** All the number of predators were releases in each micro-plot
S.d. = *Scirtothrips dorsalis*, A.c. = *Amblyseius cucumeris*

Graph 4.68
Effect of Predator; *Amblyseius cucumeris* on the population of *Scirtothrips dorsalis* in Chilli field of First Net House during the year 2014-15 & 2015-16

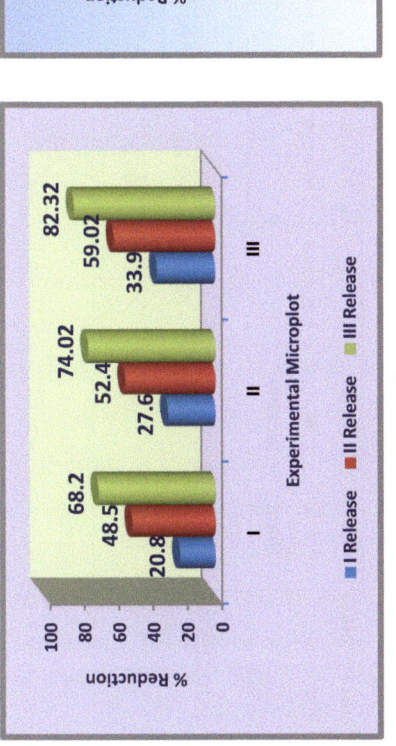

181 | Chapter 4 | Experimental Analysis

Table 4.22

Effect of Predator; *Macrotracheliella nigra* on the population of *Scirtothrips dorsalis* in Chilli field of Second Net House during the year 2014-15 & 2015-16

Experimental Microplot	2014-15					2015-16				
	Initial No. of S.d.*/ Plant	No. of M.n. release**	% reduction			Initial No. of S.d.*/ Plant	No. of M.n. release**	% reduction		
			I release	II release	III release			I release	II release	III release
I	45	45	25.2	43.7	63.01	55	55	32.4	41.8	61.49
II	45	50	32.9	50.23	77.04	55	60	35.6	49.4	78.09
III	45	55	41.1	53.6	85.8	55	65	39.9	59.04	86.1

* Mean Value, ** All the number of predators were releases in each micro-plot

S.d. = *Scirtothrips dorsalis*, M.n. = *Macrotracheliella nigra*

Graph 4.69
Effect of Predator; *Macrotracheliella nigra* on the population of *Scirtothrips dorsalis* in Chilli field of Second Net House during the year 2014-15 & 2015-16

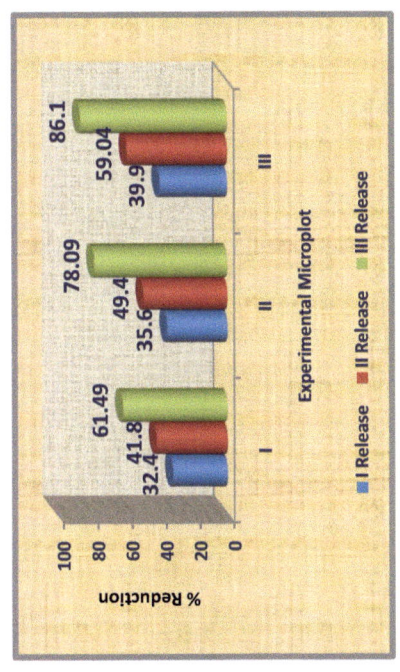

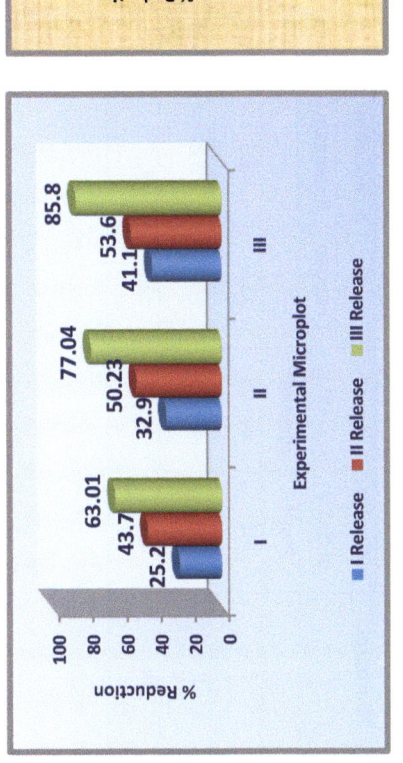

Table 4.23

Effect of Predators; *Amblyseius cucumeris* and *Macrotracheliella nigra* on the population of *Scirtothrips dorsalis* in Chilli field of Third Net House during the year 2014-15 & 2015-16

Experimental Microplot	2014-15						2015-16					
	Initial No. of S.d.*/ Plant	No. of A.c.+M.n. release**	% reduction				Initial No. of S.d.*/ Plant	No. of A.c.+M.n. release**	% reduction			
			I release	II release	III release				I release	II release	III release	
I	45	40+45	23.6	68.3	88.7		55	50+55	27.09	70.9	86.78	
II	45	45+50	26.2	75.8	92.0		55	55+60	30.65	77.05	90.8	
III	45	50+55	32.9	80.7	---		55	60+65	33.92	82.5	---	

* Mean Value, ** All the number of predators were releases in each micro-plot

S.d. = *Scirtothrips dorsalis*, A.c. = *Amblyseius cucumeris*, M.n. = *Macrotracheliella nigra*

Graph 4.70
Effect of Predators; *Amblyseius cucumeris* and *Macrotracheliella nigra* on the population of *Scirtothrips dorsalis* in Chilli field of Third Net House during the year 2014-15 & 2015-16

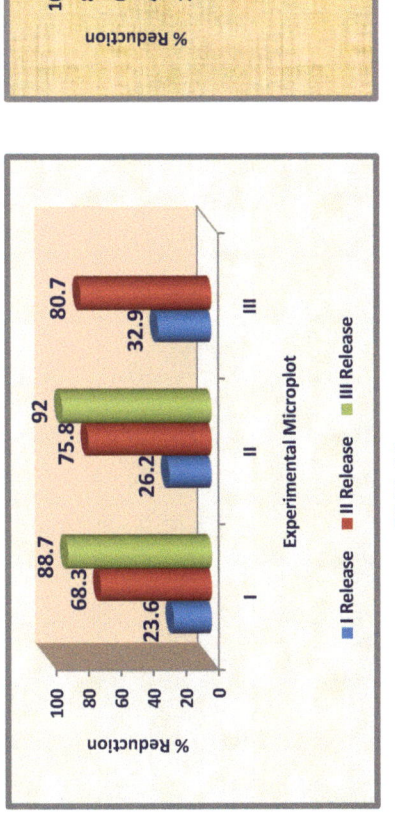

Table 4.24

Effect of Parasitoid; *Thripobius semiluteus* on the population of *Scirtothrips dorsalis* in Chilli field of Fourth Net House during the year 2014-15 & 2015-16

Experimental Microplot	2014-15						2015-16					
	Initial No. of S.d.*/ Plant	No. of T.s. release**	% reduction				Initial No. of S.d.*/ Plant	No. of T.s. release**	% reduction			
			I release	II release	III release				I release	II release	III release	
I	45	15	10.6	43.23	71.4		55	20	10.56	41.1	69.04	
II	45	20	36.68	46.7	80.7		55	25	17.6	50.15	78.8	
III	45	25	32.16	55.6	85.09		55	30	28.4	53.4	87.79	

* Mean Value, ** All the number of parasitoids were releases in each micro-plot
S.d. = *Scirtothrips dorsalis*, T.s. = *Thripobius semiluteus*

Graph 4.71
Effect of Parasitoid; *Thripobius semiluteus* **on the population of** *Scirtothrips dorsalis* **in Chilli field of Fourth Net House during the year 2014-15 & 2015-16**

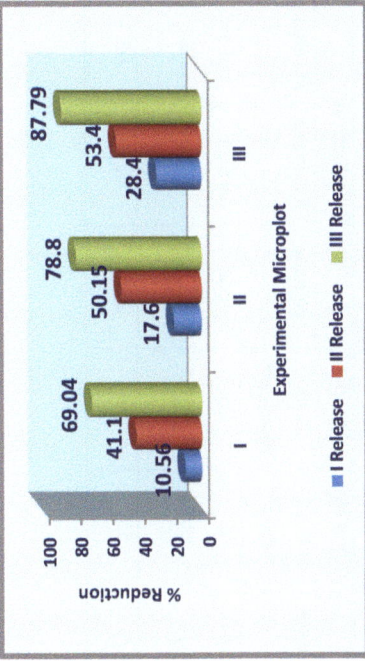

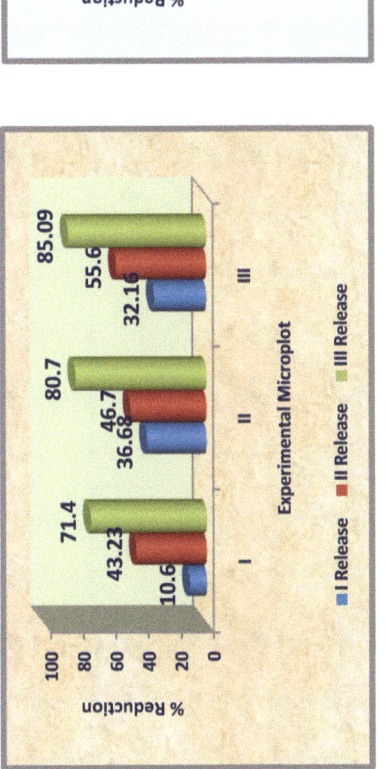

Table 4.25

Effect of Parasitoid; *Ceranisus menes* on the population of *Scirtothrips dorsalis* in Chilli field of Fifth Net House during the year 2014-15 & 2015-16

Experimental Microplot	2014-15					2015-16				
	Initial No. of S.d.*/ Plant	No. of C.m. release**	% reduction			Initial No. of S.d.*/ Plant	No. of C.m. release**	% reduction		
			I release	II release	III release			I release	II release	III release
I	45	20	11.9	52.1	68.4	55	25	15.25	45.21	65.64
II	45	25	21.02	55.3	75.01	55	30	23.58	54.11	77.3
III	45	30	27.93	64.8	78.75	55	35	28.18	67.5	83.1

* Mean Value, ** All the number of parasitoids were releases in each micro-plot

S.d. = *Scirtothrips dorsalis*, C.m. = *Ceranisus menes*

**Graph 4.72
Effect of Parasitoid; *Ceranisus menes* on the population of *Scirtothrips dorsalis* in Chilli field of Fifth Net House during the year 2014-15 & 2015-16**

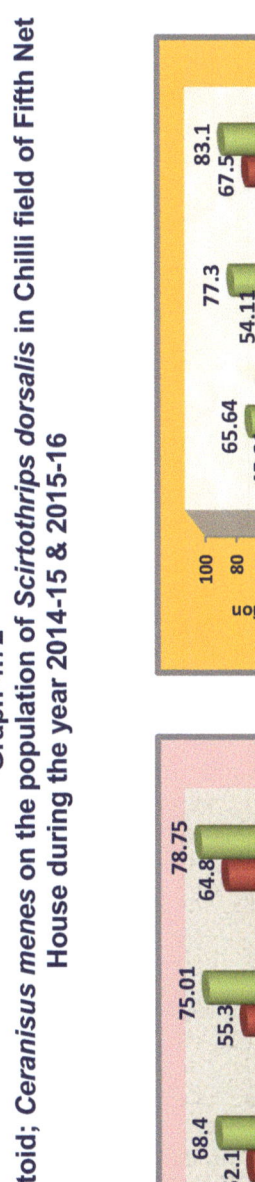
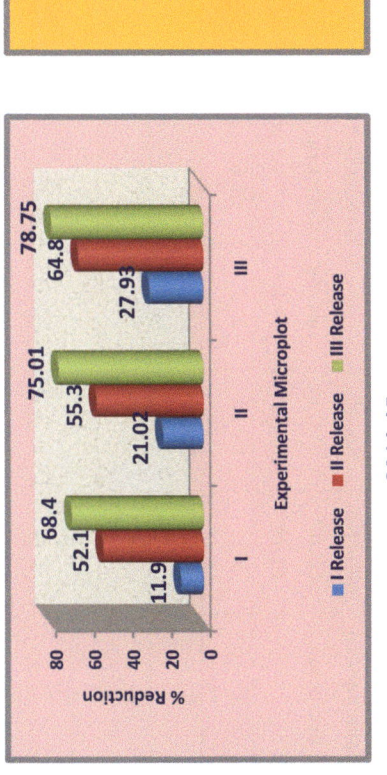

189 | Chapter 4 | Experimental Analysis

Table 4.26

Effect of Parasitoids; *Thripobius semiluteus* and *Ceranisus menes* on the population of *Scirtothrips dorsalis* in Chili field of Sixth Net House during the year 2014-15 & 2015-16

Experimental Microplot	2014-15						2015-16					
	Initial No. of S.d.*/ Plant	No. of T.s.+C.m. release**	% reduction				Initial No. of S.d.*/ Plant	No. of T.s.+C.m. release**	% reduction			
			I release	II release	III release				I release	II release	III release	
I	45	15+20	21.7	66.6	88.05		55	20+25	19.6	65.6	87.26	
II	45	20+25	33.11	83.7	--		55	25+30	33.72	82.9	--	
III	45	25+30	43.7	87.6	--		55	30+35	39.70	85.7	--	

* Mean Value, ** All the number of parasitoids were releases in each micro-plot
S.d. = *Scirtothrips dorsalis*, T.s. = *Thripobius semiluteus*, C.m. = *Ceranisus menes*

Graph 4.73
Effect of Parasitoids; *Thripobius semiluteus* and *Ceranisus menes* on the population of *Scirtothrips dorsalis* in Chill field of Sixth Net House during the year 2014-15 & 2015-16

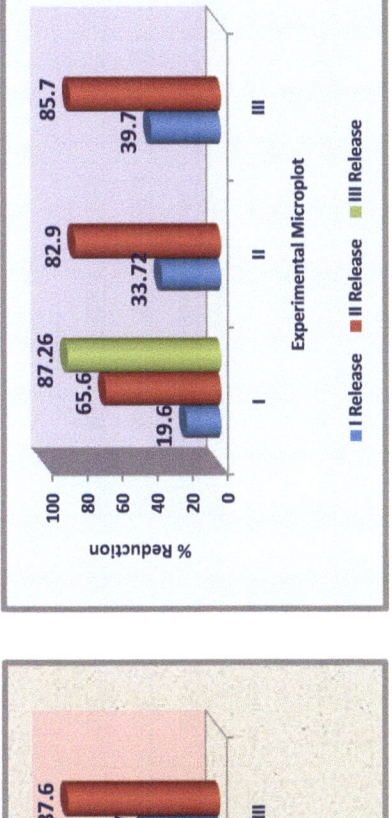

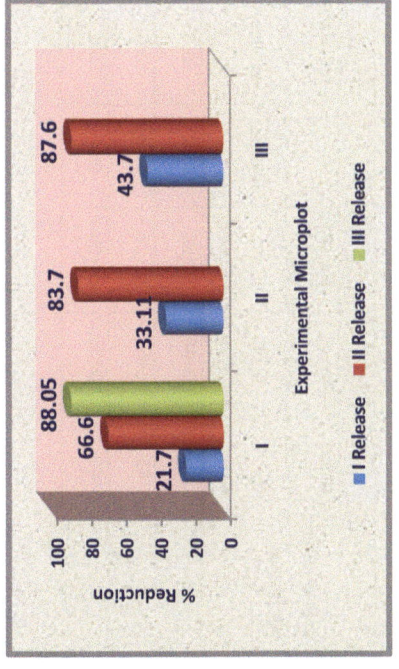

191 | Chapter 4 | Experimental Analysis

Table 4.27

Effect of Predators (*Amblyseius cucumeris* and *Macrotracheliella nigra*) and parasitoids (*Thripobius semiluteus* and *Ceranisus menes*) on the population of thrips *Scirtothrips dorsalis* in Chilli Field of Seventh Net House during the year 2014-15 & 2015-16

Experimental Micro-plot	2014-15						2015-16					
	Initial No. of S.d.*/ Plant	No. of A.c.+ M.n. release**	No. of T.s. + C.m. release**	% reduction			Initial No. of S.d.*/ Plant	No. of A.c.+ M.n. release**	No. of T.s. + C.m. release**	% reduction		
				I release	II release	III release				I release	II release	III release
I	45	40+45	15+20	41.8	86.6	--	55	50+55	20+25	44.7	88.6	--
II	45	45+50	20+25	44.3	91.6	--	55	55+60	25+30	50.3	89.7	--
III	45	50+55	25+30	46.4	93.6	--	55	60+65	30+35	59.9	94.07	--

* Mean Value, ** All the number of predators and parasitoids were releases in each micro-plot

S.d. = *Scirtothrips dorsalis*, A.m. = *Amblyseius cucumeris*, M.n. = *Macrotracheliella nigra*

T.s. = *Thripobius semiluteus*, C.m. = *Ceranisus menes*

Graph 4.74
Effect of Predators (*Amblyseius cucumeris* and *Macrotracheliella nigra*) and parasitoids (*Thripobius semiluteus* and *Ceranisus menes*) on the population of thrips *Scirtothrips dorsalis* in Chilli Field of Seventh Net House during the year 2014-15 & 2015-16

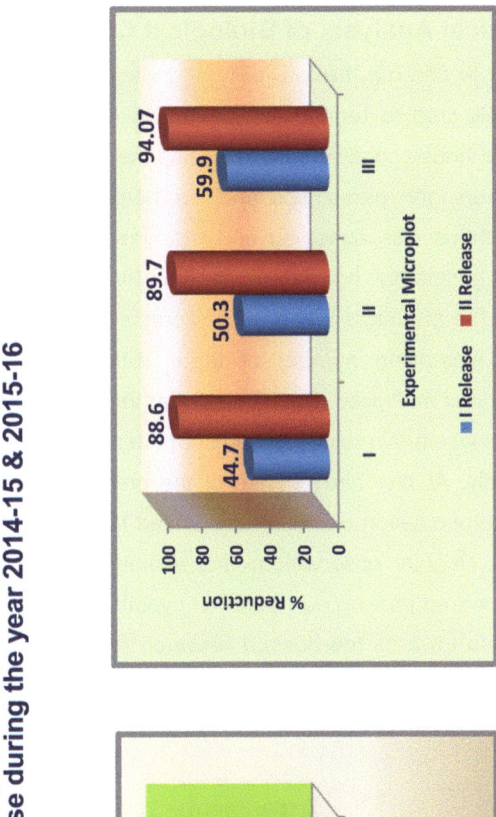

2015-16

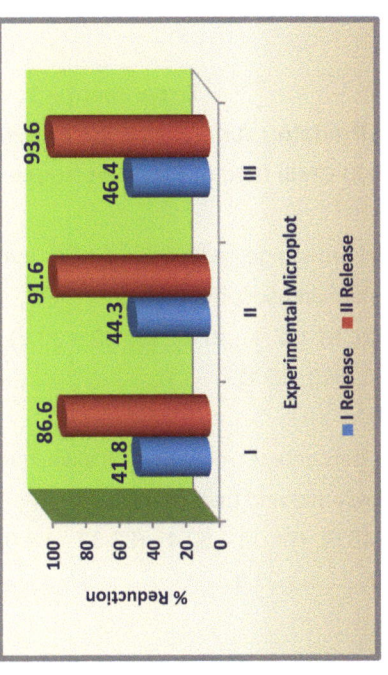

2014-15

193 | Chapter 4 | Experimental Analysis

4.5 Statistical Analysis of Biological Controlled Experiments

In order to find out the feasibility and viability of the biological control method on chilli crop to reduce the infestation of thrips, experiments were made in the net house conditions by releasing the natural enemies of thrips of course, predators and parasitoids and the percentage of reduction in the population of thrips was observed and examined. The population of thrips started reducing instantly, but after some time the infestation of thrips began to increase as the predators and parasitoids as selected in number could not consume the increasing number of thrips with as much speed as the population of thrips multiplied. So, a week after the equal number of predators and parasitoids was again released and then the population of thrips began to decrease rapidly. At the third release of the predators and parasitoids, the reduction in the population of thrips accelerated fast, and, thus, viable results were obtained. Regular observations and examinations produced significant data which appeared proving our research hypothesis. In order to confirm the progressive data towards the desired research results, statistical analysis of the data with T-test was done as in the following pages and thus the results were obtained.

Experiment – 1

Effect of Predator; *Amblyseius cucumeris* on the population of *S. dorsalis* in Chilli field of First Net House during the year 2014-15

Sample – 1

Gain in percentage reduction of alive number of *Scirtothrips dorsalis* during first release and second release.

(48.5 – 20.8), (52.4 – 27.6), (59.02 – 33.9)

x_i = 27.7, 24.8, 25.12

Sample – 2

Gain in percentage reduction of alive number of *Scirtothrips dorsalis* during second release and third release.

(68.2 – 48.5), (74.02 – 52.4), (82.32 – 59.02)

y_i = 19.7, 21.62, 23.3

$\bar{X}$ (Mean of 1st Sample) = $\dfrac{27.7+24.8+25.12}{3}$

$\bar{X}$ = $\dfrac{77.62}{3}$

$\bar{X}$ = 25.8

$\bar{Y}$ (Mean of 2nd Sample) = $\dfrac{19.7+21.62+23.3}{3}$

$\bar{Y}$ = $\dfrac{64.62}{3}$

$\bar{Y}$ = 21.54

The Sum of Square Deviation

$\sum (x_i - \bar{X})^2$ = $(27.7 - 25.8)^2 + (24.8 - 25.8)^2 + (25.12 - 25.8)^2$

= $(1.9)^2 + (1)^2 + (0.68)^2$

$\sum (x_i - \bar{X})^2$ = 5.07

$\sum (y_i - \bar{Y})^2$ = $(19.7 - 21.54)^2 + (21.62 - 21.54)^2 + (23.3 - 21.54)^2$

= $(1.84)^2 + (0.08)^2 + (1.76)^2$

$\sum (y_i - \bar{Y})^2$ = 6.5

S = Standard Deviation

$S = \sqrt{\dfrac{\sum(x_i-\bar{X})^2 + \sum(y_i-\bar{Y})^2}{n_1+n_2-2}}$

Degree of Freedom

d.f = $n_1 + n_2 - 2$

$S = \sqrt{\dfrac{5.07+6.5}{3+3-2}}$

S = 1.7

Null Hypothesis (H_0)

Assume that difference between gain in percentage reduction of alive *Scirtothrips dorsalis* during two successive releases is not significantly different.

$$|t| = \frac{\bar{X} - \bar{Y}}{S\sqrt{\frac{1}{n_1} + \frac{1}{n_2}}}$$

$$|t| = \frac{(25.8 - 21.54)}{1.7\sqrt{\frac{1}{3} + \frac{1}{3}}}$$

$$|t| = 3.08$$

Conclusion

But tabulated $|t|_{0.05}$ for Degree of Freedom 4 is 2.78 which is smaller than 3.08. Hence, Null Hypothesis (H_0) is accepted at 5% level of significance.

∴The gain in % reduction of the two samples is not significantly different and it is accepted at 5% level.

Thus, we conclude statistically that the percentage gain in *Scirtothrips dorsalis* reduction is affected by the release of predators; *Amblyseius cucumeris*, during two successive releases.

Experiment – 2

Effect of Predator; *Macrotracheliella nigra* on the population of *S. dorsalis* in Chilli field of Second Net House during the year 2014-15

Sample – 1

Gain in percentage reduction of alive number of *Scirtothrips dorsalis* during first release and second release.

x_i = 18.5, 17.33, 12.5

Sample – 2

Gain in percentage reduction of alive number of *Scirtothrips dorsalis* during second release and third release.

y_i = 19.31, 26.81, 32.2

$\bar{X}$(Mean of 1st Sample) = 16.11

$\bar{Y}$(Mean of 2nd Sample) = 26.17

The Sum of Square Deviation

$\sum (x_i - \bar{X})^2$ = 20.23

$\sum (y_i - \bar{Y})^2$ = 83.86

S = Standard Deviation

S = 5.10   d.f = 4

Null Hypothesis (H_0)

Assume that difference between gain in percentage reduction of alive *Scirtothrips dorsalis* during two successive releases is not significantly different.

$|t|$ = 2.43

Conclusion

But tabulated $|t|_{0.05}$ for Degree of Freedom 4 is 2.78 which is equal to the calculated value 2.43. Hence, Null Hypothesis (H_0) is accepted at 5% level of significance.

∴ The gain in % reduction of the two samples is not significantly different and it is accepted at 5% level.

Thus, we conclude statistically that the percentage gain in *Scirtothrips dorsalis* reduction is affected by the release of predators; *Macrotracheliella nigra*, during two successive releases.

Experiment – 3
Effect of Predators; *Amblyseius cucumeris* and *Macrotrachelielia nigra* on the population of *S. dorsalis* in Chilli field of Third Net House during the year 2014-15

Sample – 1

Gain in percentage reduction of alive number of *Scirtothrips dorsalis* during first release and second release.

x_i = 44.7, 49.6, 47.8

Sample – 2

Gain in percentage reduction of alive number of *Scirtothrips dorsalis* during second release and third release.

$$y_i \quad = \quad 20.4, 16.2$$

$\bar{X}$(Mean of 1st Sample) = 47.37

$\bar{Y}$(Mean of 2nd Sample) = 18.3

The Sum of Square Deviation

$$\sum (x_i - \bar{X})^2 \quad = \quad 12.28$$
$$\sum (y_i - \bar{Y})^2 \quad = \quad 8.82$$

S = Standard Deviation

$$S \quad = \quad 8.72 \qquad d.f \quad = \quad 3$$

Null Hypothesis (H$_0$)

Assume that difference between gain in percentage reduction of alive *Scirtothrips dorsalis* during two successive releases is not significantly different.

$$|t| \quad = \quad 3.65$$

Conclusion

But tabulated $|t|_{0.05}$ for Degree of Freedom 3 is 3.18 which is smaller than the calculated value 2.43. Hence, Null Hypothesis (H$_0$) is accepted at 5% level of significance.

∴ The gain in % reduction of the two samples is not significantly different and it is accepted at 5% level.

Thus, we conclude statistically that the percentage gain in *Scirtothrips dorsalis* reduction is affected by the release of both the predators; *Amblyseius cucumeris* and *Macrotracheliella nigra*, during two successive releases.

Experiment – 4
Effect of Parasitoid; *Thripobius semiluteus* on the population of *S. dorsalis* in Chilli field of Fourth Net House during the year 2014-15

Sample – 1

Gain in percentage reduction of alive number of *Scirtothrips dorsalis* during first release and second release.

x_i = 32.63, 28.3, 27.1

Sample – 2

Gain in percentage reduction of alive number of *Scirtothrips dorsalis* during second release and third release.

y_i = 28.17, 34, 29.49

$\bar{X}$(Mean of 1st Sample) = 29.34

$\bar{Y}$(Mean of 2nd Sample) = 30.5

The Sum of Square Deviation

$\sum (x_i - \bar{X})^2$ = 16.89

$\sum (y_i - \bar{Y})^2$ = 18.67

S = Standard Deviation

S = 2.9   d.f = 4

Null Hypothesis (H$_0$)

Assume that difference between gain in percentage reduction of alive *Scirtothrips dorsalis* during two successive releases is not significantly different.

$|t|$ = 0.49

Conclusion

But tabulated $|t|_{0.05}$ for Degree of Freedom 4 is 2.78 which is greater than the calculated value 0.49. Hence, Null Hypothesis (H$_0$) is not accepted at 5% level of significance.

∴ The gain in % reduction of the two samples is significantly different and it is not accepted at 5% level.

Thus, we conclude statistically that the percentage gain in *Scirtothrips dorsalis* reduction is not affected by the release of parasitoids; *Thripobius semiluteus* during two successive releases.

Experiment – 5
Effect of Parasitoid; *Ceranisus menes* on the population of *S. dorsalis* in Chilli field of Fifth Net House during the year 2014-15

Sample – 1

Gain in percentage reduction of alive number of *Scirtothrips dorsalis* during first release and second release.

x_i = 40.2, 34.28, 36.87

Sample – 2

Gain in percentage reduction of alive number of *Scirtothrips dorsalis* during second release and third release.

y_i = 16.3, 19.71, 13.95

$\bar{X}$(Mean of 1st Sample) = 37.11

$\bar{Y}$(Mean of 2nd Sample) = 16.65

The Sum of Square Deviation

$\sum (x_i - \bar{X})^2$ = 17.61

$\sum (y_i - \bar{Y})^2$ = 16.77

S = Standard Deviation

S = 2.93   d.f = 4

Null Hypothesis (H₀)

Assume that difference between gain in percentage reduction of alive *Scirtothrips dorsalis* during two successive releases is not significantly different.

|t| = 8.6

Conclusion

But tabulated $|t|_{0.05}$ for Degree of Freedom 4 is 2.78 which is smaller than the calculated value 8.6. Hence, Null Hypothesis (H_0) is accepted at 5% level of significance.

∴ The gain in % reduction of the two samples is not significantly different and it is accepted at 5% level.

Thus, we conclude statistically that the percentage gain in *Scirtothrips dorsalis* reduction is affected by the release of parasitoids; *Ceranisus menes* during two successive releases.

Experiment – 6

Effect of Parasitoids; *Thripobius semiluteus* and *Ceranisus menes* on the population of *S. dorsalis* in Chilli field of Sixth Net House during the year 2014-15

Sample – 1

Gain in percentage reduction of alive number of *Scirtothrips dorsalis* during first release and second release.

x_i = 44.9, 50.59, 43.9

Sample – 2

Gain in percentage reduction of alive number of *Scirtothrips dorsalis* during second release and third release.

y_i = 21.45

$\bar{X}$(Mean of 1st Sample) = 46.46

$\bar{Y}$(Mean of 2nd Sample) = 21.45

The Sum of Square Deviation

$\sum (x_i - \bar{X})^2$ = 26.04

$\sum (y_i - \bar{Y})^2$ = 0.0

S = Standard Deviation

S = 2.61   d.f = 2

Null Hypothesis (H_0)

Assume that difference between gain in percentage reduction of alive *Scirtothrips dorsalis* during two successive releases is not significantly different.

$$|t| = 6.007$$

Conclusion

But tabulated $|t|_{0.05}$ for Degree of Freedom 2 is 4.30 which is smaller than the calculated value 6.007. Hence, Null Hypothesis (H_0) is accepted at 5% level of significance.

∴ The gain in % reduction of the two samples is not significantly different and it is accepted at 5% level.

Thus, we conclude statistically that the percentage gain in *Scirtothrips dorsalis* reduction is affected by the release of both the parasitoids; *Thripobius semiluteus* and *Ceranisus menes* during two successive releases.

Experiment – 7
Effect of Predators (*Amblyseius cucumeris* and *Macrotracheliella nigra*) and Parasitoids (*Thripobius semiluteus* and *Ceranisus menes*) on the population of *S. dorsalis* in Chilli field of Seventh Net House during the year 2014-15

Sample – 1

Gain in percentage reduction of alive number of *Scirtothrips dorsalis* during first release and second release.

$$x_i = 41.8, 44.3, 46.4$$

Sample – 2

Gain in percentage reduction of alive number of *Scirtothrips dorsalis* during second release and third release.

$$y_i = 86.6, 91.6, 93.6$$

$\bar{X}$(Mean of 1st Sample) = 44.1

$\bar{Y}$(Mean of 2nd Sample) = 90.6

The Sum of Square Deviation

$\sum (x_i - \bar{X})^2$ = 10.62

$\sum (y_i - \bar{Y})^2$ = 26

S = Standard Deviation

S = 3.02   d.f = 4

Null Hypothesis (H$_0$)

Assume that difference between gain in percentage reduction of alive *Scirtothrips dorsalis* during two successive releases is not significantly different.

$|t|$ = 18.95

Conclusion

But tabulated $|t|_{0.05}$ for Degree of Freedom 4 is 2.78 which is smaller than the calculated value 18.95. Hence, Null Hypothesis (H$_0$) is accepted at 5% level of significance.

∴ The gain in % reduction of the two samples is not significantly different and it is accepted at 5% level.

Thus, we conclude statistically that the percentage gain in *Scirtothrips dorsalis* reduction is affected by the release of both the natural enemies : predators (*Amblyseius cucumeris* and *Macrotracheliella nigra*) and parasitoids (*Thripobius semiluteus* and *Ceranisus menes*) during one successive release.

Experiment – 8

Effect of Predator; *Amblyseius cucumeris* on the population of *S. dorsalis* in Chilli field of First Net House during the year 2015-16

Sample – 1

Gain in percentage reduction of alive number of *Scirtothrips dorsalis* during first release and second release.

x_i = 28.1, 26.2, 26.6

Sample – 2

Gain in percentage reduction of alive number of *Scirtothrips dorsalis* during second release and third release.

y_i = 21.22, 24.51, 28

$\bar{X}$(Mean of 1st Sample) = 26.9

$\bar{Y}$(Mean of 2nd Sample) = 24.57

The Sum of Square Deviation

$\sum (x_i - \bar{X})^2$ = 1.93

$\sum (y_i - \bar{Y})^2$ = 23.02

S = Standard Deviation

S = 2.5   d.f = 4

Null Hypothesis (H$_0$)

Assume that difference between gain in percentage reduction of alive *Scirtothrips dorsalis* during two successive releases is not significantly different.

$|t|$ = 2.87

Conclusion

But tabulated $|t|_{0.05}$ for Degree of Freedom 4 is 2.78 which is smaller than the calculated value 2.87. Hence, Null Hypothesis (H$_0$) is accepted at 5% level of significance.

∴ The gain in % reduction of the two samples is not significantly different and it is accepted at 5% level.

Thus, we conclude statistically that the percentage gain in *Scirtothrips dorsalis* reduction is affected by the release of predators; *Amblyseius cucumeris*, during two successive releases.

Experiment – 9

Effect of Predator; *Macrotracheliella nigra* on the population of S. dorsalis in Chilli field of Second Net House during the year 2015-16

Sample – 1

Gain in percentage reduction of alive number of *Scirtothrips dorsalis* during first release and second release.

x_i = 9.4, 13.8, 19.14

Sample – 2

Gain in percentage reduction of alive number of *Scirtothrips dorsalis* during second release and third release.

y_i = 19.69, 28.69, 27.06

$\bar{X}$(Mean of 1st Sample) = 14.11

$\bar{Y}$(Mean of 2nd Sample) = 25.14

The Sum of Square Deviation

$\sum (x_i - \bar{X})^2$ = 47.69

$\sum (y_i - \bar{Y})^2$ = 45.98

S = Standard Deviation

S = 4.84   d.f = 4

Null Hypothesis (H_0)

Assume that difference between gain in percentage reduction of alive *Scirtothrips dorsalis* during two successive releases is not significantly different.

$|t|$ = 2.82

Conclusion

But tabulated $|t|_{0.05}$ for Degree of Freedom 4 is 2.78 which is smaller than the calculated value 2.82. Hence, Null Hypothesis (H_0) is accepted at 5% level of significance.

∴ The gain in % reduction of the two samples is not significantly different and it is accepted at 5% level.

Thus, we conclude statistically that the percentage gain in *Scirtothrips dorsalis* reduction is affected by the release of predators; *Macrotracheliella nigra*, during two successive releases.

Experiment – 10

Effect of Predators; *Amblyseius cucumeris* and *Macrotracheliella nigra* on the population of *S. dorsalis* in Chilli field of Third Net House during the year 2015-16

Sample – 1

Gain in percentage reduction of alive number of *Scirtothrips dorsalis* during first release and second release.

x_i = 43.81, 46.4, 48.58

Sample – 2

Gain in percentage reduction of alive number of *Scirtothrips dorsalis* during second release and third release.

y_i = 15.9, 13.75

$\bar{X}$(Mean of 1st Sample) = 46.3

$\bar{Y}$(Mean of 2nd Sample) = 14.9

The Sum of Square Deviation

$\sum (x_i - \bar{X})^2$ = 11.4

$\sum (y_i - \bar{Y})^2$ = 2.32

S = Standard Deviation

$$S = 4.6 \qquad d.f = 3$$

Null Hypothesis (H_0)

Assume that difference between gain in percentage reduction of alive *Scirtothrips dorsalis* during two successive releases is not significantly different.

$$|t| = 7.49$$

Conclusion

But tabulated $|t|_{0.05}$ for Degree of Freedom 3 is 3.18 which is smaller than the calculated value 7.49. Hence, Null Hypothesis (H_0) is accepted at 5% level of significance.

∴ The gain in % reduction of the two samples is not significantly different and it is accepted at 5% level.

Thus, we conclude statistically that the percentage gain in *Scirtothrips dorsalis* reduction is affected by the release of both the predators; *Amblyseius cucumeris* and *Macrotracheliella nigra*, during two successive releases.

Experiment – 11
Effect of Parasitoid; *Thripobius semiluteus* on the population of *S. dorsalis* in Chilli field of Fourth Net House during the year 2015-16

Sample – 1

Gain in percentage reduction of alive number of *Scirtothrips dorsalis* during first release and second release.

$$x_i = 30.54, 32.6, 32.55$$

Sample – 2

Gain in percentage reduction of alive number of *Scirtothrips dorsalis* during second release and third release.

$$y_i = 27.94, 28.65, 34.39$$

$\bar{X}$(Mean of 1st Sample) = 31.9

$\bar{Y}$(Mean of 2nd Sample) = 30.3

The Sum of Square Deviation
$$\sum (x_i - \bar{X})^2 = 2.75$$
$$\sum (y_i - \bar{Y})^2 = 25.06$$

S = Standard Deviation
S = 2.63   d.f = 4

Null Hypothesis (H$_0$)
Assume that difference between gain in percentage reduction of alive *Scirtothrips dorsalis* during two successive releases is not significantly different.
$$|t| = 0.72$$

Conclusion
But tabulated $|t|_{0.05}$ for Degree of Freedom 4 is 2.78 which is greater than the calculated value 0.72. Hence, Null Hypothesis (H$_0$) is not accepted at 5% level of significance.

∴ The gain in % reduction of the two samples is significantly different and it is not accepted at 5% level.

Thus, we conclude statistically that the percentage gain in *Scirtothrips dorsalis* reduction is not affected by the release of parasitoids; *Thripobius semiluteus* during two successive releases.

Experiment – 12
Effect of Parasitoid; *Ceranisus menes* on the population of *S. dorsalis* in Chilli field of Fifth Net House during the year 2015-16

Sample – 1
Gain in percentage reduction of alive number of *Scirtothrips dorsalis* during first release and second release.

x$_i$ = 29.96, 30.53, 39.32

Sample – 2

Gain in percentage reduction of alive number of *Scirtothrips dorsalis* during second release and third release.

y_i = 20.43, 23.19, 15.6

$\bar{X}$(Mean of 1st Sample) = 33.27

$\bar{Y}$(Mean of 2nd Sample) = 19.74

The Sum of Square Deviation

$\sum (x_i - \bar{X})^2$ = 55

$\sum (y_i - \bar{Y})^2$ = 29.47

S = Standard Deviation

S = 4.6   d.f = 4

Null Hypothesis (H$_0$)

Assume that difference between gain in percentage reduction of alive *Scirtothrips dorsalis* during two successive releases is not significantly different.

$|t|$ = 16.5

Conclusion

But tabulated $|t|_{0.05}$ for Degree of Freedom 4 is 2.78 which is smaller than the calculated value 16.5. Hence, Null Hypothesis (H$_0$) is accepted at 5% level of significance.

∴ The gain in % reduction of the two samples is not significantly different and it is accepted at 5% level.

Thus, we conclude statistically that the percentage gain in *Scirtothrips dorsalis* reduction is affected by the release of parasitoids; *Ceranisus menes* during two successive releases.

Experiment – 13

Effect of Parasitoids; *Thripobius semiluteus* and *Ceranisus menes* on the population of *S. dorsalis* in Chilli field of Sixth Net House during the year 2015-16

Sample – 1

Gain in percentage reduction of alive number of *Scirtothrips dorsalis* during first release and second release.

x_i = 46, 49.18, 46

Sample – 2

Gain in percentage reduction of alive number of *Scirtothrips dorsalis* during second release and third release.

y_i = 21.7

$\bar{X}$(Mean of 1st Sample) = 47.06

$\bar{Y}$(Mean of 2nd Sample) = 21.7

The Sum of Square Deviation

$\sum (x_i - \bar{X})^2$ = 6.74

$\sum (y_i - \bar{Y})^2$ = 0.0

S = Standard Deviation

S = 3.37   d.f = 2

Null Hypothesis (H_0)

Assume that difference between gain in percentage reduction of alive *Scirtothrips dorsalis* during two successive releases is not significantly different.

$|t|$ = 5.64

Conclusion

But tabulated $|t|_{0.05}$ for Degree of Freedom 2 is 4.30 which is smaller than the calculated value 5.64. Hence, Null Hypothesis (H_0) is accepted at 5% level of significance.

∴ The gain in % reduction of the two samples is not significantly different and it is accepted at 5% level.

Thus, we conclude statistically that the percentage gain in *Scirtothrips dorsalis* reduction is affected by the release of both the parasitoids; *Thripobius semiluteus* and *Ceranisus menes* during two successive releases.

Experiment – 14

Effect of Predators (*Amblyseius cucumeris* and *Macrotracheliella nigra*) and Parasitoids (*Thripobius semiluteus* and *Ceranisus menes*) on the population of *S. dorsalis* in Chilli field of Seventh Net House during the year 2015-16

Sample – 1

Gain in percentage reduction of alive number of *Scirtothrips dorsalis* during first release and second release.

x_i = 44.7, 50.3, 59.9

Sample – 2

Gain in percentage reduction of alive number of *Scirtothrips dorsalis* during second release and third release.

y_i = 88.6, 89.7, 94.07

$\bar{X}$(**Mean of 1st Sample**) = 51.6

$\bar{Y}$(**Mean of 2nd Sample**) = 90.8

The Sum of Square Deviation

$\sum (x_i - \bar{X})^2$ = 118.19

$\sum (y_i - \bar{Y})^2$ = 16.75

S = Standard Deviation

S = 5.81   d.f = 4

Null Hypothesis (H_0)

Assume that difference between gain in percentage reduction of alive *Scirtothrips dorsalis* during two successive releases is not significantly different.

$$|t| = 8.17$$

Conclusion

But tabulated $|t|_{0.05}$ for Degree of Freedom 4 is 2.78 which is smaller than the calculated value 8.17. Hence, Null Hypothesis (H_0) is accepted at 5% level of significance.

∴ The gain in % reduction of the two samples is not significantly different and it is accepted at 5% level.

Thus, we conclude statistically that the percentage gain in *Scirtothrips dorsalis* reduction is affected by the release of both the natural enemies : predators (*Amblyseius cucumeris* and *Macrotracheliella nigra*) and parasitoids (*Thripobius semiluteus* and *Ceranisus menes*) during one successive release.

4.6 Discussion

Population Fluctuation of Thrips

From Locality Talib Nagar : During the year 2014, the maximum population of thrips was recorded in the first and second half of the month October. It was recorded 583 in the first half and 390 in the second half of this month. But in the month of December 2014(First half), it was observed that the minimum abundance in thrips population was recorded. Our regular fortnightly survey in the year 2015, it was recorded that the minimum population of thrips was in the month of February and it was recorded as 3 in the first half and 11 in the second half. But, the maximum abundance in thrips population was recorded in the month of April and May (Summer season). It was 460 in the second half of the month April and 510 in the first half of the month May. There was no enhancement in the growth of the population of thrips in the month of December 2014 (II Half – Winter Season) and January 2015.

From Locality Jalali : During the year 2014, the maximum population of thrips was recorded in the first and second half of the month October. It was recorded 465 in the first half and 473 in the second half of this month. But in the month of December 2014 (First half), it was observed that the minimum abundance in thrips population (2) was recorded. During our regular fortnightly survey in the year 2015, it was recorded that the minimum population of thrips was in the month of February and it was recorded as 2 in the first half and 13 in the second half. But, the maximum abundance in thrips population was recorded in the month of April and May (Summer season). It was 465 in the second half of the month April and 515 in the first half of the month May. There was no enhancement in the growth of the population of thrips in the month of December 2014 (II Half – Winter Season) and January 2015.

From Locality Tappal : During the year 2014, the maximum population of thrips was recorded in the first and second half of the month October. It was recorded 544 in the first half and 339 in the second half of this month. But in the month of December 2014 (First half), the minimum abundance in thrips population (4) was recorded. In our regular fortnightly survey in the year 2015, it was recorded that the minimum population of thrips was in the month of February and it was recorded as 0 in the first half and 17 in the second half. But, the maximum abundance in thrips population was recorded in the month of April and May (Summer season). It was 389 in the second half of the month April and 384 in the first half of May. There was no enhancement in the growth of the population of thrips in the month of December 2014 (II Half – Winter Season) and January 2015.

From Locality Kayamganj : During the year 2014, the maximum population of thrips was recorded in the first and second half of October. It was recorded 451 in the first half and 416 in the second half of this month. But in the month of December 2014 (First half), the minimum abundance in thrips population (2) was recorded. In our regular fortnightly survey in the year 2015, it was recorded that the minimum population of thrips was in the month of February and it was recorded as 0 in the first half and 10 in the second half. But, the maximum abundance in thrips population was recorded in the month of April and May (Summer season). It was 410 in the first half of the month

April and 408 in the second half of this month. There was no enhancement in the growth of the population of thrips in the month of December 2014 (II Half – Winter Season) and January 2015.

From Locality Sumera : During the year 2014, the maximum population of thrips was recorded in the first and second half of October. It was recorded 603 in the first half and 480 in the second half of this month. But in the month of December 2014 (First half), the minimum abundance in thrips population (3) was recorded. In our regular fortnightly survey in the year 2015, it was recorded that the minimum population of thrips was in the month of February and it was recorded as 0 in the first half and 5 in the second half. But, the maximum abundance in thrips population was recorded in the month of April and May (Summer season). It was 445 in the second half of the month April and 375 in the first half of the month May. There was no enhancement in the growth of the population of thrips in the month of December 2014 (II Half – Winter Season) and January 2015.

Collection of Thrips – Natural Enemies

From Locality Talib Nagar : In this locality outfields, the highest population of selected predator *Amblyseius cucumeris* was recorded 10 & 14 on first and second half of the month of October during the year 2014. The maximum abundance of another predator *Macrotracheliella nigra* was recorded at 8 and 11 in first and second half of the month October 2014. While during the year 2015, the maximum abundance of the predator *Amblyseius cucumeris* was recorded 18 and 20 in the first and second half of the month of May. The maximum count of another predator *Macrotracheliella nigra* was recorded 19 and 16 in first and second half of the month of May. On the other hand, the evaluated research work was done with the help of parasitoids. In this locality, the presence of selected parasitoid *Thripobius semiluteus* was recorded 4 and 5 in first and second half of the month of October 2014. The presence of another selected parasitoid *Ceranisus menes* was also recorded. It was recorded 5 and 7 in first half and second half of the month October. During regular research and sampling survey, in the year 2015, it was recorded that the natural presence of parasitoids *Thripobius semiluteus* was 6 and 7 in first and second half of the month of May. In the same manner, the sampling of another parasitoid *Ceranisus menes* was done and it was recorded 6 in first half of the month of April and 7 in first half of the month of May. During our research survey throughout the years of 2014 and 2015, it was recorded that there was no enhancement in the population of thrips natural enemies in the month of December 2014 and in the month of January 2015.

From Locality Tappal : In this locality outfields, the highest population of selected predator *Amblyseius cucumeris* was recorded 12 & 16 in first and second half of the month of October during the year 2014. The maximum abundance of another predator *Macrotracheliella nigra* was recorded at 15 and 13 in first and second half of the month October 2014. While during the year 2015, the maximum abundance of the predator *Amblyseius cucumeris* was recorded 17 and 15 in the first and second half of the month April. The maximum count of another predator *Macrotracheliella nigra* was recorded 17 and 19 in first and second half of the month of May. On the other hand, the evaluation of parasitoids was done. In this locality, the presence of selected parasitoid *Thripobius semiluteus* was recorded 4 and 3 in second half of the month of October, 2014 and in the first half of the month of November, 2014 respectively. The presence of another selected parasitoid *Ceranisus menes* was also recorded. It was recorded 8 in the second half of the month of October. During regular research and sampling survey in the year 2015. The natural presence of parasitoids *Thripobius semiluteus* was 8 in first half of the month of May. In the same manner, the sampling of another parasitoid *Ceranisus menes* was done in 2015 also. It was recorded 7 in the second half of the month of April. There was no enhancement recorded in the population of thrips natural enemies in the month of December, 2014 and in the month of January 2015.

From Locality Sumera : In this locality outfields, the highest population of selected predator *Amblyseius cucumeris* was recorded 14 in the second half of the month of October during the year, 2014. The maximum abundance of another predator *Macrotracheliella nigra* was recorded at 13 in the second half of the month of October, 2014. While during the year, 2015, the maximum abundance of the predator *Amblyseius cucumeris* was recorded 16 and 14 in the first and second half of the month of May. The maximum count of another predator *Macrotracheliella nigra* was recorded 19 in the second half of the month of May. On the other hand, the evaluated research work was done with the help of parasitoids. In this locality, the presence of selected parasitoid *Thripobius semiluteus* was recorded 3 on second half of the month of October 2014. The presence was recorded of another selected parasitoid *Ceranisus menes*. It was recorded 4 on second half of the month October. During regular research and sampling survey in the year 2015, it was recorded that the natural presence of parasitoids *Thripobius semiluteus* was 6 on second half of the month April. In this manner, the sampling of another parasitoid *Ceranisus menes* was done. And, it was recorded 9 in the second half of the month of April. There was no enhancement recorded in the population of thrips natural enemies in the month of December, 2014 and in the month of January, 2015.

From Locality Jalali : In this locality outfields, the highest population of selected predator *Amblyseius cucumeris* was recorded 13 and 15 in the first and second half of the month of October during the year 2014. The maximum abundance of another predator *Macrotracheliella nigra* was recorded 6 and 7 in the first and second half of month October 2014. While during the year 2015, the maximum abundance of the predator *Amblyseius cucumeris* was recorded 18 and 16 in first half of April and first half of the month of May respectively. The maximum count of another predator *Macrotracheliella nigra* was recorded 10 & 9 in the first and second half of the month of April. On the other hand, the evaluation of parasitoids was done. In this locality, the presence of selected parasitoid *Thripobius semiluteus* was recorded 3 in the second half of the month of November, 2014. The natural presence of another selected parasitoid *Ceranisus menes* was recorded. It was recorded 3 in the second half of the month of September, 2014. During regular research and sampling survey in the year 2015. The natural presence of parasitoids *Thripobius semiluteus* was 5 and 6 in the first half of the month of April and second half of the month of June respectively. In the same manner, the sampling of another parasitoid *Ceranisus menes* was done in 2015 also. It was recorded 5 in the second half of the month of April. There was no enhancement recorded in the population of thrips natural enemies in the month of December, 2014 and January, 2015.

From Locality Kayamganj : In this locality outfields, the highest population of selected predator *Amblyseius cucumeris* was recorded 10 and 17 in the first and second half of the month of October during the year 2014. The maximum abundance of another predator *Macrotracheliella nigra* was recorded at 7 and 8 in the first and second half of the month of October, 2014. While during the year 2015, the maximum abundance of the predator *Amblyseius cucumeris* was recorded 15 and 16 in the first and second half of April and 16 in the second half of the month of May respectively. The maximum count of another predator *Macrotracheliella nigra* was recorded 11 in the first half of the month of April. On the other hand, the evaluation of research work was done. In this locality, the presence of selected parasitoid *Thripobius semiluteus* was recorded 3 and 3 in the second and first half of the month of October and November in the year 2014. The presence was also recorded of another selected parasitoid *Ceranisus menes*. It was recorded 2 in the first half of the month of October, 2014. During regular research and sampling survey in the year 2015, it was recorded the natural presence of parasitoids *Thripobius semiluteus*. It was 5 in the second half of the month of March. In the same manner, the sampling of another parasitoid *Ceranisus menes* was done, and, it was recorded maximum in the month of March, April and in the month of May. There was no enhancement recorded in the population of thrips natural enemies in the month of December 2014 and January 2015.

4.7 Discussion of Biological Controlled Experiments

The experiment was done to see the effect of biological control after the release of the predators (*Amblyseius cucumeris* and *Macrotrcheliella nigra*) and parasitoids (*Thripobius semiluteus* and *Ceranisus menes*) in controlled conditions during the years 2014 and 2015.

All the experiments were done in 21 different sets / microplots of seven net houses. Each treated and replicated with three times. It was in first half of October, 2014 that selected number of thrips were released and the natural enemies (predators and parasitoids) were introduced in selected numbers.

Experiment 1

In the first net house of first microplot the initial number of *S. dorsalis* was 45. In order to find out the effect of biological control, the predators in number 40 were released. A day after, it was found that the average number of alive S.d. was 35.60, and, thus, the reduction percentage was recorded at 20.8%. The plot was supervised properly and second release of predators was done after one week at the rate of 40. At that time the remaining number of *Scirtothrips dorsalis* was 70.4. After one day, it was further examined and on examination it was found that the percentage reduction was 48.5% for the average number of alive *Scirtothrips dorsalis* i.e. 36.5. In the same microplot, the third release was done after two weeks of the first release and 40 predators were again released to control the remaining number of *Scirtothrips dorsalis* i.e. 36.5 per plant. At the close of the third week, the average number of alive *Scirtothrips dorsalis* was examined by proper methods and it was found 11.60 at the reduction percent of 68.2%.

In the second microplot of first net house, the initial number of *Scirtothrips dorsalis* was the same (45) of *Scirtothrips dorsalis* as in the first microplot. But, the rate of the release predators was increased. It was made 45 in number in order to see its effectiveness in a better way. A day after of the release of the predators, it was recorded that the average number of alive *Scirtothrips dorsalis* was 32.54. The reduction rate was at 27.6%. The second release of predators was done after one week as it was done in the first microplot. The rate of release was again the same i.e. 45. On examination the remaining number of *Scirtothrips dorsalis* was 63.5. When the effect was examined, it was found that the number of *Scirtothrips dorsalis* after one day was reduced by 52.4% as the average number of alive *Scirtothrips dorsalis* was found 30.2. In the third release the remaining number of thrips, *Scirtothrips dorsalis* was found 33.50 after examination. For this the release rate of predators was 45. In our examination after one day, the average number of alive *Scirtothrips dorsalis* was calculated as 8.70 and the rate of reduction of thrips was reduced at 74.02%.

The same examination was done at the same net house in the third microplot with the release of more number of predators i.e. 50. The initial number of *Scirtothrips dorsalis* was same as the first and second microplot i.e.

45. A day after observation it was found that the average number of alive *Scirtothrips dorsalis* was 29.71 and in this manner the rate of reduction was recorded at 33.9%. At the time of the second release (after 1 week) of the predators the remaining number of *Scirtothrips dorsalis* was 57.6. In the second release of the predators was also at the rate of 50. When the result was examined over *Scirtothrips dorsalis* population, it was found that the average number of alive *Scirtothrips dorsalis* reduced to 23.6 at the rate of 59.02%. The third release was done after one week of the second release again at the same rate of predators i.e. 50. At this time the remaining number of *Scirtothrips dorsalis* was recorded as 30.6. When the effect was seen and examined, it was found that the average of alive *Scirtothrips dorsalis* reduced considerably. It was only 5.41 per plant. In this manner, the total reduction rate came to 82.32% of the first net house.

Experiment 2

In the second net house of the first microplot the release rate of selected predators; *Macrotracheliella nigra* for the effective control on thrips population was 45. The initial number of *Scirtothrips dorsalis* was 45, same as it was in the first experimental net house. In observation a day after the release, it was found that the average number of alive thrips, *Scirtothrips dorsalis* was 33.64, and, thus, the rate of reduction was at 25.2%. Second release of the predators was done after a week. The remaining number of *Scirtothrips dorsalis* at that time was 70.4. 45 predators were again released and the result was the average number of alive *Scirtothrips dorsalis* after one day was 39.6. So, the reduction rate in thrips population was at 43.7%. In the same microplot the third release of 45 predators was done. At that time the remaining number of *Scirtothrips dorsalis* was 36.5. When examination after 1 day was done, the average number of *Scirtothrips dorsalis* was 13.50, and, thus, the total reduction rate of the first microplot was at 63.01%.

In the second microplot the initial number of *Scirtothrips dorsalis* was 45, the same as it was in the first microplot. But the number of predators was increased to 50. A day after of release of the predators it was recorded that the average number of alive *Scirtothrips dorsalis* was 30.19, and, it was reduced at the rate of 32.9%. After a week, the second release of the selected predators in equal number (50) was done. At that time the remaining number of *Scirtothrips dorsalis* was 63.5. A day after it was found that the average number of alive *Scirtothrips dorsalis* was 31.6 and the reduction rate was at 50.23%. The third release of the predators in equal number (50) was done again. At the calculated time the number of remaining *Scirtothrips dorsalis* was 33.50. When observation was done after one day it was found that the average number of alive *Scirtothrips dorsalis* was 7.69, and, thus, the percentage reduction in alive thrips was at 77.04%.

In the third experimental microplot the initial number of *Scirtothrips dorsalis* was again 45. But the number of release predators was 55. After examination at the scheduled time, the average number of alive *Scirtothrips dorsalis* was 26.50, and, the percentage reduction was at 41.1%. A week after, the equal number (55) of predators was again released. At that time, the remaining number of *Scirtothrips dorsalis* was 57.6. A day after it was found that the average number of alive *Scirtothrips dorsalis* was 26.7, and, thus, reduction rate was recorded at 53.6%. In the third release of the predators (55) in number, the number of remaining *Scirtothrips dorsalis* was 30.6. After one day of observation, it was found that the average number of alive *Scirtothrips dorsalis* was 4.34 and the gain in percentage reduction in thrips population was at 85.8%.

Experiment 3

In the third net house both types of predators (*Amblyseius cucumeris* and *Macrotracheliella nigra*) were released simultaneously. The initial number of *Scirtothrips dorsalis* in the first microplot was 45. The predators *Amblyseius cucumeris* 40 and *Macrotracheliella nigra,* 45 were released. After one day the average number of alive *Scirtothrips dorsalis* was recorded 34.36. Thus, the reduction in thrips population was at 23.6%. One week after, the second release of the predators was made in the same number, as it was done at the time of the first release i.e. (40+45). At that time the number of remaining *Scirtothrips dorsalis* was 70.4. Proper examination was made after one day, and, it was found that the average number of *Scirtothrips dorsalis* was 22.26. Thus, the reduction in thrips population was at 68.3%. At the time of the third release of the predators in equal number (40+45). The number of remaining *Scirtothrips dorsalis* was 36.5. A day after of observation it was found that the average number of alive *Scirtothrips dorsalis* was 4.10 and the reduction rate was calculated at 88.7%.

In the second experimental microplot the initial number of *Scirtothrips dorsalis* was the same as in the first microplot i.e. 45. But increasing number of predators (45+50) was released in order to see the impact over thrips population. A day after the average number of alive *Scirtothrips dorsalis* reduced to 33.20 and in this manner the reduction rate was calculated at 26.2%. A week after another group of same predators with equal numbers (45+50) was released. At that time the remaining number of *Scirtothrips dorsalis* in the second microplot was 63.5. One day after the data was collected and it was observed that the average number of *Scirtothrips dorsalis* was 15.36. Thus, the reduction rate was calculated at 75.8%. The third release of the same predators (45+50) was made at the scheduled time, the remaining number of *Scirtothrips dorsalis,* 33.50. A day after, the average number of *Scirtothrips dorsalis* was 2.67 and the reduction rate in thrips population was at 92.0%.

In the third experimental microplot the initial number of *Scirtothrips dorsalis* was 45. But, the number of selected predators was increased. It was made (50+55). On examination after a day it was found, that the number of alive *Scirtothrips dorsalis* was 30.16 and the reduction rate was claculated at 32.9%. At the time of second release of the same predators (50+55) in the same number for the remaining number of *Scirtothrips dorsalis* was 57.6. One day after, the average number of *Scirtothrips dorsalis* was found 11.10 which was at the reduction rate of 80.7%. Since, the significant level of reduction in the population of thrips was obtained at this time and therefore, the third release of the predators was not necessitated.

Experiment 4

In the first microplot of the fourth net house, the initial number of *Scirtothrips dorsalis* was 45, as, it was in the first, second and third experimental nethouses. In order to find out the effect of parasitoids over thrips population 15 parasitoids were released. It was found after 3 days interval the average number of alive *Scirtothrips dorsalis* was 40.19, and, thus, the reduction percentage was calculated at 10.6%. The microplot was examined properly and, the second release of the parasitoids was done again after one week at the number of 15. At that time the remaining number of *Scirtothrips dorsalis* was 70.4. Three days after observation was done, and, it was found that the average number of alive *Scirtothrips dorsalis* was 39.96, and, the reduction rate in thrips population was at 43.23%. In the same microplot the third release of parasitoids was done at scheduled time i.e. 2 weeks after first release and 15 parasitoids were again released to control the remaining number of *Scirtothrips dorsalis* i.e. 36.5. At the close of the third week of the third release the average number of alive *Scirtothrips dorsalis* was 10.41 and it was examined after three days of released parasitoids. At that time the percentage reduction of alive thrips was at 71.4%.

In the second microplot of the fourth net house, the initial number of *Scirtothrips dorsalis* was 45, as in the first microplot. But, the rate of release parasitoids was increased. It was made 20 in number, in order to see the better efficacy in a better way. Three days after of release parasitoids, it was recorded that the average number of alive *Scirtothrips dorsalis* was 36.68. At that time, the reduction in population was at 18.4%. The second release of the same parasitoids (20) was done after one week. The efficacy of these parasitoids was examined over the remaining number of *Scirtothrips dorsalis* i.e. 63.5. Three days after, it was found that the average of alive *Scirtothrips dorsalis* was 33.80, and, the reduced percentage in thrips population was at 46.7%. In the third release, the remaining number of *Scirtothrips dorsalis* was 33.50. For this, the release rate of selected parasitoids was 20. Three days after examination, the average number of alive *Scirtothrips dorsalis* was calculated as 6.45 and the reduction in thrips population was reduced at the rate of 80.7%.

The same examination was done at the same exprimental net house in the third microplot. The initial number of *Scirtothrips dorsalis* was 45. For this, the release rate of parasitoids was increased i.e. 25. Three days after observation, it was found that the average number of alive *Scirtothrips dorsalis* was 32.16 and in this manner, the rate of reduction was recorded at 28.5%. At the time of second release (after 1 week) of the parasitoids, the remaining number of *Scirtothrips dorsalis* was 57.6. For second release, the release rate of parasitoids was also at the rate of 25. When the result was examined after the efficacy over *Scirtothrips dorsalis,* it was found that the average number of alive *Scirtothrips dorsalis* (after 3 days) was 25.56, and, the rate of reduction was recorded at 55.6%. The third release of parasitoids was done after two weeks of the first release, to obtain the better efficacy over *Scirtothrips dorsalis* population. At this time the remaining number of *Scirtothrips dorsalis* was recorded at 30.6. When the effect was seen and examined, it was found that the average of alive *Scirtothrips dorsalis* was 4.56 per plant. In this manner, the total rate of reduction in the population of thrips was came to 85.09% of the fourth net house.

Experiment 5

In the first microplot of the fifth net house, the release rate of selected parasitoids; *Ceranisus menes* for the effective control on thrips population was 20. The initial number of *Scirtothrips dorsalis* was 45, as, it was in the previous experimental net houses. In observation after three days of the release, it was found that the average number of alive *Scirtothrips dorsalis* was 39.61, and, the reduction rate was calculated at 11.9%. In the second release of the same microplot, the remaining number of *Scirtothrips dorsalis* was 70.4. It was done after a week of the first release. 20 parasitoids were again released to control over it. At that time, the average number of alive *Scirtothrips dorsalis* was 33.7, and, thus, the reduction rate was calculated at 52.1%. The third release was also done in the same microplot. The release of the selected parasitoids was again 20 to control the remaining number of *Scirtothrips dorsalis,* and, it was 36.5. When examination was done after 3 days interval, the average number of alive *Scirtothrips dorsalis* was found at 11.5, and, thus, the rate of reduction was recorded at 68.4%.

In the second experimental microplot of the same net house the initial number of *Scirtothrips dorsalis* was 45, the same as, it was in the first microplot. But the number of parasitoids was increased to 25. Three days after of the release of the parasitoids, it was recorded that the average number of alive *Scirtothrips dorsalis* was 35.54, and, it was reduced at the rate of 21.02%. After a week the second release of the parasitoids in equal number (25) was done. At that time the remaining number of *Scirtothrips dorsalis* was 63.5. After three days, it was found that the average number of alive *Scirtothrips dorsalis* was 28.34, and, the rate of reduction was at 55.3%. The third release of the parasitoids in equal number (25) was done. At the calculated time, the number of remaining number of *Scirtothrips dorsalis* was 33.50. When observation was done at three days after, it was found that the average number of alive *Scirtothrips dorsalis* was 8.37. And, thus the percentage reduction in thrips population was at 75.01%.

In the third experimental microplot, the initial number of *Scirtothrips dorsalis* was again 45. But the number of released parasitoids was 30. After examination at the scheduled and calculated time (after 3 days), the average number of alive *Scirtothrips dorsalis* was 32.43, and, the rate of reduction was at 27.93%. A week after, the equal number (30) of parasitoids was again released. At that time, the remaining number of *Scirtothrips dorsalis* was 57.6. Three days after, it was found that the average number of alive *Scirtothrips dorsalis* was 20.26 and the percentage reduction was recorded at 64.8%. In the third release of the same microplot the released number of parasitoids was 30. At that time, the remaining number of *Scirtothrips dorsalis* was 30.6. Three days after, it was found that the average number of alive *Scirtothrips dorsalis* was 6.50, and, the total gain after this experiment in the percentage reduction in *Scirtothrips* population was at 78.75%.

Experiment 6

In the sixth experimental net house both types of parasitoids (*Thripobius semiluteus* and *Ceranisus menes*) were released simultaneously. The initial number of *Scirtothrips dorsalis* in the first microplot was 45. The parasitoids i.e. 15+20 for T.s. & C.m. were released. After three days the average number of alive *Scirtothrips dorsalis* was recorded at 35.20 per plant. So, the reduction in the thrips population was at 21.7%. One week after, the second release of the parasitoids was made in the same number, as, it was done at the time of first release (15+20). At that time the remaining number of *Scirtothrips dorsalis* was recorded 70.4. Proper examination was made after three days interval, and, it was found that the average number of alive *Scirtothrips dorsalis* was 23.46 and thus, the rate of reduction in thrips population was recorded at 66.6%. In this manner, at the time of third release of the parasitoids in equal number (15+20), the number of remaining *Scirtothrips dorsalis* was recorded 36.5. Three days after the average number of alive *Scirtothrips dorsalis* was found 4.36 and the reduction rate in thrips population was calculated at 88.05%.

In the second microplot of this experiment, the initial number of *Scirtothrips dorsalis* was the same as in the first, microplot i.e. 45. But increasing number of parasitoids (20+25) for T.s. and C.m. was released, in order to see the impact of biological controlled parameters over thrips population. Three days after the average number of alive *Scirtothrips dorsalis* was reduced to 30.10 and in this manner, the reduction rate was 33.11%. A week after, another group of the same parasitoids with equal numbers (20+25) was released. At that time the remaining number of *Scirtothrips dorsalis* in the experimental microplot was 63.5. Three days after, the data was claculated and found the average number of alive *Scirtothrips dorsalis* was 10.30. The reduction rate in thrips population was calculated at 83.7%. Since, the significant level of percentage reduction in the population of thrips, *Scirtothrips dorsalis* was obtained in the second release of the second microplot. So, the third release of the parasitoids was not needed.

In the third experimental microplot, the initial number of *Scirtothrips dorsalis* was the same as previous experiments. i.e. 45. But, the number of selected parasitoids was increased. It was made (25+30). On examination after three days, it was found that the number of alive *Scirtothrips dorsalis* was 25.30 and thus, the percentage reduction was recorded at 43.7%. At the time of second release of the same group of parasitoids in the same number; (25+30). At that time, the remaining number of *Scirtothrips dorsalis* was 57.6. Three days after on the basis of observation, it was recorded that the average number of alive *Scirtothrips dorsalis* was 7.10. In this manner, the reduced percentage in thrips population was recorded at 87.6%. Since, at the time of third release the level of significant in thrips population was obtained in the second release. Therefore, at this time the third release of the parasitoids was not necessitated.

Experiment 7

In the seventh experimental net house both predators (A.c. + M.n.)* and parasitoids (T.s. + C.m.)* were released simultaneously. In the first microplot the initial number of *Scirtothrips dorsalis* was 45, in order to find out the biological parameters over thrips population. The release number of predators (A.c. + M.n.) was (40+45) and the release number of parasitoids was (15+20). A day after, it was found that the average number of alive

Scirtothrips dorsalis was 33.04 and after four days of the released of natural enemies, it was found that the average number of alive *Scirtothrips dorsalis* was 26.15, and, thus, the reduction percentage was recorded at 41.8%. The microplot was supervised properly and the second release of natural enemies was done after one week, in which the release rate of selected predators was again (40+45) and the release rate of selected parasitoids was (15+20), same as in the first release. After one day of the release of natural enemies, the average number of alive *Scirtothrips dorsalis* was 15.05 and after 4 days it became 9.37. Thus, the reduction percentage of thrips population was recorded at 86.6%. After the second release the significant data of reduction in the thrips population was obtained and therefore, there was no need to carry the experiment to the third release in the first microplot.

In the second experimental microplot of seventh net house, the initial number of *Scirtothrips dorsalis* was 45. But the released rate of selected natural enemies (predators and parasitoids) was increased. It was made (45+50) for predators and (20+25) for parasitoids. After four days of release of these natural enemies, the number of alive *Scirtothrips dorsalis* was 25.05. At that time, the reduction rate was recorded at 44.3%. The second release of both natural enemies was done after one week as it was done in the first release of second microplot i.e. (45+50) for predators and (20+25) for parasitoids. On examination, the remaining number of *Scirtothrips dorsalis* was 63.5. When the efficacy was examined, it was found that the number of *Scirtothrips dorsalis* after four days was reduced at 91.6%, when the average number of alive *Scirtothrips dorsalis* was found 5.30. Since, the significant level of *Scirtothrips dorsalis* population was obtained and therefore, there was no need to carry the experiment to the third stage with third release in the second microplot.

In the third experimental microplot, the initial number of *Scirtothrips dorsalis* was again 45. But the number of released predators and parasitoids, was (50+55) for selected predators (A.c. + M.n.) and for selected parasitoids (T.s. + C.m.) it was (25+30). After observation at the scheduled time (after 4 days), the average number of alive *Scirtothrips dorsalis* was observed 24.10 and the reduction rate was recorded at 46.4%. A week after the equal number of natural enemies was again released i.e. (50+55) (A.c. + M.n.) predators and (25+30) (T.s. + C.m.) parasitoids. At that time the remaining number of *Scirtothrips dorsalis* was 57.6. Four days after, it was found that the average number of alive *Scirtothrips dorsalis* was 3.67, and, the reduction rate was recorded at 93.6%. After the second release of the third microplot, the significant data of reduction in the population of thrips was obtained and therefore, there was no need to carry the experiment to the third stage with third microplot.

Year 2015-16

The experiments based on biological management of thrips on chilli plants as done in the year 2014-15, were again done in the year 2015-16 in order to find out the continuity and sustainability of the methods adopted therein. For the purpose of gaining concrete results it was necessary to confirm the effectiveness of the biological parameters adopted in the scientific method. The repeated confirmation of the results sets is thinking about the practical application of these methods in getting proper and healthy yields of chilli in the farms, if economic threshold is also obtained.

Experiment 8

In the first experimental net house of the first microplot, the release rate of selected predators; *Amblyseius cucumeris* for the better efficacy over thrips population was 50. The initial number of *Scirtothrips dorsalis* was 55. In observation a day after the release, it was found that the average number of alive thrips was 45.16, and, thus, the rate of reduction was calculated at 17.8%. Second release of the predators was done after a week. The remaining number of *Scirtothrips dorsalis* was 73.3. 50 predators were again released. At that time, the average number of alive *Scirtothrips dorsalis* was 39.6. Thus, the reduction rate in thrips population was at 45.9%. In the same microplot for the third release the number of predators was again 50 and the remaining number of *Scirtothrips dorsalis* was 39.73. When examination after one day was done the average number of *Scirtothrips dorsalis* was 13.06 and the total reduction rate of first microplot was recorded at 67.12%.

The same examination was done at the same net house of second micropot with the release of more number of predators i.e. 55. The initial number of *Scirtothrips dorsalis* was same as the first microplot. A day after observation it was found that the average number of alive *Scirtothrips dorsalis* was 41.36 and in this manner, the rate of reduction was recorded at 24.8%. At the time of second release (after 1 week) of the predators the remaining number of *Scirtothrips dorsalis* was 65.6. In the second release of the predators was also at the same rate of first release and i.e. 55. When the result was examined over *Scirtothrips dorsalis* population, it was found that the average number of alive *Scirtothrips dorsalis* reduced to 32.14 at the rate of 51.0%. The third release was done after one week again at the same rate of predators i.e. 55. At this time the remaining number of *Scirtothrips dorsalis* was recorded as 34.06. When the effect was seen and examined, it was found that the average of alive *Scirtothrips dorsalis* reduced considerably. It was 8.34 per plant and the reduction rate was calculate at 75.51%.

In the third microplot of first net house the initial number of *Scirtothrips dorsalis* was 55 as in the first and second experimental microplot. But, the rate of the release predators was increased. It was made 60 in number, in order to see the effectiveness in a better way. A day after of release predators it was recorded that the average number of alive *Scirtothrips dorsalis* was 38.34%. The reduction rate was at 30.2%. The second release of predators was done after one week as it was done in the first release of the predators i.e. 60. On examination the remaining number of *Scirtothrips dorsalis* was 56.5. When the effect was examined it was found that the average number of alive *Scirtothrips dorsalis* after one day was calculated 24.36 per plant and thus, the reduction rate was recorded at 56.8%. In the third release the remaining number of *Scirtothrips dorsalis* was found 29.16. For this the release rate of predators was again 60. The examination was doe after one day. At that time, the average number of alive *Scirtothrips dorsalis* was calculated only 4.43 per plant. In this manner, the total reduction rate came to 84.80% of the first net house.

Experiment 9

In the second net house of the first microplot, the initial number of *Scirtothrips dorsalis* was 55. In order to find out the biological parameters on thrips population, the release number of predators was 55. A day after examination the average number of alive *Scirtothrips dorsalis* was 37.17, and, thus, the reduction percentage was recorded at 32.4%. The second release of predators was done after one week at the rate of 55. At that time, the remaining number of *Scirtothrips dorsalis* was 73.3. After one day, it was further examined and on examination it was found that the average number of alive *Scirtothrips dorsalis* was 42.63 and the rate of reduction was recorded at 41.8%. In the same microplot the third release was done after two weeks of the first release and 55 predators were again released to control over the thrips population. At that time the remaining number of *Scirtothrips dorsalis* was 39.73. After examination at scheduled time, the average number of alive *Scirtothrips dorsalis* was recorded 15.30 and the percentage reduction was 61.49%.

In the second microplot of second net house the initial number of *Scirtothrips dorsalis* was same as of *Scirtothrips dorsalis* as in the first microplot i.e. 55. But, the rate of the release predators was increased. It was made 60 in number, in order to see the effectiveness in a better way. A day after of the release of the predators it was recorded that the average number of alive *Scirtothrips dorsalis* was 35.40. The reduction rate was recorded at 35.6%. The second release of the predators was done after one week as it was done in the first release of second microplot. The rate of release again the same and it was made 60 in number. On examination the remaining number of *Scirtothrips dorsalis* was 65.6. When the effect was examined it was found that the number of *Scirtothrips dorsalis* after one day was reduced by 40.4% as the average number of alive *Scirtothrips dorsalis* was found 33.16. In the third release the remaining number of *Scirtothrips dorsalis* was found 34.06 after observation. For this the release rate of the selected predators was 60. In the search results the average number of alive *Scirtothrips dorsalis* after one day was recorded 7.46. Thus, the rate of reduction was recorded at 78.09%.

In the third experimental microplot the initial number of *Scirtothrips dorsalis* was again 55. But the increased number of released predators was 65. In the search results after one day, the average number of alive *Scirtothrips dorsalis* was 33.04 and the reduction percentage of thrips population at 39.9%.

After a week the second release of the predators in equal number (65) was done. At that time the remaining number of *Scirtothrips dorsalis* was 56.5.

A day after it was found that the average number of alive *Scirtothrips dorsalis* was 23.14 and the reduction rate was 59.04%. The third release of the predators in equal number was done again i.e. 65. At the calculated time the number of remaining *Scirtothrips dorsalis* was 29.16. When observation was done after one day of release, it was found that the average number of alive *Scirtothrips dorsalis* was 4.03. And, thus the gain in the percentage reduction was at 86.1%.

Experiment 10

In the third experimental net house both types of predators (*Amblyseius cucumeris* and *Macrotracheliella nigra*) were released simultaneously. The initial number of *Scirtothrips dorsalis* in the first microplot was 55. The predators *Amblyseius cucumeris* 50 and *Macrotracheliella nigra* 55 were released. After one day, the average number of alive *Scirtothrips dorsalis* was recorded 40.10. Thus, the reduction rate was recorded at 27.09%. One week after second release of both the predators was made in the same number as it was done at the time of first release i.e. 50+55. At that time, the number of remaining *Scirtothrips dorsalis* was 73.3. Proper examination was made after one day, and, it was found that the average number of alive *Scirtothrips dorsalis* was 21.26, and, thus, the reduction in the thrips population was at 70.9%. At the time of the third release of the predators in equal number (50+55), the number of remaining *Scirtothrips dorsalis* was 39.73. A day after, the average number of alive *Scirtothrips dorsalis* was found 5.25, and, the reduction rate was calculated at 86.78%.

In the second experimental microplot the initial number of *Scirtothrips dorsalis* was 55. But the increasing number of predators (55+60) was released, in order to see the efficacy on thrips population. A day after, the average number of alive *Scirtothrips dorsalis* reduced to 38.14 and in this manner, the reduction rate was at 30.65%. A week after, another group of the same predators with equal numbers (55+60) was released. At that time, the remaining number of *Scirtothrips dorsalis* in the microplot was 65.6. One day after the data was collected and calculated, at that time the average number of alive *Scirtothrips dorsalis* was 15.05. The reduction rate was calculated at 77.05%. The third release of the same predators (55+60) was made at the scheduled time, the remaining number of *Scirtothrips dorsalis* was 34.06. A day after the average number of alive *Scirtothrips dorsalis* was 3.13 and the reduction rate of thrips population was recorded at 90.8%.

In the third experimental microplot the initial number of *Scirtothrips dorsalis* was the same as (55) as it was in the first and second microplots. But, the number of selected predators was increased. It was made 60+65. On examination after a day it was found that the average number of alive *Scirtothrips dorsalis* was 36.34, and, thus, the rate of reduction in thrips population was recorded at 33.92%. At the time of second release of the same predators in the same number (60+65), the remaining number of *Scirtothrips dorsalis* was 56.5. One day after the average number of alive *Scirtothrips dorsalis* was found 9.87 which was at the reduction rate 82.5%. Since, the significant level of reduction in the population of thrips was obtained at this time, and, therefore, the third release of the predators was not necessitated.

Experiment 11

In the fourth experimental net house of the first microplot, the initial number of *Scirtothrips dorsalis* was 55 as it was in the first, second and third experimental net houses. In order to find out the effect of parasitoids on thrips population, 20 parasitoids were released. It was found after 3 days interval that theaverage number of alive *Scirtothrips dorsalis* was 49.19, and, thus, the reduction percentage was calculated at 10.56%. The microplot was examined properly and the second release of the parasitoids was done again after one week at the number of 20. At that time the remaining number of *Scirtothrips dorsalis* was 73.3. Three days after observation was done and it was found that the average number of alive *Scirtothrips dorsalis* was 43.16 and the reduction rate was at 41.1%. In the same microplot, the third release was done at scheduled time i.e. 2 weeks after the first release and 20 parasitoids were again released to control the remaining number of *Scirtothrips dorsalis* i.e. 39.73. At the close of the third week of the third release the average number of alive *Scirtothrips dorsalis* was 12.30 and it was examined after three days of the release of parastioids. At that time, the percentage reduction of alive thrips was at 69.04%.

In the second microplot, the initial number of *Scirtothrips dorsalis* was was 55. But, the release of the parasitoids was increased. It was made 25 in number in order to see the better efficacy in a better way. Three days after the release of the parasitoids, it was recorded that the average number of alive *Scirtothrips dorsalis* was 45.30. At that time, the reduction in the thrips population was 17.6%. The second release of the same parasitoids i.e. 25 was done after one week. The efficacy of these parasitoids was examined over the remaining number i.e. 65.6. Three days after, it was found that the average number of alive *Scirtothrips dorsalis* was 32.70 and the reduced percent of thrips population was 50.15%. In the third release, the remaining number of *Scirtothrips dorsalis* was 34.06. For this the release rate of selected parasitoids was 25. Three days after, it was examined that the average number of alive *Scirtothrips dorsalis* was 7.19 and the reduction in the thrips population was reduced at the rate of 78.8%.

The same examination was done at the same experimental net house. In the third microplot, the initial number of *Scirtothrips dorsalis* was 55. For this, the release rate of parasitoids was increased i.e. 30. Three days after, it was found that the average number of alive *Scirtothrips dorsalis* was 39.36 and thus, the rate of reduction was recorded at 28.4%. At the time of second release (after 1 week) of the parasitoids, the remaining number of *Scirtothrips dorsalis* was 56.5. For the second release, the release rate of selected parasitoids was also at the rate of 30. When the result was examined after three days in order to see the efficacy over thrips population, it was found that the average number of alive *Scirtothrips dorsalis* was 26.30 and the rate of reduction was recorded at 53.4%. The third release of parasitoids i.e. 30 in number was done after two weeks of the first release in order to obtain the better efficacy over *Scirtothrips dorsalis* population. At that time, the remaining number of *Scirtothrips dorsalis* was recorded at 29.16. When the effect was seen and examined, it was found that the average number of alive *Scirtothrips dorsalis* was 3.56 per plant. In this manner, the total rate of reduction came to 87.79% of the fourth experimental nethouse.

Experiment 12

In the fifth experimental net house of the first microplot the release rate of selected parasitoids for the effective control over thrips population was 25.

The initial number of *Scirtothrips dorsalis* was 55 as it was in the previous experimental net houses. In observation after three days of the release, it was found that the average number of alive *Scirtothrips dorsalis* was 46.61 and the reduction rate was calculated at 15.25%. In the second release of the same microplot the remaining number of *Scirtothrips dorsalis* was 73.3. It was observed after a week of the first release. 25 parasitoids were again release to control over it and it was found that after 3 days, the average number of alive *Scirtothrips dorsalis* was 40.16. So, the reduction rate was recorded at 45.21%. In the third release of selected parasitoids i.e. 25 in number to control the remaining number of *Scirtothrips dorsalis* it was 39.73 per plant. In the search results after 3 days it was found that the average of alive *Scirtothrips dorsalis* was 13.65 and thus, the rate of reduction was observed at 65.64%.

In the second experimental microplot of the same net house the initial number of *Scirtothrips dorsalis* was 55, the same as it was in the first microplot. But the number of parasitoids was increased to 30. Three days after of the release of selected parasitoids, it was recorded that the average number of alive *Scirtothrips dorsalis* was 42.03 and it was reduced at 23.58%. After a week, the second release of the parasitoids in equal number (30) was done. At that time the remaining number of *Scirtothrips dorsalis* was 65.6. After three days it was found that the average number of alive *Scirtothrips dorsalis* was 30.10 and the rate of reduction was at 54.11%. The third release of the parasitoids in equal number i.e. 30 was done. At the calculated time the number of remaining number of *Scirtothrips dorsalis* was 34.06. When observation was done at three days after it was found that the average number of alive *Scirtothrips dorsalis* was 7.70. And, thus the percentage reduction of thrips population was at 77.3%.

In the third experimental microplot the initial number of *Scirtothrips dorsalis* was again 55. But the number of release parasitoids was 35. After examination at the scheduled time (after 3 days), the average number of alive *Scirtothrips dorsalis* was 39.50 and the rate of reduction was at 28.18%. A week after the equal number (35) of parasitoids was again released. At that time the remaining number of *Scirtothrips dorsalis* was 56.5. Three days after it was found that the average number of alive *Scirtothrips dorsalis* was 18.34 and the percentage reduction was recorded at 67.5%. In the third release of the same microplot the release number of parasitoids was 35. At that time, the remaining number of *Scirtothrips dorsalis* was 29.16. Three days after it was found that the average number of alive *Scirtothrips dorsalis* was 4.90 and the total percentage reduction was at 83.1%.

Experiment 13

In the six experimental net house both types of parasitoids (*Thripobius semiluteus* and *Ceranisus menes*) were released simultaneously. The initial number of *Scirtothrips dorsalis* in the first microplot was 55. The parasitoids i.e. 20+25 in number for *Triphobius semiluteus* and *Ceranisus menes* were released. After three days, the average number of alive *Scirtothrips dorsalis* was recorded at 44.20 per plant. So, the reduction was 19.6%. One week after, the second release of the parasitoids was made in the same number as it was done at the time of first release (20+25). At that time the remaining number of *Scirtothrips dorsalis* was recorded 73.3. Proper examination was made after three days and it was found that the average number of alive *Scirtothrips dorsalis* was 25.16 and thus, the rate of reductionin thrips population was at 65.6%. In this manner, at the time of the third release of the parasitoids was (20+25) for the remaining number of *Scirtothrips dorsalis* i.e. 39.73. Three days after the average number of alive *Scirtothrips dorsalis* was found 5.06 and the reduction rate of thrips population was calculated at 87.26%.

In the second experimental microplot of this experiment, the initial number of *Scirtothrips dorsalis* was the same as in the first experimental microplot i.e. 55. But increasing number of parasitoids (25+30 for T.s. and C.m.) was released in order to see the impact of biological controlled parameters over thrips population. Three days after the average number of *Scirtothrips dorsalis* was reduced to 36.45, and, in this manner the reduction rate was recorded at 33.72%. A week after another group of the same parasitoids with equal number i.e. (25+30) was released. At that time the remaining number of *Scirtothrips dorsalis* was 65.6. One day after the average number of alive *Scirtothrips dorsalis* was 11.19. At that time, the reduction rate was calculated at 82.9%. Since, the significant level of percentage reduction in the population of thrips, *Scirtothrips dorsalis* was obtained in the second release of the second microplot. So, the third release of the parasitoids was not needed.

In the third experimental microplot the initial number of *Scirtothrips dorsalis* was the same as previous experiments i.e. 55. But the number of selected parasitoids was increased. It was made 30+35. On examination after three days it was found that the average number of alive *Scirtothrips dorsalis* was 33.16 and thus, the percentage reduction was recorded at 39.70%. At the time of second release, the same group of parasitoids were released in the same number as first release (30+35). At that time, the remaining number of *Scirtothrips dorsalis* was 56.5. Three days after on the basis of observation, it was recorded that the average number of alive *Scirtothrips dorsalis* was 8.06 perplant. In this manner, the reduced percentage of thrips population was recorded at 85.7%. Since, at the time of third release the level of significance in the thrips population was obtained. Therefore, at this time the release of the parasitoids was not necessitated.

Experiment 14

In the seventh experimental net house both predators (A.c. + M.n.) and parasitoids (T.s. + C.m.) were released simultaneously. In the first microplot, the initial number of *Scirtothrips dorsalis* was 55, in order to find out the biological parameters over thrips population. The release number of predators (A.c. + M.n.) was (50+55) and the release number of parasitoids was (20+25). A day after, it was found that the average number of alive *Scirtothrips dorsalis* was 36.10 and after four days of the release of natural enemies, it was found that the average number of alive *Scirtothrips dorsalis* was 30.39 and thus the reduction percentage was recorded at 44.7%. The microplot was supervised properly and second release of natural enemies was done after one week in which the release rate of selected predators was again (50+55) and the release rate of selected parasitoids was (20+25), same as in the first release. After one day of the release of natural enemies, the average number of alive *Scirtothrips dorsalis* was 14.10 and after 4 days it became 8.30. Thus, the reduction percentage of thrips population was recorded at 88.6%. After the second release the significant data of reduction in the thrips population was obtained and therefore, there was no need to carry the experiment to the third release in the first microplot.

In the second experimental microplot of seventh net house the initial number of *Scirtothrips dorsalis* was 55. But the released rate of selected natural enemies (predators and parasitoids) was increased. It was made (55+60) for predators and (25+30) for parasitoids. After four days of release of these natural enemies, the number of alive *Scirtothrips dorsalis* was 30.36. At that time, the reduction rate was recorded at 50.3%. The second release of both natural enemies was done after one week as it was done in the first release of second microplot i.e. (55+60) for predators and (25+30) for parasitoids. On examination the remaining number of *Scirtothrips dorsalis* was 65.6. When the efficacy was examined it was found that the number of *Scirtothrips dorsalis* after four days was reduced at 89.7% and the average number of alive *Scirtothrips dorsalis* was found 6.75. Since, the significant level of *Scirtothrips dorsalis* population was obtained and therefore, there was no need to carry the experiment to the third stage with third release in the second microplot.

In the third experimental net house of the first microplot, the initial number of *Scirtothrips dorsalis* was again 55. But the number of released predators and parasitoids was (60+65) for selected predators (A.c. + M.n.) and for selected parasitoids (T.s. + C.m.) it was (30+35). After observation at the scheduled time (after 4 days) the average number of alive *Scirtothrips dorsalis* was recorded 22.03 and the reduction rate was recorded at 59.9%. A week after the equal number of natural enemies was again released i.e. (60+65) (A.c. + M.n.) predators and (30+35) (T.s. + C.m.) parasitoids. At that time the remaining number of *Scirtothrips dorsalis* was 56.5. Four days after observation, it was found that the average number of alive *Scirtothrips dorsalis* was 3.35 and the reduction rate was recorded at 94.07%. After the second release, the significant data of reduction in the population of thrips was obtained and therefore, there was no need to carry the experiment to the third stage with third microplot.

Chapter 4: Experimental Analysis

Number of Figures: 03

Number of Graphs: 74

Number of Tables: 27

Chapter-5
Observation

5.1 Thysanopterans on Chilli Crop in District Aligarh of Western Uttar Pradesh
5.2 Occurrence and Damage of Thrips on Chilli Crop
5.3 Pest Investigation by the Monitoring Observation on Chilli Crop
5.4 Weather Influences on Insects
5.5 Taxonomy of *Scirtothrips dorsalis*
5.6 Biology of *Scirtothrips dorsalis*
5.7 Life Cycle of *S. dorsalis*
5.8 Taxonomy and Biology of Natural Enemies of Thrips

Chapter-5
Observation

5.1 Thysanopterans on Chilli Crop in District Aligarh of Western Uttar Pradesh

In the field of agriculture, an insect may be understood as a pest if the damage caused by it reduces the yield of the host crop quantitatively and qualitatively by an amount that is unacceptable to the farmers. Pest activities cause injury on the plants and consequently damage to the crop as a whole results in significant loss. In the field of agriculture chilli has occupied an important position due to its large scale use in vegetables, spices and daily meals. Many insects attack this crop. The insects included under the order, Thysanoptera are popularly known as "Thrips" or "Bladder-footed insects." They are one of the smallest pterygotes. They range from 0.5mm to 1.0mm. They possess remarkable structural peculiarities like fringe wings, assymetrical feeding apparatus and a protrusible bladder. They received little attention in the past among entomologists possibly because of its minute size and unattractive colouration. They have assumed considerable importance as pests of many crops of agriculture and horticulture. Some species of Thrips are vectors of viral diseases. But, majority of the species are phytophagous. Thrips are found in different types of vegetation and are considered to be very susceptible to environmental changes and can thrive only under particular climatic and microclimatic situation. The Biological diversity of Thysanopterans possess remarkable structural diversity. The structural diversity stimulated possibly due to fluctuations in habitat and internal environment. Most of the Thrips species are unicolorous and bright, some are bicolorous and tricolorous. Cuticle shows striking patterns of sculptures, irregular corrugations and transverse straie or granulations.

In a regular monitoring survey of chilli crop in the selected outfields during favourable environmental conditions, we observed that the thrips breeding population infested chilli plant and almost failed to produce any fruits or produced very minimum number of deformed fruits. It was also observed that thrips were feeding on meristems, terminals and other tender plant parts of the host plant. Thrips feeding was examined on those parts which were present above the soil surface. This pest preferred young plant tissues and was not observed to feed on mature host tissues. During infestation, the colour of damaged tissues changed from silvery to brown or black. The infestation usually started from seedling stage although severe infestation appeared first in the vegetative stages of the chilli plants and then in the flowering and fruiting stages. Thrips population were mostly confined in the ventral surface of the chilli leaf more than what they were in the upper leaf surface. Thrips nymphs and adults sucked the cell sap mostly from the ventral surface of the leaf. This resulted in the infested leaf, which lost its vitality and vigour leaves became curled or twisted. We also observed at that time along with leaf both nymphs and adults sucked the cell sap from the twigs and stem and finally infested and affected twigs which were distorted due to loss of vitality. The indirect damage on host plant also occurred in the form of photosynthetic reduction. During regular monitoring it was also observed that under favourable conditions, thrips reproduced at a faster rate and caused damage upto 35-40% of total chilli production.

Among many pests, thrips is one of the most important constraints in chilli production. Feeding injury by thrips can reduce the leaf size affecting photosynthetic activity of plant and eventually can result in significant yield loss. The thrips may constitute a substantial part of insect fauna associated with chilli fields. In our obversation it was found that chilli plants are very sensitive to direct cosmetic damage caused by thrips on the fruits and it is more injurious to the yield loss. On the basis of our observation we found that the short time periods of high densities of thrips resulted in immediate loss in fruit quality which would affect the economic return.

Figure 5.1 : (a) Pest preferred young plant tissues; (b) Thrips feeding on meristems, terminals and tender parts of the chilli plant; (c) The colour of damaged tissue changes from silvery to brown or black; (d) Plant loses their vitality; (e) Thrips : Vector of viral disease; (f) Chilli plant produced minimum number of deformed fruits; (g) Thrips : Mostly confined in the Ventral Surface

5.2 Occurrence and Damage of Thrips on Chilli Crop
Direct Damage

It is by their mouth parts that thrips feed upon the plant tissues. They pierce into it and suck out the cell contents of the their host plant. In this manner, tissue scarification on the leaves of the plants are visible. The result is that the natural resources for the growth of the plants become deficient leaving it in diseased form. This pest prefers young plant tissue. It was seen during our observation that on mature tissues they fed less then they had already done on the young ones. The infestation of thrips reduced the value of the crop directly. The leaves of the plants were used as oviposition site and food for their survival and growth was also used from there. In this manner, damage to the plant was done to such an extent that deformation in the leaves, flowers and fruits were clearly visible. Silver patches and flecking on expanded leaves were also seen. Chilli thrips can feed on the wide range of plant species. But its real growth and heavy infestation can be found among those plants which provide safer place for their reproduction besides provisioning food and shelter. Plants infested with thrips showed the damaged symptoms; leaf surface was spotted with silvery patches. Leaf lamina thickened in linear manner and in some cases the senescence and abscission of leaves occurred alongwith brown frass markings on the infested leaves. Damage included feeding scars and leaf-distortion. They caused damage such as streaking and scarring of petals, distortion of chilli flowers and flower buds. Feeding on chilli plant caused hazy streaks of spots on the chilli fruit. It was found that thrips fed on the fruist calyx and turned its colour and created ghost spotting. In this manner, heavy infestation reduced the ability of the host plant to photosynthesize. Thus, the infested plant became dwarfed with its leaves beginning to detach from the stem. It caused defoliation in a few plants. We observed that the abundance of thrips population was not on growth during rainy season. In the dry weather, it had better growth. Heavy infestations on chilli plants caused by thrips resulted in the changes in plant appearance. In some plants where thrips inserted their eggs, we found discoloured spots and it was due to the damage to the plant cells in the event of thrips feeding on them. Deformation of leaves and shoots and discolouring and dwarfing of fruits were the next symptoms resulting out of thrips feeding. It was seen that thrips deposited their greenish black facal on the leaves while feeding on them. It was also observed that one thrips adult and its pathogen could infect a plant after feeding on it for about 30 minutes. In this manner it is understood that it is highly polyphagous pest which causes serious infestation in the form of totally defoliated plant.

Figure 5.2 : (a) Silver patches on expanded leaves; (b) Brown frass marking on the infected leaf; (c) Frass marking on infected fruit; (d) Flower deformation; (e) Ghost spotting by thrips feeding on chilli fruit; (f) Linear thickening of the leaf lamina; (g) Deformed chilli plant growth; (h) Tissue scarification on chilli plant

Chilli thrips fed on succulent leaves and young fruits. Feeding on young leaves caused irregular bronzing or scarring on both upper and lower sides of the leaf. It was observed that discolouration was typically concentrated along the midrib and lateral leaf veins. It also appeared in scattered patches between veins as population increased. Rapid breeding of thrips caused premature leaf drop. As fruits grew, this early feeding became apparent as scabby or leathery brown scars expanded across the skin. Thrips scarring is sometimes called **'Alligator skin'**. During survey, the mechanical injury or abrasionh as the blows of strong winds, also caused fruit scarring that could be confused with thrips injury. Chilli thrips prefer to feed and lay eggs in the succulent leaves. They move to young fruit when leaves harden. We observed that major damage occurred when fruit was 0.2-0.6 inch (5-5mm) long. Although fruit was susceptible to thrips feeding when they attained the growth about 2 inches (5 cm) in length. The feeding only caused scars on leaves as well as on fruits, when they were less than about ¾ inch (19mm) long. Prolonged feeding by thrips is recognized as **'Mudra Disease'**. We also observed that thrips feed in high densities and insufficiently dry climates. The process results in the desiccation and death of their host plant. Yield loss solely dedicated to *S. dorsalis* damage can range between 61-74%.

Figure 5.3 : (a) Desiccation and Death of the chilli leaf; (b) Prolonged feeding : Recognized as 'Murda Disease'; (c) Scattered patches between veins; (d) Discolouration : Concentrated along the midrib and lateral leaf veins

Chilli Leaf Curl Disease caused by *Scirtothrips dorsalis*

Thrips is a general plant feeder. Plant provides for thrips the food, a protective habitat and the site of oviposition. Chilli thrips has a potential to cause damage to the leaves and reduce the photosynthetic capacity of the host plant. Heavy feeding reduces the physiobiological reaction of the host plant. Thereafter, infestation caused by Chilli thrips on chilli plant changes in the appearance of plant and it is termed **"Chilli leaf curl"**. It is due to the major damage factor of this pest on the chilli plant that this pest is generally known as **"Chilli thrips"**.

Figure 5.4 :   Chilli Leaf Curl caused by Chilli Thrips

Figure 5.5 :   Reduced Photosynthetic Capacity and Physiobiological Reaction of the Chilli Plant

Indirect Damage

As noted earlier, thrips cause direct damage to the plants as twisting of leaves, decolouring of fruits and dwarfing of the plant as whole. Besides this, it causes indirect damage to affect other plants in the surrounding region. It spreads vectors indirectly in the form of several viruses. Following plant viruses have been identified: (1) Chilli Leaf Curl (CLC), (2) Peanut Necrosis Virus (PBNV), (3) Peanut Yellow Spot Virus (PYSV), (4) Tobacco Streak Virus (TSV), (5) Watermelon Silver Mottle Virus (WSMOV).

5.3 Pest Investigation by the Monitoring Observation on Chilli Crop

During our survey, we observed newly flushed leaves of chilli plant by the sampling and tapping method to get a clue about the availability of thrips to make a problem and examined them till the young fruits appeared. A biological monitoring was conducted in growing season of chilli crop, when the production of the chilli crop was high. Monitoring observation on thrips was done before young fruits started coming and continued our monitoring till fruits were full grown. During observation survey succulent leaves were marked and inspected with the help of magnifying glass lens. On inspection leaves were found in light green colour to reddish brown. In our examination we avoided those leaves that were fully hardened and dark green in colour. We also avoided those leaves that touched fruit and were very close to flower. In this manner, we monitored young fruits and pinched stem and then examined the entire fruit surface. We knew that thrips developed significantly in cool temperature and dry weather. During monitoring, population typically began increasing in early summer and spring, when chilli thrips fed on young leaves. Abundance of thrips reached its peak in spring and early summer days. It was the time when fruits were young. Thrips get better opportunity for growth from foliage to young fruits. Thrips population was suppressed by Rainy and Winter conditions and when most fruits were large and ready to be harvested and no long susceptible to new damage.

It was necessary for us to monitor thrips population so that the related problem in the chilli crop might be understood in the right perspective for adopting appropriate control actions. Early detection of the problem is necessary because the elimination of thrips can be and should be done at the early stage. There is no use controlling the thrips if the damage to the plants had already been done. Moreover, it is easier to control small thrips in early stages in comparison to larger ones. In our experiments and observations it was difficult to count thrips on the plants. It was time consuming as well as cost effective. So we assessed the presence of thrips on sampled basis. In the selected samples, we monitored thrips alternatively with sticky traps and the tapping method. Having tapped the leaves and flowers of a selected sample on the sheet of white paper, we dislodged thrips and made them visible. Such a tapping of plants was used to find out where thrips presence was there or not. In this manner, we could get a rough idea and estimate of the number of thrips per plant.

Figure 5.6 : Monitoring : (a) For Chilli Plants; (b) Monitoring of Newly flushed leaves; (c) Defoliation of Flower; (d) Reduced of photosynthetic activity; (e) Necrosis tissues of chilli plant; (f) Distortion of chilli plant; (g) Discolouration of chilli plant; (h) Dislodged Thrips population

For more precision, sticky cards were used. We observed the presence of thrips on sticky traps when they were placed near the host plant. Yellow sticky card was used to monitor the presence of thrips and it was an important feature during our survey.

During our monitoring survey some species of thrips other than *S. dorsalis* were also found on chilli crop of different selected localities namely:

- *Thrips tabaci*
- *Frankiniella occidentalis*
- *Frankniella schultzei*

Infestation caused by these thrips species on chilli resulted in the form of curling of leaves, necrosis tissues, silver to brown scars on fruits, complete defoliation of buds and flowers, distortion and discolouration of plant and finally yield loss. We observed thrips feeding was high and in dry climates and it resulted in the form of desication and death of the host plant.

✦ *Thrips tabaci* (Commonly called Onion Thrips)
Key Characters

1. They have seven segmented antennae.
2. They do not possess elongated anterior setae on the pronotum.
3. Its Pleurotergites contain rows of fine microtrichia.
4. They have pigmented ocelli which are grey in colour in comparison to red pigmented ocelli in other thrips species.
5. During development period, the body colours vary with temperature and are seen from yellow to brown in colour.

Figure 5.7: *Thrips tabaci*

✦ *Frankliniella occidentalis* (Commonly called Western Flower Thrips)
Key Characters

1. Western Flower Thrips is yellow to brown in colour and it has eight antenal segments.

2. Third and fourth segments of antennae have forked sensorium. The fourth, postocular setae is more pronounced.
3. With its long and irregular setae the tergite VIII comb is present in it.

Figure 5.8: *Frankliniella occidentalis*

✦ *Frankliniella schultzei* (Commonly called Tomato Thrips – Blossom Thrips)

Key Characters
1. It is found in two colours – yellow with a little brownish colour and dark brown.
2. Eight segmented antennae is also found in it.

Figure 5.9: *Frankliniella schultzei*

3. Setae originate along the marginal line connecting the front edges. Its front edges are connected with marginal lines where from setae originates. It has two hind ocelli.
4. Its anteroangular setae is longer than its prenatal anteromarginal.
5. Absence of tergite VIII can also be marked in it.

5.4 Weather Influences on Insects

Thrips are minute insects in average of 1mm in length. They show diverse life features and habits. Majority of these species feeds upon plant materials like leaves, flowers and stem tissues. During observation it was seen that weather conditions significantly affect the proliferation rate. Warm dry conditions are most suitable for the growth of thrips pests. They can hardly survive where temperature goes down 4^0C. Sometimes adaptation on high temperature was observed within thrips population. Drought stress conditions could increase the population development of thrips. Direct effect of temperature can be seen not only on the growth of pests but also on its natural enemies – Predators and Parasitoids. If extreme winter reduces thrips population, it also reduced the number of natural enemies. Rainfall and wind also affected the density of the pest population. Prolonged rainy season did not affect the thrips population as much. The population remained most active towards growth during long dry seasons. Pest attacks fluctuated widely throughout the year because of changing weather patterns on the one hand and on the other hand due to the activity of predators and parasitoids. Some natural enemies; predatory mites, minute pirate bugs etc. were much effective predators for chilli thrips. The presence of *S. dorsalis* is known to foreage on several plant species, but it grows more rapidly on chilli plants because its leaves, stems and flowers provide better support to their reproduction in addition to food and shelter.

5.5 Taxonomy of *Scirtothrips dorsalis*

The genus *Scirtothrips* is related to more than 100 species in the tropics and sub-tropic regions. *Scirtothrips dorsalis* is commonly known as the Assam thrips, Castor thrips, Chilli thrips, Berry thrips and Yellow tea thrips. It is a highly polyphagous adventive pest. The morphological traits of taxonomic importance for identification of *S. dorsalis* are well defined in the literature. With slide mount images, we illustrated a taxonomic identification key of *S. dorsalis*. We noted that only 2 of the 21 species of *Scirtothrips* have microtrichial folds extending fully across the sternites. The taxonomic traits of *S. dorsalis* with thrips slide mount images are very helpful. The accurate and rapid identification of this invasive and potentially devastating pest is essential to implement for an effective management.

- **Scientific Classification**

    Kingdom : Animalia
    Phylum : Arthropoda
    Class : Insecta
    Order : Thysanoptera
    Family : Thripidae
    Sub-family : Thripinae
    Genus : *Scirtothrips*
    Species : *dorsalis*

Figure 5.10: *Scirtothrips dorsalis*

- **Binomial Name** : *Scirtothrips dorsalis*

- **Synonyms** : *Heliothrips minutissimus*
    Anaphothrips andreae
    Neophysopus fragariae
    Scirtothrips padmae

S. *dorsalis* belongs to subfamily Thripinae under family Thripidae. Members of this family have a saw-like ovipositor curving downwards. They have narrow wings with two veins, and antennae of 6-10 antennomeres with stiletto- like forked sense cones on antennal segments (III) and (IV) Thripidae family belongs to suborder terebrantia under order Thysanoptera.

The following characters of *S. dorsalis* were given by Vivek Kumar, University of Florida for positive identification and morphological analysis.

1. Body of adult *S. dorsalis* is pale yellow in colour bearing dark brown antecostal ridge (AR) on tergites and sternites.

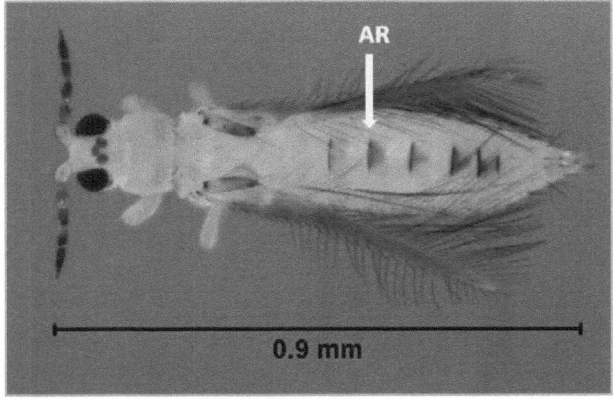

Source : Vivek Kumar, University of Florida

Figure 5.11 :   *S. dorsalis* **adult presenting dark brown antecostal ridge (AR) on tergites (Dorsal View)**

2. Head wider than long, bearing closely spaced lineation and a pair of eight segmented antennae with third and fourth segment each presents a forked sensorium.

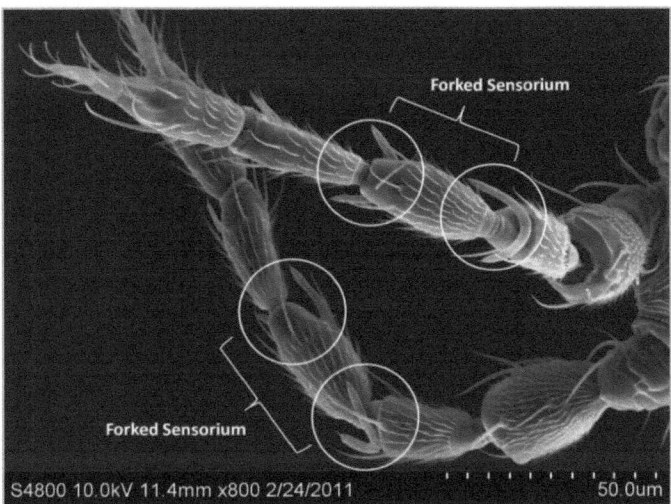

Source : Vivek Kumar, University of Florida
Figure 5.12 :   **Eight segmented antenna III and IV segments : presenting a forked sensorium**

251 | Chapter 5 | Observation

3. Of the three pairs of ocellar setae, the third pair, also known as interocellar setae (IOS), arises between the 2 hind ocelli (HO) and is nearly the same size as the two pairs of post ocellar setae (POS) on the head.

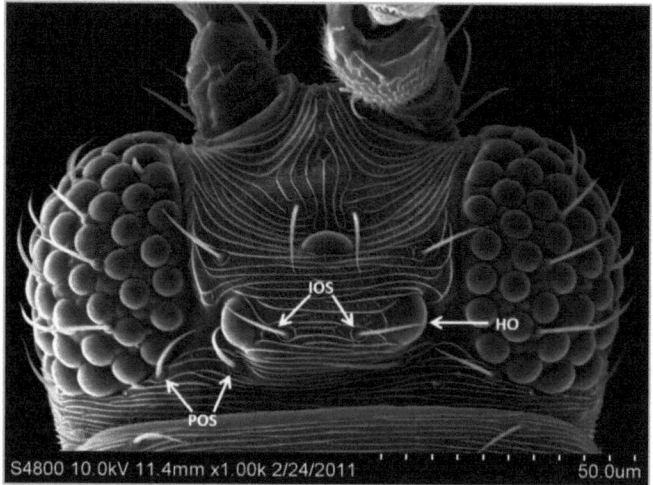

Source : Vivek Kumar, University of Florida

Figure 5.13 : *S. dorsalis* head with ocellar triangle, interocellar setae (IOS), hind ocelli (HO) and postocular setae (POS) – (Dorsal view)

4. Pronotum presents closely spaced transverse lineation.

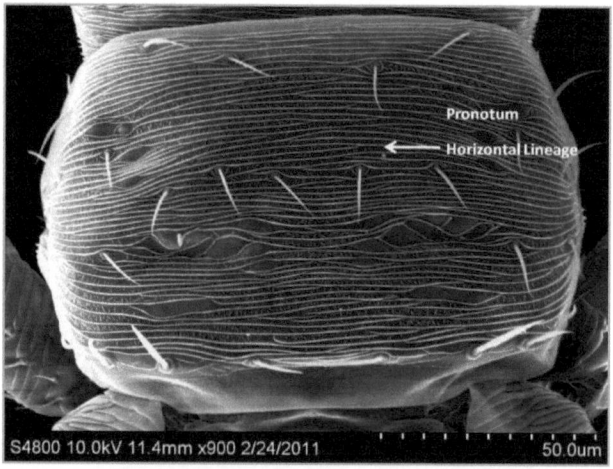

Source : Vivek Kumar, University of Florida
Figure 5.14 : Pronotum of S. dorsalis presenting horizontal closely spaced sculpture lines

5. Pronotal setae (anteroangular, antero marginal and discal setae) are short and approximately equal in length posteromarginal seta-II is broader and 1.5 times longer than posteromarginal setae-I and III. Posterior half of the metanothum presents longitudinal striations; medially located metanotal setae arise behind anterior margin, companiform sensilla are absent.

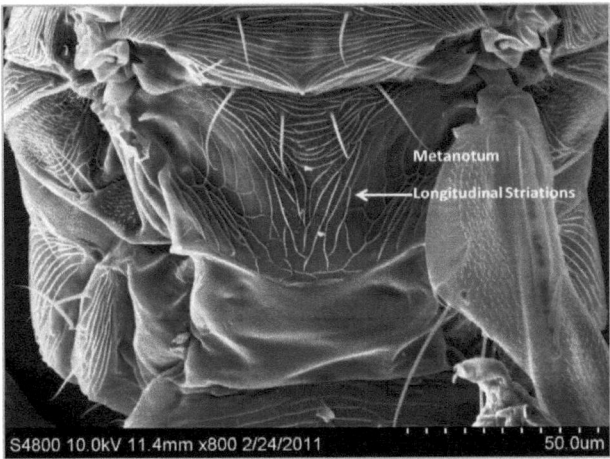

Source : Vivek Kumar, University of Florida
Figure 5.15 : Metanotum (Posterior half) presents longitudinal striations; medially located metanotal setae arise behind anterior margin, companiform sensilla are absent

6. Forewings are distally linght in colour with posteromarginal straight cilia, on distal half, first and second veins bear 3 and 2 widely spaced setae, respectively.

Source : Vivek Kumar, University of Florida
Figure 5.16 : Shaded forewing of *S. dorsalis* is light in colour distally with first and second vein: presenting 3 and 2 widely spaced setae

7. Abdominal tergites III to VI, each present a pair of small medially located setae.

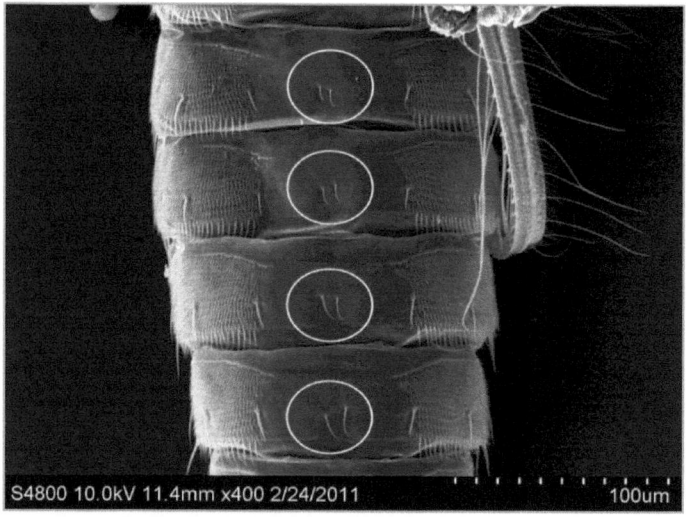

Source : Vivek Kumar, University of Florida
Figure 5.17 :   Abdominal tergites III to VI of *S. dorsalis* present small setae medially situated close to each other

8. The posteromarginal comb on segment VIII is complete, tergite IX of female presents medially located discal microtrichia.

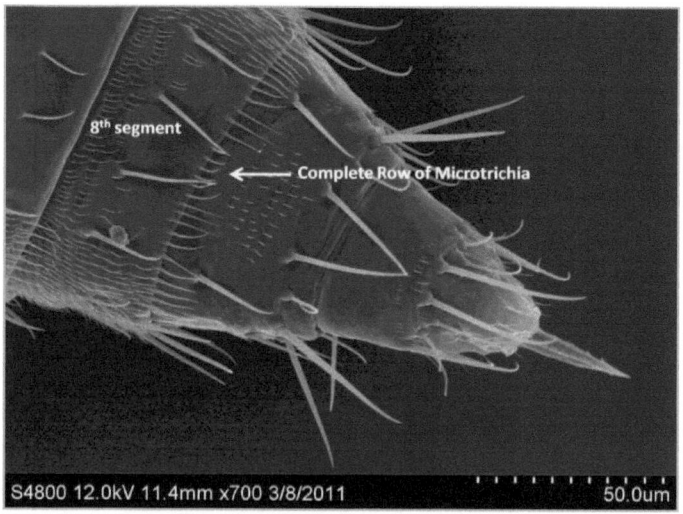

Source : Vivek Kumar, University of Florida
Figure 5.18 :   The posteromarginal comb (row of microtrichia) on segment VIII is complete

9. Discal setae absent on sternites, sternites covered with rows of microtrichia exlcuding on the antero-medial region i.e. a complete band of microtrichia transverse the posterior half of each sternite.

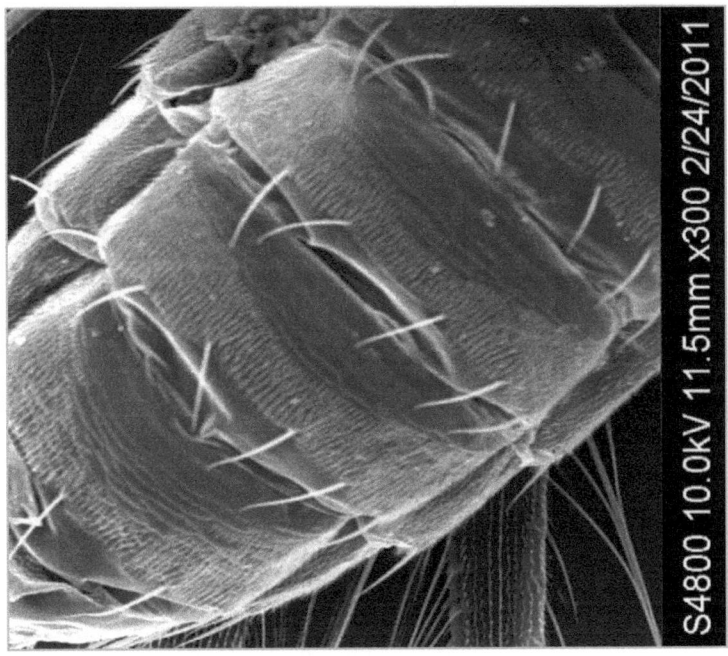

Source : Vivek Kumar, University of Florida
Figure 5.19 : Discal setae absent on sternites, posterior half of sternite presents a continuous band of microtrichia, but microtrichia are absent in the antero-medial region

+ **Taxonomic Key for *Scirtothrips dorsalis***

Identification of *Scirtothrips* spp. is based on male or female adults, both of which are winged. They are pale in colour and minute, and cleared specimens on microscopic slides are needed for identification. For identification of thrips a magnification factor between 100 and 600 is necessary.

Following key characters that allow the identification down to the genus *Scirtothrips* are shown in Table.

Key for the Identification of adults of the genus *Scirtothrips*

Abdominal segment X usually conical, not tubular, serrated ovipositor present; wing surface with microtrichia	Terabrantia
Ovipositor downturned at the apex; abdominal sternite VIII not developed; sense cones on antennal segments III & IV emergent, each more than twice as long as wide	Thripidae
Head and legs not strongly reticulately, sculptured, abdominal tergites may be laterally sculptured; antennal segments III and IV usually with microtrichia; terminal antennal segments rarely elongate; meso and/ or metathoracie furcae with or without spinula; forewing first vein not fused to costa	Thripinae
Abdominal tergites covered with numerous microtrichia body often clear yellow, 8 antennal segments, 3 ocellar setae, posteromarginal pronotal setae 2, usually elongate, pronotum transversely striate, regular without dark internal apodeme.	*Scirtothrips*

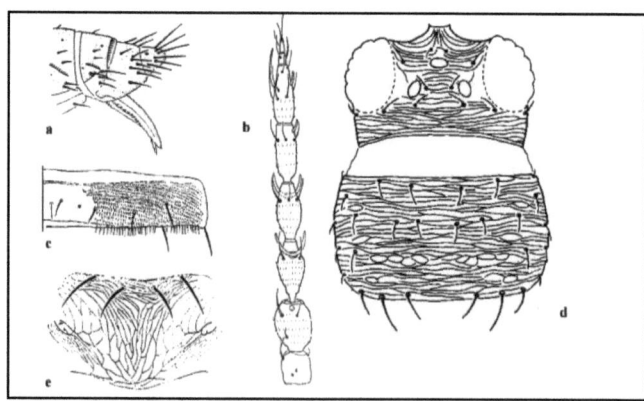

Figure 5.20 : (a) Ovipositor of *S. dorsalis*; (b) Antenna of *S. dorsalis*; (c) Right half tergite V of *S. dorsalis*; (d) Head and pronotum of *S. dorsalis*; (e) Metanotum of *S. dorsalis*

5.6 Biology of *Scirtothrips dorsalis*

Like most members within the sub-order Terebrantia and the family, Thripidae *Scirtothrips dorsalis* develops in several stages – two nymphal and two pupal stages. The average size of thrips is about 1.2 mm. In structural appearances, thrips are not as attractive as many other insects are. Generally *Scirtothrips dorsalis* are plant feeders. They suck plant cells with their mouthparts. They develop at six stages. The egg is their first stage and then they pass through two larval stages. Thereafter we find inactive and non-feeding prepupal stage. Then they pass on to pupal stage and finally become adults either male or female. Under favourable conditions the adult stage is attained within 7 to 10 days. Sometimes, a few individuals take two weeks time in their full growth depending on conditions. It depends on an individual's access to nourishing contents and favourable or unfavourable conditions of temperature. However, it can be concluded that chilli thrips is the fastest spreading thrips. Adult thrips are slender and can be noticed with naked eyes, if viewed carefully. But these thrips flying in the experimental net house are hard to be noticed. Through microscopic studies, it can be seen that adult thrips develop a few distinctive featurs on their external bodies. Two pairs of feathers may be seen in their long but narrow wings. Their wings are marked with fringes of fine hairs. These hairs are visible only when they flutter their wings in the act of flying. When they are at rest, their wings lie parallel to their back. In immature thrips these wings are not viisble. In the external features the colours of the thrips vary from creamish to yellow and even to brownish. In their biological needs they require plant cells which they suck by piercing plant tissues. In this process the plant is damaged. Leaves are studded with greenish black fecal of thrips while they feed on them. Initially, these spots on the leaves caused by thrips feeding were considered as "Murda disease". It was only at the later stage that this change of the colours on the twisted and deformed leaves was known to be the cause of the thrips infestation. When these chilli thrips feed on the plants for a longer duration, flowers and fruits of chilli begin to change colour from bronze to black and thus the plant material becomes unmarketable. When thrips feed in high densities, or in sufficiently dry climates, this process results in the eventual desication and death of the plant. Sometimes low densities of thrips can contribute to the decline in fruit production and plant health.

5.7 Life Cycle of *S. dorsalis*

Stage	Approx. Duration
Egg	2-4 days at 68 – 98^{0}F
1st Instar	1-2 days
2nd Instar	2-4 days
Prepupal stage	1-2 days
pupal stage	1-3 days
Adults	12-15 days

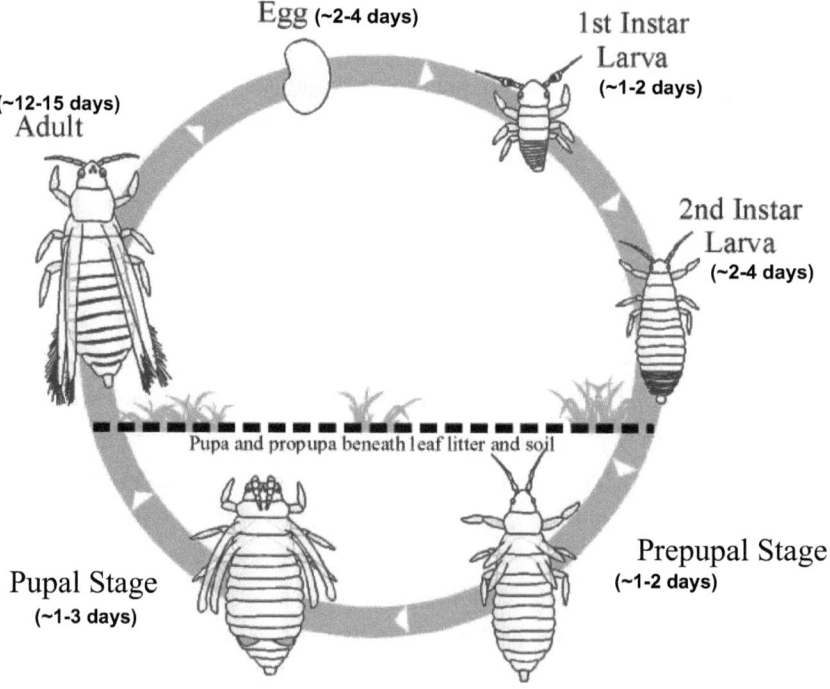

Figure 5.21 :   Life Cycle of *Scirtothrips dorsalis* under controlled conditions

Under the green house conditions, it was seen that most of the thrips were females and hence their population grows fast. In thrips, the process of reproduction without fertilization is a frequent occurrence. In our survey, the body of thrips was observed as heavily sclerotized. Thrips were gregarious in nature. The growth and longevity of the life of thrips is directly affected by humidity and temperature. The thrips development could continue uninterrupted throughout the year depending on the variety of temperature. When the weather conditions are warm and dry thrips swarm like anythings. Thrips had a rapid life cycle and showed a thigmotactic behaviour. Due to its thigmotactic behaviour and several biological characteristics, control over it was extremely difficult in our examination. Thrips produce eggs and insert them into leaves and petal tissues. The eggs hatched into larval. In this manner, thrips insects passed through two larval stages. These stages were fed under protected areas where thy showed a high reproductive rate. Sometimes the adult thrips spread with wind currents. The females started laying the eggs (3-5 days) after their emergence. Eggs were deposited on soft and succulant parts of the host plant.

Longevity

Longevity of the adult female varies from 21-24 days at a constant temperature of 27 ± 1^0C in the laboratory.

Oviposition

The eggs are deposited on the surface of the plant parts and start laying eggs in 3-5 days after their emergence. Each female lays 15-40 eggs per day and the total number of laid eggs ranges from 70-200. Eggs are heavily filled with yolk. These eggs are produced sexually as well as parthenogentically.

Egg

The minute bean-shaped eggs (0.075mm long and 0.070 mm wide) are inserted in the tender plant tissues, including leaf, flower, stem and fruit by a tiny, saw like ovipositor. In capsicum, egg hatched on leaves and have a speckled appearance. After an incubation period of 6-8 days, the egg ruptures at the anterior end. In this manner, the lid gets shifted to one side and the first instar larva wriggles out. Antennae comes out and then it is first followed by the head and thereafter, the rest of the body. In thrips life cycle the development intermediates hemimetabolism and holometabolism. Temperature and weather conditions directly affect the reproductive potential of *Scirtothrips dorsalis,* their preoviposition, fecundity and longevity.

All species are haploid (male) – diploid (female). Thrips shows a facultative parthenogenesis. Male offprings grow from unfertilized females and female offsprings come from fertilized females and this phenomenon is known as arrhenotokous.

First Instar Larva

A fleshy emerged larva is creamish colour with red-pigmentation on the lateral sides of the body. Nymphs feed on plant leaves and suck plant cells and when their feeding finishes, they either fall on the ground or they come in sedate posture on the leaves of the plant and grow into the next stage of their development.

Second Instar Larva

General body structure of second Instar larva is quite similar to first instar larva, except they are bigger in size. It is creamish yellow in colour, with red pigments. These pigments are scattered all over the body. Both immature forms of thrips are wingless. We observed that extreme winter restrained the population growth of the thrips and when the warmer days of the year came, they began to grow in numbers. Actually their abundance and thickness depended on the availability of their food and shelther. Thrips population was readily destroyed by heavily rains. They were most numerous in arid and semi arid climates.

 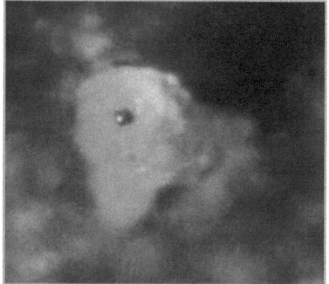

Figure 5.22 : Ist and 2nd Instar (Nymph) of *S. dorsalis*

Prepupa

It is the short duration life stage of thrips life cycle. They do not exhibit any marked and characteristic feature. In this stage, the wing buds are not distinct. The duration of life span of prepupal stage is at 1-1.5 days.

It is due to the availability of chilli crop throughout the year that the thrips also continue to develop throughout the year with rapid, less rapid or slow growth rate. During warmer periods in late afternoon the thrips sometimes shows swarming activity. They are caught by wind currents. In this manner, they are dispersed over a wide area.

Pupa

It is generally creamish in colour with red pigmentation over the body. Pupa is robust. They are non-feeding and quite inactive. Pupae of chill thrips could be seen growing on the axial of leaves and leaf litter and under the calyxes of flowers and fruits.

Small size (< 2mm) of *S. dorsalis* life stages and its rapid movement under controlled conditions made it less detectable in fresh vegetation. The thigmotactic behaviour of thrips may increase the chance of transportation through internal trade of fresh plant material. Thrips are opportunistic and they are exploiting intermittently in occurring environment.

Figure 5.23 : Prepupal and Pupal Stage of *S. dorsalis*

Adult

Under favourable conditions, the young attain full growth i.e. adult stage in 7 to 10 days. Adult species are as large as 1-2mm long and can be noticed by open eyes without the use of microscope. But they evade normal human range of visibility when they come in the act of flying. In thrips life cycle, the most distinctive external features can be seen on the full grown thrips. They develop haired feather in their long and narrow wings. The adults, while flying in the air and then resting on leaves for food and shelter may continue their lives about 30 to 40 days. Reproduction occurs with or without mating. There are one to many generations of thrips in a year. Hibernation most commonly occurs in this stage. They can hibernate in fresh vegetation and in their curled leaves. The process of migration is hardly seen, if sufficient food is available for the adult thrips. They are weak fliers and pass their time in getting food from the leaves and other plant tissues.

Adult Female

The adult female of thrips is deeply pigmented. They are pale yellow in colour. They are found under protected areas of the plant such as flowers and terminals. Antennae are light segmented with different shades of pale yellow (Segment I and II) to the (Segment III – VIII). Prothoracic setae are dark. Head is pale in colour longer than broad. Wings are well developed and devoid of microtrichia, Mouth cone is triangular in shape and acutely pointed at the end.

Adult Male

This species (male) is reproduced by arrhenotokous parthenogenesis (producing males from unfertilized eggs).

Figure 5.24 : *S. dorsalis* : Young attain full growth

5.8 Taxonomy and Biology of Natural Enemies of Thrips

Predators

We selected two species of predator to the measured control on thrips species. To evaluate the applied measures, we performed their taxonomy and biological study under laboratory conditions with the help of taxonomic experts.

(1) *Amblyseius cucumeris* **or** *Neoseilus cucumeris*

Amblyseius is a large genus of predatory mites belonging to the family; Phytoseiidae under superfamily Phytosioidea of class, Arachnida. Many members of this genus feed on the population of other mites and also on thrips.

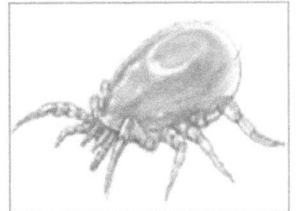

Figure 5.25 : *Amblyseius cucumeris*

Scientific Classification

Kingdom	:	Animalia
Phylum	:	Arthropoda
Class	:	Arachnida
Subclass	:	Acari
Order	:	Mesostigmata
Super family	:	Phytosioideam
Family	:	Phytoseiidae
Genus	:	*Amblyseius*
Species	:	*cucumeris*

Habit and Habitat

Mites are not parasitic but they are free-living and have predatory habit. They can be recognized by the single pair of spiracles positioned laterally on the body.

Cucumeris is a predator better known for thrips control. The colour is hazy and tan. They are found on the underside of chilli leaves along the veins or inside mature flowers of chilli plant. Mites are most effective at preventing thrips build up, when we apply them during the early days of growing season of chilli. Adult had 4 pairs of legs. The size of adult is ~0.3 – 0.5 mm and we observed that females of predatory mites were bigger in size than males.

They had elongated, pear shaped body. The colour of mite was completely dependent on what they ate. They can vary from dark red, to purple and white to light yellow. But, when they preyed on thrips, the colour tended a kind of light orange. They lacked the true head. The mite body is conspicuous and did not show segmentation. So, the body of *A. cucumeris* is a whole unit.

Predatory mites have distinctive features. It has relatively few hairs on its back, 20 pairs of hairs at the most. The palps of this mite was noted using feeding and handling the food. Adult had got typical long legs, which enabled them to move quickly. Each leg was consisted of the coxa, trochanter, femur, genu, tibia, tarsus and apotile.

Biology of *Amblyseius cucumeris*

In a chilli crop, the predatory mites lived in the microclimate and were found in the layer of air tight which was next to the leaf surface. Female of *A. cucumeris* laid 4-5 eggs per day, with an average of 40 eggs during her life cycle. Eggs of mite are oval and transparent. Eggs were deposited on hairs on the underside of leaves. They were hatched after 3 days. The eggs hatched and developed into many developmental stages such as:

Egg → Larva → Protonymph → Deutonymph → Adult.

Egg – adult	11 days
Sex – ratio (% Female)	65%
Eggs/ Female	40 eggs
Life cycle at 20^0C (68^0F)	
Longevity	3 Weeks
Consumption / day	6 thrips larvae/day

Mode of Action

It was observed that the nymphs and adults of *Amblyseius cucumeris* fed on young thrips larvae (mainly 1st instar).

Newly hatched larvae did not feed until they moulted within 2 days. The larvae, nymphs and adults are droplet shaped. The optimum temperature for *A. cucumeris* lies between 25 - 28 degree celsius. *A. cucumeris* is not able to survive at high temperatures, if sufficient food was available. In *A. cucumeris* the development from egg to the adult phase took only five to six days. The predatory mite did not go into diapause (dormancy) in response to shorter days of lower temperatures. This means that the *A. cucumeris* was also active on shorter days and showed much better efficacy on thrips population. An adult of *A. cucumeris* lived for about 3 weeks. The young larvae that emerged had only six legs. They do not feed. A nymph looked like a smaller adult, so there was found no metamorphosis. Adults pierced their prey and then fed over it.

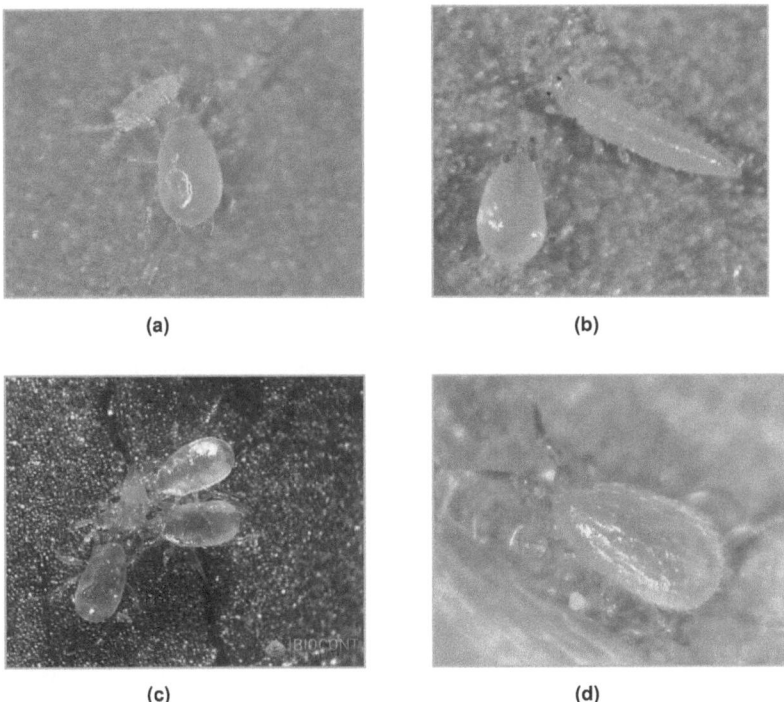

Figure 5.26 : *Amblyseius cucumeris* : (a) Feeding on 1st Instar larvae; (b) Feeding on Thrips pupa; (c) Active on thrips population; (d) Searching for their prey

Life Cycle

The life cycle of *A. cucumeris* begins with small white eggs. There are two nymphal stages which last 7 days and the adult stage which lasts upto 30 days and they feed upon immature stages of thrips. An adult can eat an average of 1 thrips per day.

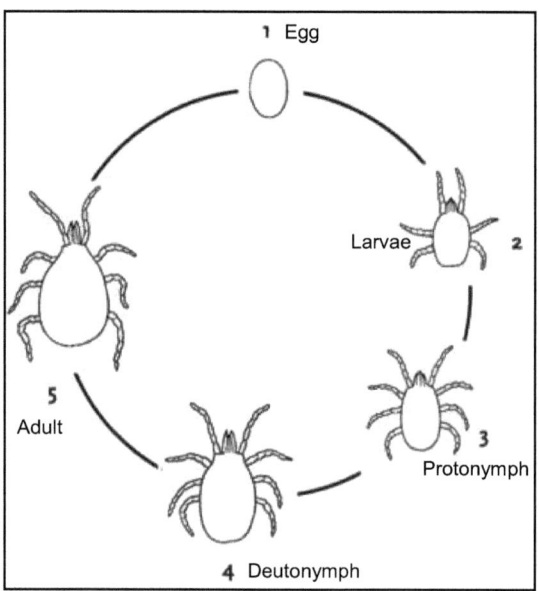

Figure 5.27 : Life Cycle of *Amblyseius cucumeris* (Developmental Stages from Egg to Adult : Under Controlled conditions) (1. Egg; 2. Larvae; 3. Protonymph; 4. Deutonymph; 5. Adult)

(2) *Macrotracheliella nigra*: Commonly known as Minute Pirate Bug

Taxonomy

Anthocorids have long been recognized as beneficial predators. They are considered to be of economic importance to man. They are phytophagous in nature. The adult male length is 2.24 mm and the width 0.77mm. The length of head is 0.45 mm. The adult female length is 2.52-2.66 nm, and 0.91-0.98mm in width. They are more robust They are dark reddish brown in colour. *Macrotracheliella nigra* belongs to the Family Anthocoridae under superfamily Cimicoidae.

Figure 5.28 : *Macrotracheliella nigra*

These bugs belong to the order Hemiptera. The Hemipterans – are most often known as "true bugs". The Heteropterans can be diagnosed with the help of their mouthparts. The typical hemipterans have mandibular stylets concentric and surrounding maxillary stylets. The labium is inserted anteriorly on the head to form a "beak" or "rostrum", called proboscis. This proboscis is able to pierce plant tissues. Rostrum extends to anterior coxae. The antennae of *Macrotracheliella nigra* are typically segmented. The second antennal segment of this bug is about 0.28 mm long and black in colour. But terminal segments are reddish brown in colour. The pronotum of this bug is 0.73mm wide at base, black in colour as well as shiny. The body is oval to triangular. They have shield body. So, they are called Hemelytran. This species is characterized by the long neck and the shape of their pronotum. The genital clasper is similar to that found in other species; *Orius*. The osteolar canal is dinstinctive in feature.

Figure 5.29 :   *Macrotracheliella nigra* preyed on their host : thrips

Scientific Classification

Class	:	Insecta
Subclass	:	Pterygota
Infraclass	:	Neoptera
Superorder	:	Paraneoptera
Order	:	Hemiptera
Suborder	:	Heteroptera
Superfamily	:	Cimicoidea
Family	:	Anthocoridae
Sub-family	:	Anthocorinae
Genus	:	*Macrotracheliella*
Species	:	*nigra*

They are recognized as beneficial predators. They have predatory nature and depend on insects and other arthropods, thrips for food. During our observation they mostly preyed on thrips eggs. When the supplied food exhausted in one habitat, they sought other areas for food and shelter. The habitats of prey provided excellent hiding place for this predator.

Biology and Life Cycle of *Macrotracheliella nigra*

Macrotracheliella nigra was found in the population of chilli thrips in natural conditions. They are a potential predator which feed on the different species of thrips. They might vary from 1.07 – 4.9 mm in length. The general appearance of anthocorids is similar to the mirids except that anthocorids have ocelli. They are flattened and glabrous or pubescent. The wings of this predator are Macropterous or Brachypterous. After spawning, the duration of the developmental stages takes place depending on the varied temperature. Only within a few minutes hatching takes place. The nymphs of *Macrotracheliella nigra* are almost similar to the adults. *Macrotracheliella nigra* did not show metamorphosis during growth after hatching. They were only driven by shedding or the process ecdysis. The main characteristic of this predator is *Fledging* in which they accompany the development of their structure. Even the wings and sexual organs develop at this stage. During their life cycle, the nymphs develop in five stages. On examination, it was found that their sexual organs were visible externally. In the fifth instar of this predator, the ovipositor of the female could be seen apparently. The other sources of this predator other than arthropod insects were noticed during our observation such as forbs, shrubs and trees etc. Nymphal food was as much useful to them as the adults. They preferred only individuals for their food and chose them on the basis of their size. During the nymphal stage of a single individual several hundred thrips were consumed. It seemed that at least one meal was needed between two molts, and the most hearty meal preceded for the phenomenon of ecdysis. The egg of this predator was ovoid and appeared pale sculptured chorion. Fertilization tubes or micropyles were not present in them. However, numerous pseudo micropyles were present in them. Oviposition of this predator varied on the basis of different developmental levels of the ovipositor of the females. Predatory habits of most Anthocorids are well recognized during our research examination.

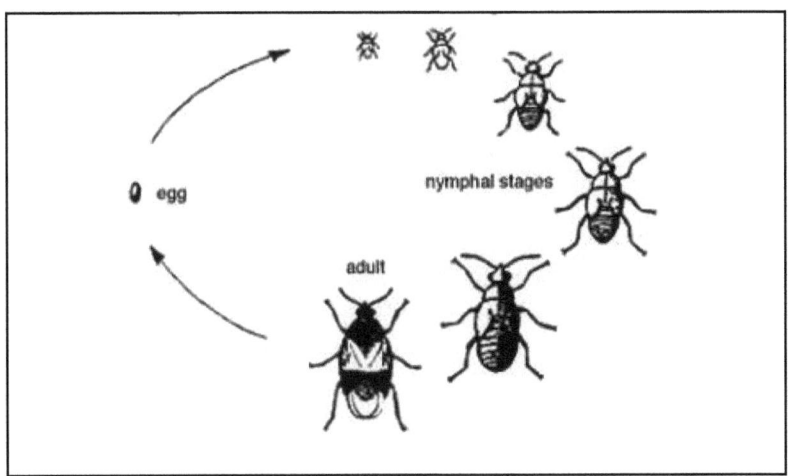

Figure 5.30 : Developmental Stages (From Egg to Adult) *Macrotracheliella nigra*

Parasitoids

Introduction : Thrips known to be parasitized by many hymenopterans which belongs to the Family; Eulophidae. Parasitoids are mostly host specific and appeared more effective to prevent the population of thrips at low densities. It was because parasitoids have inability to switch on to another host. Some factors such as, Reproductive Strategies and fecundity limits could influence the efficiency of the parasitoids. In order to assess their role as biocontrol agents for the biological management over *Scirtothrips dorsalis*, we evaluated the informatio on aspects such as, the biology, reproductive behaviour of selected parasitoids and their bioefficacy. An attempt was made in detail with reference to two parasitoids; *Thripobius semiluteus* and *Ceranisus menes*.

Hymenoptera is found as a singnificant order in all the entomophagous insects. Over two-third insects and pests are biologically controlled by hymenopterous parasites. Hymenopterous parasitoids of thrips are related to the superfamily Chalcidoidea. There are several extremely interesting biological adaptations among the Hymenopterans, in which one is the ovipositor (specialized organ or egg laying). It is composed of long interlocking chitinous stylets. Egg passes in Hymenopterans through this stylet. The ovipositor acts as a drill to pierce the host or the material surrounding the host. In many cases parasitoids serves as a hypotermic (inject paralyzing venom into the host). In some species it is used to secret and form a feeding tube. This feeding tube helps adult parasitoids suck the host body fluid in order to obtain protein for continued egg production. Most of Hymenopterans are solitary endoparasitoids of larvae (Eulophidae). Ovipositor of Hymenopterans lacks muscles and it is supplied with nerves which extends its length to the tip. Ovipositor bears highly sensitive sense organs that can be discerned by chemical stimuli whether a host is suitable or not. The well-known method of reproduction of female production is from fertilized eggs. But the production of male shows arrhenotoky (production from unfertilized eggs). Hymenopterans shows the characteristic of biparental. But this characteristic is supplemented in some species in which uniparental reproduction and parthenogenesis occurred. The phenomenon of parthenogenesis is called Thelytoky. In this process, females give rise to females and males are either lost or if they surrive they became capable of killing the pest.

Parasites may have one generation and it is called univoltine. If in certain cases, these parasites have two or more generation it is called Multivoltine to one of the host. The Life cycle of hymenopteran parasites are generally short, of course, they live only between ten days and thirty days. In warm and hot weather, it is only for about ten days and in cold weather, it can go upto 30 days. In general, **parasitoids have great potential rate of increase.**

A common parasitoids has two distinctions such as Ectoparasitoids and Endoparasitoids. Ectoparasitoids feeds externally upon the host, but in case of Endoparasitoids, the development is internal within the host. Certain species of parasitoids may start their life as endoparasitoids and later emerge from their host externally for continuous feeding on it. Some others may start as ectoparasites and then bear into the host.

During our monitoring observation, ectoparasitoids most frequently occured in hosts that live in some protected areas – a larva in a leaf mine, a pupa in soil and an armored insect under a wax shield. It was because they were likely to be dislodged and lost their host. Many of them parasitoids stang and paralyzed the host prior to oviposition. On the other hand, endoparasitoids were usually well protected within the host. So, we selected for our research work those endoparasitoids which acted more dominantly over the population of thrips species. In this manner, we observed a special adaptation in those endoparasitoids. They respired in a liquid or semi-liquid medium. Some endoparasitoids larvae could directly obtain oxygen from the host's body fluids (either it was through the entire integument or through a posterior vesicle). Many endoparasitoids **'Mummifed'** their hosts upon completion of their own larval development. This was especially noticeable in parasitized thrips, in which the integument becames distended and hard. It is a biological characteristic that most insects parasitize upon other insects; they are protelean parasites. They parasitize only in their larval (immature) stages, and prefer free lives as soon as they become adults. They feed upon their victims body material first and then begin eating pupates.

(1)　　*Thripobius semiluteus*

Synonyms

- *Thripobius hirticornis*
- *Thripobius semiluteus*

Scientific Classification

Figure 5.31 :
Thripobius semiluteus :
A Parasitic Wasp

Kingdom	:	Animalia
Subkingdom	:	Eumetazoa
Phylum	:	Arthropoda
Subphylum	:	Hexapoda
Class	:	Insecta
Order	:	Hymenoptera
Suborder	:	Apocrita
Superfamily	:	Chalcidoidea
Family	:	Eulophidae
Subfamily	:	Entedoninae
Genus	:	*Thripobius*
Species	:	*semiluteus*

Taxonomy

Thripobius semiluteus, a larval endoparasitoid belongs to the order Hymenoptera of the family Eulophidae. It is solitary in habit and shows a uniparental characteristic. In preference it parasites the first and early second larvae of *S. dorsalis*. Females of this species are 0.6mm in size. Head and thorax are purple black in colour and in the lateral view it seems sharply convex in shape. Their wings are almost transparent and are called hyaline. They have slightly curved subcubital vein. Their antennae are pale yellow in colour. The gaster of *T. semiluteus* has 1-2 small dark spots which is sublateral in position. The male of *T. semiluteus* was not known during our research examination.

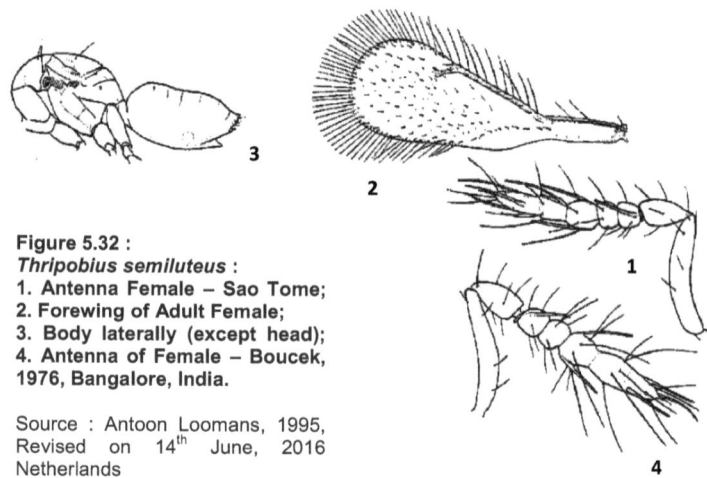

Figure 5.32 :
Thripobius semiluteus :
1. Antenna Female – Sao Tome;
2. Forewing of Adult Female;
3. Body laterally (except head);
4. Antenna of Female – Boucek, 1976, Bangalore, India.

Source : Antoon Loomans, 1995, Revised on 14th June, 2016 Netherlands

Biology

T. semiluteus is uniparental (only females are produced). During searching for their hosts, the female of this species walks slowly at sideways over the leaf surface. In the beginning they feed upon larvae of first and early stages. However, the parasitoids were found less comfortable in feeding upon the second stage larvae and the result was that about half of the pupae could not develop due to want of food. The prepupal and pupal stages of thrips could not provide sufficent food for parasitoids. The parasitoid killed mature larva for its food and having eaten upon it and it transformed into black pupa.

This pupa is present on the surface where the parasitized thrips larva has been feeding. The Developmental time of *T. semiluteus* averages 22-25 days at 23°C during our research examination.

(2) *Ceranisus menes*

Synonyms

- *Ceranisus*
- *Thripoctenus*
- *Epomphale*
- *Cryptomphale*

The C.m. is solitary in habit. It acts as internal parasitoids of the larval stages. But, sometimes the prepupa and pupa may be attacked.

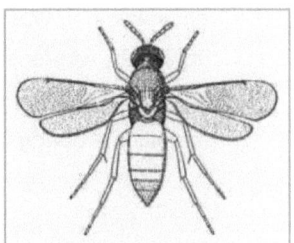

Figure 5.33 :
Ceranisus menes :
A parasitic wasp

Scientific Classification

Kingdom	:	Animalia
Subkingdom	:	Eumetazoa
Phylum	:	Arthropoda
Subphylum	:	Hexapoda
Class	:	Insecta
Order	:	Hymenoptera
Suborder	:	Apocrita
Superfamily	:	Chalcidoidea
Family	:	Eulophidae
Subfamily	:	Entedoninae
Genus	:	*Ceranisus*
Species	:	*menes*

Taxonomy

Ceranisus menes belongs to the subfamily Entedoninae under the family Eulophidae. Hymenoptera is a large order of insects that is represented by several hundred thousand different species. Head and mesosoma of adult female was observed as dark brown in colour. Females are 0.66mm till 1.06mm in size. Ovipositor sheath infuscata if viewed apically of *C. menes* is characterized as head and thorax. All legs are yellow in colour. The hyaline wings are a distnguished feature of this species. The antennae of male and female have marked differences and it is on this basis that they are distinguished. Further, the male has truncate brownish abdomen. In many records, colour features of this species are not mentioned. But on the basis of observation, it was found that the colour of the abdomen of the females was yellowish, whereas in males we found brownish abdomen. The only known males of *C. menes* are also associated with females having a brown metastoma during our taxonomical examination.

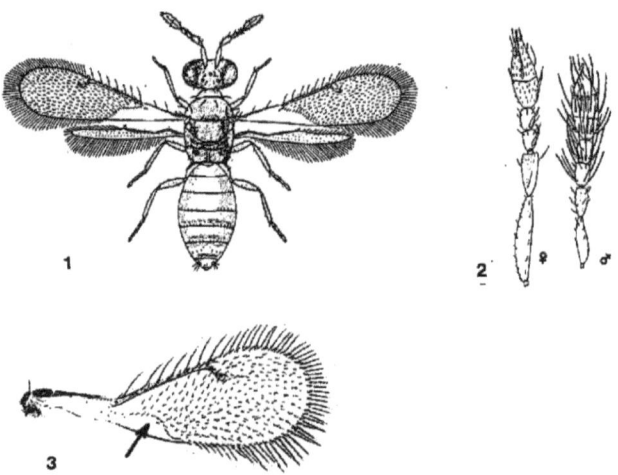

Source : Antoon Loomans, 1995, Revised on 14th June, 2016 Netherlands

Figure 5.34 :   *Ceranisus menes* : 1. Adult female; 2. Antenna of Female and Male; 2. Front Wing (showing Sinuate subcubital vein – arrow)

Biology and Life Cycle of *C. menes*

C. menes is a very effective parasitoid of thrips larvae. Female inserted its ovipositor into the abdomen of their host; *S. dorsalis* and deposited a single egg inside the abdomen of the paralyzed larva of thrips. Both the males and females reproduce by arrhenotokous and thelytokous parthenogensis. During our observation, larvae of this species slowly recovered within a few seconds. The larvae of *C. menes* consumed the entire body of the larval host inside its empty shell. They search for their host by handling behaviour and it is similar for various host parasitoid combinations. The incubation period of *C. menes* is about 1-3 days. Larvae moulted in pupa within 3-4 days and the pupa converted to adult in 5-6 days. Hence, the complete life cycle of *C. menes* was about 18-22 days under optimum temperature.

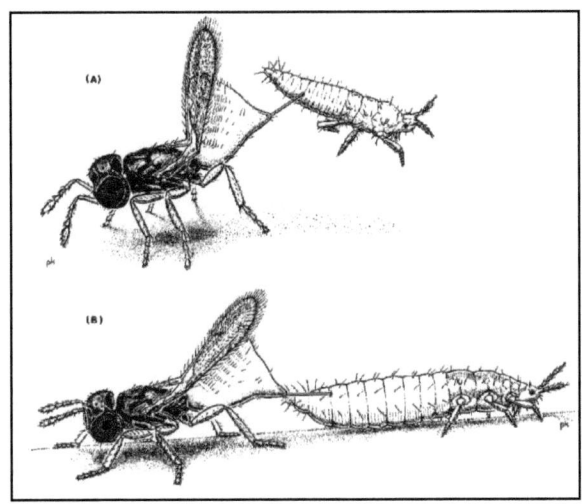

Source : Antoon Loomans, 1995, Revised on 14th June, 2016 Netherlands
Figure 5.35 : *Ceranisus menes* : Oviposition postures : (a) Lifting; (b) Tailing or Dragging (Original)

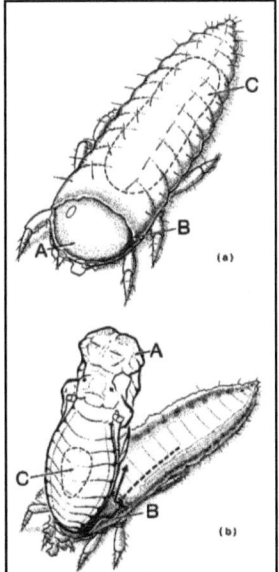

Source : Antoon Loomans, 1995, Revised on 14th June, 2016 Netherlands
Figure 5.35 : *Ceranisus menes* : (a) Prepupa at the movement of pupation; (b) Emerged pupa of *Ceranisus menes*. A = pupa, B = host skin, C = orange/red central spot. The way by which the pupa emerged from the host skin, is indicated by the arrow (original).

Chapter 5: Observation
Number of Figures: 35; Number of Charts: 0; Number of Tables: 0

Chapter-6
Summary and Conclusion

6.1 Summary and Conclusion
6.2 Recommendations
6.3 Future Scope
6.4 Popularization : May Reach at Gross Root Level Need
6.5 General Suggestions : Through Our Entire Research
6.6 Limitations : Over Restrain the Growth of Harmful Pests
6.7 The "Pros and Cons" of Biological Control

Chapter-6
Summary and Conclusion

6.1 Summary and Conclusion

From decades, agriculture has an extensive background. Its association with needful food crops is of vital significance. Agriculture plays a critical role in the economy of the country. It also provides great opportunities in the field of self employment and labour employment. We find the origin of the word, "Agriculture" in Latin language. 'Ager', 'culturo', the two words gave birth to the meaning of agriculture. 'Ager' is related to field and 'culturo' means cultivation or growing. Agriculture contributes a lot in the production of commodities, including food, fibre, forest and horticultural crops maintain the quality of life. Agriculture deserves the second rank worldwide on the basis of the farm input and other life forms for food, bio-fuel, medicinal products etc. In India, agricultural production prepares the backbone of the country and place an important role in its economy. Its share in the Gross Domestic Products prepares the economic health of the country. In the developing countries agriculture is the main source of livelihood in rural areas. Today, the people of the world are conscious about eco-system and healthy environment endangered by pollution and other damages.

The chilli is being cultivated throughout the country for thousands of years. Its use is still relevant despite many changes in the food habits of the people. The genus *Capsicum* represents a diverse plant group which is famous for its intense pungent taste. From the Taxonomic view point, the Kingdom of Chilli is Plantae, the Division, Magnoliophyta, the Class Magnoliopsida, the Order Solanales, the Family Solanaceae, the Genus *Capsicum* and the species *annuum*.

Its popularity among its increasing number of users has made it a crop of economic importance. Besides its nutritional value it is also used in medicines. Unripe fruits are green in colour, but after the process of ripening, they attain the shade of red colour. Green chillies are the excellent source of natural colour and wide range of spectrum of vitamins. These are also rich with group of compounds like Castonoids, Capsaicinoids and Phenolic. From the morphological point of view the chilli plant is a white coloured flower with dark-green leaves that grow up to 1.5 in height. The pungency of chilli is due to its active constitutent namely, *Capsaicin*. It is recognized as a powerful stimulant with no narcotic effect. Chilli is definitely a good item of food for the benefit of human health. They serve as anticancer, anti-ulcer, analgesic and anti-inflammatory, anti-epileptic and anti-haemorrhoidal agents. Chemical analysis of chilli represent the dried pod yields of 100 gm contains 160 calories of energy in which we find carbohydrates 36 gm., protein 18 gm, fat 16 gm, iron 31 mg, niacin 2.5 mg, vitamin A 640 I.U. and vitamin C 40 mg. Chilli is considered as commercial vegetable-cum-spice crops. We need it in our daily life and enjoy its taste from mild to tingling and even explosively hot.

Chilli is widely cultivated throughout the warm temperate, tropical and sub-tropical countries. Chilli is known as a favourite item in the use of cooking, curries, breads and appetizers. Although India is one of the biggest producers of chilli but the productivity rates and quality are not yet satisfactory. In order to compete in the international market effective measures are to be taken to increase its productivity and quality level. In India, the production of chilli is popular in all its corners, but there are several states which do not produce as much chilli as they consume. The production of chilli is found at large scale in States, where climate is generally hot and dry. Its pungency level is better realized in such states of tropical areas. The important growing states of chilli in India are; Karnataka, Andhra Pradesh, Orissa, Rajasthan, Maharashtra, West Bengal, Tamil Nadu and Uttar Pradesh.

Western Uttar Pradesh, which is famous for its fertile soil, abundant water, and varied climate, cold, wet and hot. In Western Uttar Pradesh, agriculture is known as vital source of 'State Wealth' as far as the production of chilli is concerned. It contributes 10% of the total output of chilli in India. our observation areas are the villages around the district Aligarh namely; Tappal, Jalali, Talib Nagar, Sumera and Kayamganj. In those areas chilli is grown at a large scale. These areas contribute a lot in the chilli production of the Western Uttar Pradesh. Cosmetic damage largely affects the chilli fruits, which is directly proportional to yield loss. On the other hand, pests cause indirect damage in the form of several deformities on twigs, leaves, buds, flowers and affect the growth of the plant. High densities of pests population cause an immediate loss in quality in a short period of time. Anyone engaged in agricultural, horticultural or medicinal entomology needs to know something about the insects which commonly affect plants, animals or humans.

There are several studies in environmental sciences where we find the description of the varieties of insects and their patterns of growing in behaviour. Regular studies have enable the scientists to find out certain methods to check the proliferation of insects. The crop ecosystem needs proper management of insect pests and it is on this basis that several methods to control insects have been brought in use. Certain insects are harmful to the crop, but at the same time there are a few insects which are useful for the proper growth of the crop. In the Insect Pest Management all these aspects have to be considered. A better side of the knowledge is related to the relative value of the beneficial insects and it is increasingly recognized. In this way, from the agricultural view point an insect which does not cause damage to crops as a whole is kept out from the list of pests animals. So many insects belong to generally accepted pest species but the individual populations are not always counted as pests and not recognized as economic pests. The 'pest' is a living being in the ecosystem and its presence and rapid breeding causes annoyance to crops in the way of its healthy growth. In this way, from the agricultural view point an insect which does not cause damage to crops as a whole is kept out from the list of pests animals.

From among the numerous pests, we have to recognize certain pests which cause substantial damage to the crop and on the basis of that damage we would like to call them pest species and their value is placed upon these consequences by human society. Large complex of Insects and their species is found in chilli ecosystems. Wherever the crop is grown a few pests are of economic significance. It is reported that the large number of insects creates major pest fauna of chilli including thrips, whiteflies, aphids, mites, caterpillars etc.

There are homogenous group of insects and they are recognized as opportunistic species. These insects have distinctive wings and long fringe in them. There is less nervature in them. Thysanopeteran adults are very small with only a few mm length. They are at ease in flying and migrating into the crops areas. Their body has pecularity of its own. Its mouth parts can easily pierce and suck the tissue contents. It has protursible bladder at the apex of tarsal of leg.

Minute insects i.e. Thrips fall in order – Thysanoptera and Family Thripidae. They show exploiting behaviour intermittently. It plays an active role as a pest of many ornamental, vegetable and agricultural crops. The word 'Thrips' has been derived from a Greek word which means 'Woodworm'. Thrips also refers to 'Fringed Wing Insects, Bladder Footed insects, Storm flies, Thunder Flies, Thunder blights and Corn lice." Large number of thrips species is considered as pests, and affect the plants in commercial value. Thrips have their primary and independent status. Systematic studies on these groups have been extensive.

In all the Asian countries where chilli is grown, we find the spread of *S. dorsalis*. For example, Indonesia, Pakistan, Bangladesh, Korea, Japan, Sri Lanka, China, Taiwan, Phillipines and Thailand are such countries where we find the infestation of *S. dorsalis* profugely. Different scientists express different views about the origin of chilli, but one thing is concluded for certain that the crop of chilli became popular in Asian countries in sixteenth century when Spanish and Portuguese adventurers came to Asia. On the basis of recent studies it is concluded that the origin of chilli is in Indonesia, where it was in use much before Spanish and Portuguese invasion.

There are uncertainities about the original host of *S. dorsalis*. In India, this species is found in plants of castor, pepper, cotton, tea and even in mango and peanuts. This pest is found in the new world also and has been traced in more than 100 plants of about forty families. Thrips are recognized as serious pests on the basis of the extent of damage they cause to the host plant. The presence of thrips creates a kind of cosmetic infestation on the leaves of the host plants. They cause damage to the leaves by feeding on the plant tissues by their specific mouth parts. The presence of thrips causes tissue scarification and depletion of the host plants. Generally, thrips spread on young plants. Their direct feeding on leaves, flowers and fruits causes sufficient damage to the host plants. Thrips are phytophagous. They also show the characteristics of Omnivorous herbivores. Thrips like to spread on the host plants and they grow in numbers constantly by oviposition of eggs. After sometime their infestation becomes thicker and thicker.

Chilli thrips; *S. dorsalis* does not feed on mature host tissues, it feeds on young fruit calyx and leaves. Thrips feeding on delicate plant organs can create silvery scarring on leaves. This early damage to plant growth can lead to malformation of the fruits. It is the result of feeding injury caused by *S. dorsalis* on the host plant that the photosynthetic activity is also affected and the final result is significant yield loss. The infestation usually starts from seedling stage although the severe infestation appears in the stage of vegetation. Thrips possess piercing and sucking mouthparts by which, they feeds and cause damage. They suck the epidermal cells of the host plant and cause necrosis and distortion of leaves, buds, flowers and fruits. Most of the thrips attacking their host plants. Thrips are not very impressive due to its morphological nature in contrast to our better known insects.

Most of the thrips attacking their host plants have a simple life history, which may be enumerated like this - first step is egg. Second and third stages are larval which are non-feeding. The forth stage is prepupal. The fifth is pupate stage and the sixth and the final stage is adult through the phenomenon of metamorphosis. Under favourable environment the young attain full growth in 7-10 days. The growing importance of chilli thrips warrants more quantitative research, not only to provide broader prospective on the biological and economic problems posed by this pest species but one recent study reported that severally infested thrips host plant almost failed to produce any fruits or produced very minimum number of deformed fruits. Thrips are opportunistic species for many scientists, agriculturists and entomologists, exploring intermittently and occurring in environment. The population of thrips grows rapidly during the warmer parts of the year. In this manner, several generations of thrips overlap during the growing season. The population of thrips does not grow in the rainy season or during chilly winters with less than 4^0C temperature. Warm and dry season is the appropriate season for abundant infestation of thrips. The feeding areas caused by thrips are not restricted to leaves alone but they also extend to almost all parts of their host plant. They readily fly and can easily migrate into and around the greenhouses.

In India, chemical pesticides are used to curb the harmful effects of insects and pests. But, the immediate gain in this process has adverse effect on the environment in the long run. Regular use of chemicals leads to insecticide resistance. Then, biodiversity is distributed by pest resurgence and pesticide residues. There are several problems caused to human life by the use of chemicals. Residual toxicity in fruits and vegetables is the immediate effect of chemicals. In the long run it causes pollution of water also. It affects the total scheme of plantation by causing harm to the predators and parasitoids – which are beneficial to the crop. The use of chemicals eliminates the natural enemies of the harmful pests and in this manner there comes the next problem of the outbreak of secondary pests. The recent scientific studies have proved that the pesticides in them toxic heavy metals. It is true that these chemicals have harmful effects on the human life in general in the long run. However, as yields started increasing, the use of pesticide started and becoming widespread. Their adverse effects on the environment and human health also soon became apparent. Now, it is realized that the continuous use of pesticides leads to altering the genetic make up of pests also and they become pesticide resistant in the long run.

In this manner, biological control of pesticides is a safe alternative. This method may replace pesticides and can bring good results both ways – at economic threshold and on environmental issues. It is most cost-effective and environmentally safest way of controlling the pest and its management. The population of thrips is reduced by its natural enemies and in this manner the crop is saved from the harmful effects. In the scheme, nature is used against nature in order to safeguard nature itself. It is a very successful management approach. International Biological Programme has defined biological control in the simplest way, "Using Biota to Reduce Biota." This approach is beneficial in all the ways. Farmers are helped as their crop is saved. Further, it is safer to the environment. In this manner, this approach is cost-effective as well as safer for the environment. That is why this approach is getting popularity among agricultural scientists. Now a days, it has become an interesting area of entomological research. Really, it saves ecosystem from further injury and becomes beneficial for the economics of the man-made society.

In order to make biological control management a success, we must have accurate knowledge of the pest behaviour of the harmful pests as well as the behaviour of those insects which act as their natural enemies. Biological control is a sub-discipline of applied ecology. To adequately practice it, one should have a firm understanding of population and behavioural ecology. Without any systematic and proper identification of insect pests, and their associations of natural enemies, biological control as a science would fail to function.

There are three basic approaches of biological control strategies : (i) Importation; sometimes called Classical Biological Control; (ii) Augmentation; (iii) Conservation. So, biological control of arthropods can be defined as, "The study and uses of Parasites, Predators and Pathogens for the regulation of host (pest) densities."

In Augmentation, we count the supplemental release of natural enemies. With the help of this approach we can boost the population of the natural enemies.

The Predators, Parasites and Pathogens of pests are a large component of world's biodiversity in biological control management. In the scheme of nature, we find the theory of killing. One species lives and prolongs its life by feeding upon some other species. Thrips are attacked by the natural enemies like parasites, predators or pathogens. In this manner, the injurious pests are kept at very low levels by the natural working of predators and parasites. In this manner, crop can be saved from economic hazards. Harmful insects are naturally controlled by these natural enemies. All plant and animal species are the objects of attack by some or other natural agents for their natural habits of attaining food for the perpetuation of their lives. The impact of these natural enemies ranges from a temporary or minor effect to the death of the host or prey.

Predators and parasites are animals that feed on other animals. They show the entomopathogenic nature. Effective biological control may be achieved through their utilization. **The term predation is composed of a broad base of animal → plant (herbivore) – predation supporting the complex web of animal → animal (carnivore) – predation.**

The objectives of our evaluated research work were in favour of useful flora and fauna. Through these objectives we obtained more and more useful yield of the chilli crop. We applied the biological control experiments and methods to control the infestation of chilli thrips, *Scirtothrips dorsalis*. Our objectives were :

(i) Find out the impacts of chilli thrips, *Scirtothrips dorsalis* and their useful natural enemies.

(ii) Evaluated the infestation and damage level of thrips, *Scirtothrips dorsalis* in chilli, *Capsicum annuum*, under biological controlled conditions.

(iii) Study the Biology of thrips and their natural enemies by the way of exploration.

(iv) Evaluate the seasonal fluctuation of thrips.

(v) Provided the biological control management techniques by the use of natural enemies against thrips, *Scirtothrips dorsalis* under controlled climatic conditions.

(vi) Popularized the bio-control technology among the farmers of district Aligarh.

All experiments had done on the host plant chilli, *Capsicum annuum*, which grown throughout the world as a main agricultural crop and known to their useful substance in various forms. Various aspects and legislative approaches have been considered to accomplish our research work. Various insect pests are known to cause damage to chilli crop. But during certain circumstances this crop species are heavily infected with *Scirtothrips dorsalis*. Natural enemies were examined are used as a consultant in the form of various predators and parasitoids. They prey and parasite over the adults of thrips population and significantly controlled. Eating (trophic) relationships often link several species in a community through herbivory, predation and/ or parasitism. This is a version of : Predators eat herbivores, and herbivores eat plants i.e. Tritrophic Interaction. Owing to the need to feed the world's population in a sustainable and enviromentally fashion, it is necessary to consider the third trophic level (natural enemies of herbibvorous insects), when studying insect-plant interactions, not only can this help improve ecological understanding but it is enable us to improve biological control.

In the present research, all the experiments were successfully carried out on *S. dorsalis* with their host plant and their natural enemies. Predators and parasitoids were selected for biological control of thrips. Predators *Amblyseius cucumeris* and *Macrotracheliella nigra* and parasitoids *Thripobius semiluteus* and *Ceranisus menes* were selected for *Scirtothrips dorsalis*. Therefore to reduce the attack of thrips, attempts were made under this research entitled **"Impact of Chilli Thrips;** *Scirtothrips dorsalis*, **Thysanoptera : Thripidae, On Their Host Plant Chilli;** *Capsicum annuum* **with Special Reference of their Biological Control in District Aligarh."**

In this manner, for the evaluation of our scientific, we established a research laboratory in the department of Zoology, D.S. College, Aligarh. However, a experimental net hosue was arranged for the cultivation of chilli plants and for biological management against chilli thrips, by the way of exploration. All biological experiments were done by the augmentative release of selected predators and parasitoids. Some scientific tools i.e. BOD incubator and insect rearing cages, plexiglass containers, petridishes, insect collecting boxes, Binocular microscope and camera Lucida, glasswares, mounting chemicals etc. These scientific tools were arranged to perform the morphological characterization of thrips and their natural enemies and the biology of thrips and their natural enemies.

To accomplish our research hypothesis, five experimental sites, Talib Nagar, Sumera, Jalali, Tappal and Kayamganj were selected for survey and sampling of thrips and its natural enemies in District Aligarh of Western Uttar Pradesh. This extensive survey was made to collect the population of *S. dorsalis* on chilli crop with regard to tritrophic interaction: plant, pest and their natural enemies. Natural enemies namely, *Amblyseius cucumeris* and *Macrotracheliella nigra* in the form of predators and *Thripobius semiluteus* and *Ceranisus menes* in the form of parasitoids were sampled and, then, collected during our survey. For sampling of insects, five plants were selected randomly, four in each direction and fifth was selected from the central area of the field. The survey was made on the basis of 15 days (fortnightly) interval during the year 2014-15. The sampled insect specimens were transferred in vials contained alcohol and, then, brought to our research laboratory for further evaluation. In the research labratory, the BOD incubator was used against rearing the sampled insects. In the same manner, the Binocular Digital Microscope was used against the study of biology and morphology of thrips and their natural enemies. The camera Lucida was also applied to draw the sketch diagrams of sampled specimen. Plastic vials with alcohol were also used to preserve the different life stages of thrips namely : eggs, larvae, pupa and adults. Insect collecting boxes were used to the preservation of insects and to maintain the record of arthropod complex. The ventilated glass containers with lids were also used for rearing of sampled specimens. The culture of thrips was also preserved in the plexi glass containers for further studies.

In order to fulfil our research purpose, regarding biological management of thrips on chilli crop, we also checked the natural presence of natural enemies. In the regular survey, we used the method of sampling and tapping method for sampled insects. The survey was done in the morning hours. It was done from August 2014 to August 2015. The population fluctuation count of sampled insects was calculated in the experimental units on the basis of simple tabular analysis. The presence of natural enemies was examined at that time, when they preyed and parasitized on the thrips population. We collected these natural enemies for our research experiments and used them as consultants. Regular observations and examinations at regular intervals produced significant findings for our rsearch work. These findings are summarized in the following manner:

From Locality Talib Nagar : During the year 2014, the maximum population of thrips was recorded in the first and second half of the month of October. It was recorded 583 in the first half and 300 in the second half. But, in the month of November and December, 2014, the minimum abundance in thrips population was recorded. It was 6 in the second half of the month of November and was 3 in the first half of the month of December. During our regular survey in the year 2015, it was recorded that the minimum abundance in thrips population was in the month of February. It was recorded, 3 in the first half and 11 in the second half respectively. But, the maximum abundance in thrips population was recorded in the month of April. It was 460 in the first half and 510 in the second half. There was no enhancement recorded in the growth of the population of thrips in the month of December 2014 (second half) and January 2015.

From Locality Jalali: During the year 2014, it was recorded that the maximum population in thrips was in the first and second half of the month of October. It was recorded, 465 and 473 respectively. On the other hand, in the month of November and December, the minimum abundance in thrips population was recorded. It was 9 in the second half of the month of November and was 2 in the first half of the month of December. During our regular survey in the year 2015, it was recorded that the minimum count in the population of thrips was in the month of February. It was calculated 2 and 13 for the first and second half respectively. In the same year, the highest population was recorded in the month of April and in the month of May. It was, 465 in the second half of the month of April and 515 in the first half of the month of May. There was no enhancement recorded in the growth of the population of thrips in the month of December (second half) and January.

From Locality Tappal: During the year 2014, it was recorded that the maximum population in thrips was in the first and second half of the month of October. It was recorded, 544 and 339 respectively. On the other hand, the minimum abundance in thrips population was recorded in the month of November and December. It was 5 in the second half of the month of November and was 4 in the first half of the month of December. During our regular survey in the year 2015, it was recorded that the highest population in the month of April and in the month of May. It was, 389 in the second half of the month of April and 384 in the first half of the month of May. There was no enhancement recorded in the growth of the population of thrips in the month of December (second half) and January.

From Locality Kayamganj: During the year 2014, it was recorded that the maximum population in thrips was in the first and second half of the month of October. It was recorded, 451 and 416 respectively. On the other hand, the minimum abundance in thrips population was recorded in the month of November and December. It was 3 in the second half of the month of November and was 2 in the first half of the month of December. During our regular survey in the year 2015, it was recorded that the highest population in the month of April and in the month of May. It was, 410 in the first half of the month of April and 393 in the first half of the month of May. There was no enhancement recorded in the growth of the population of thrips in the month of December, January and February (first half).

From Locality Sumera: During the year 2014, it was recorded that the maximum population in thrips was in the first and second half of the month of October. It was recorded, 603 and 480 respectively. On the other hand, the minimum abundance in thrips population was recorded in the month of November and December. It was 7 in the second half of the month of November and was 3 in the first half of the month of December. During our regular survey in the year 2015, it was recorded that the highest population in the month of April and in the month of May. It was, 445 in the second half of the month of April and 375 in the first half of the month of May. There was no enhancement recorded in the growth of the population of thrips in the month of December (second half) and January.

Similar sampling method at different outfields of selected localities was applied during our research survey for thrips natural enemies, predators and parasitoids. On the basis of random sampling, we determined the natural occurrence and abundance of natural enemies. Survey and sampling were made for the study of *S. dorsalis* on chilli crop with regard to tritrophic interaction between plants, pests and their natural enemies. Two predator species *Amblyseius cucumeris* and *Macrotracheliella nigra* and two species of parasitoids namely, *Thripobius semiluteus* and *Ceranisus menes* were sampled during survey. Adults predators and parasitoids were directly hand-picked from the sampled chilli plants with the aid of a fine hair brush, few species of parasitoids were emerged from the infected or field collected eggs and larvae of thrips collected from the account during the process of rearing which were transferred to the research laboratory. Afterthat, they were placed in BOD incuabtor for their emerging out. The obtained and recorded results of different selected outfields were done in the following manner:

From Locality Talib Nagar: In this locality outfields, the highest population of selected predator, *Amblyseius cucumeris* was recorded 10 & 14 in the first and second half of the month of October during the year, 2014. The maximum abundance of another selected thrips predator, *Macrotracheliella nigra* was recorded 8 and 11 in number in the first and second half of the month of October, 2014. While during the year 2015, the maximum abundance of the predator *Amblyseius cucumeris* was recorded 18 and 20 in the first and second half of the month of May. In the same manner, the maximum count of another predator, *Macrotracheliella nigra* was recorded 19 and 16 in the first and second half of the month of May respectively. In the same manner, the evaluated research work was done with the help of parasitoids. In this locality the presence of selected parasitoid, *Thripobius semiluteus* was recorded 4 and 5 in the first and second half of the month of October, 2014. In the same manner, the presence was recorded of another selected parasitoid, *Ceranisus menes*. It was recorded 5 and 7 in the first half and second half of the month of October. During regular research and sampling survey in the year 2015, it was recorded that the natural presence of parasitoids, *Thripobius semiluteus* was 6 and 7 in the first and second half of the month of May. In the same manner, the samplings of another parasitoid, *Ceranisus menes* was done, and, it was recorded 6 in the first half of the month of April and 7 in the first half of the month of May. There was no enhancement recorded in the population of natural enemies in the month of December, 2014 and January, 2015.

From Locality Tappal: In this locality outfields, the highest population of selected predator, *Amblyseius cucumeris* was recorded 12 & 16 in the first and second half of the month of October during the year, 2014. The maximum abundance of another selected thrips predator, *Macrotracheliella nigra* was recorded 5 and 13 in number in the first and second half of the month of October, 2014. While during the year 2015, the maximum abundance of the predator *Amblyseius cucumeris* was recorded 17 and 15 in the first and second half of the month of April. In the same manner, the maximum count of another predator, *Macrotracheliella nigra* was recorded 17 and 19 in the first and second half of the month of May respectively. In the same manner, the evaluated research work was done with the help of parasitoids. In this locality the presence of selected parasitoid, *Thripobius semiluteus* was recorded 4 in the second half of the month of October and 3 in the first half of the month of November, 2014. In the same manner, the presence was recorded of another selected parasitoid, *Ceranisus menes*. It was recorded 3 in the second half of the month of October and 2 in the first half of the month of November respectively. During regular research and sampling survey in the year 2015, it was recorded that the natural presence of parasitoids, *Thripobius semiluteus* was 6 in the first half of the month of April and 8 in the first half of the month of May. In the same manner, the samplings of another parasitoid, *Ceranisus menes* was done, and, it was recorded 5 and 7 in the first half and second half of the month of April respectively. There was no enhancement recorded in the population of natural enemies in the month of December, 2014 and January, 2015.

From Locality Sumera: In this locality outfields, the highest population of selected predator, *Amblyseius cucumeris* was recorded 14 in the second half of the month of October and 12 in the first half of the month of November during the year, 2014. The maximum abundance of another selected thrips predator, *Macrotracheliella nigra* was recorded 13 in number in the first half of the month of September, 2014 and 13 in the second half of the month of October, 2014. While during the year 2015, the maximum abundance of the predator *Amblyseius cucumeris* was recorded 16 and 14 in the first and second half of the month of May. In the same manner, the maximum count of another predator, *Macrotracheliella nigra* was recorded 13 and 19 in the first and second half of the month of May respectively. The help of parasitoids was taken for the evaluated research work. The presence of selected parasitoids was observed in the selected locality. It was found that *Thripobius semiluteus* was in no. 3 on the selected plants in the month of October. *Ceranisus menes*, a parasitoid was also found in no. 4 on the selected plants in the second half of the month of October. It was recorded 4 in the second half of the month of October. During regular research and sampling survey in the year 2015, it was recorded that the natural presence of parasitoids, *Thripobius semiluteus* was 6 in the second half of the month of April and 5 in the second half of the month of May. In the same manner, the samplings of another parasitoid, *Ceranisus menes* was done, and, it was recorded 9 in the second half of the month of April. There was no enhancement recorded in the population of natural enemies in the month of December, 2014, January, February, June, July and August, 2015.

From Locality Jalali: In this locality outfields, the highest population of selected predator, *Amblyseius cucumeris* was recorded 13 & 15 in the first and second half of the month of October during the year, 2014. The maximum abundance of another selected thrips predator, *Macrotracheliella nigra* was recorded 6 and 7 in number in the first and second half of the month of October, 2014. While during the year 2015, the maximum abundance of the predator *Amblyseius cucumeris* was recorded 18 in the first half of the month of April and 16 in the first half of the month of May. In the same manner, the maximum count of another predator, *Macrotracheliella nigra* was recorded 10 and 9 in the first and second half of the month of April respectively. In the same manner, the evaluated research work was done with the help of parasitoids. In this locality the presence of selected parasitoid, *Thripobius semiluteus* was recorded 2 in the second half of the month of October and 3 in the second half of the month of November, 2014. In the same manner, the presence was recorded of another selected parasitoid, *Ceranisus menes*. It was recorded 3 in the second half of the month of September and 2 in the second half of the month of November. During regular research and sampling survey in the year 2015, it was recorded that the natural presence of parasitoids, *Thripobius semiluteus* was 5 in the first half of the month of April and 6 in the first half of the month of June. In the same manner, the samplings of another parasitoid, *Ceranisus menes* was done, and, it was recorded 4 and 5 in the first half and second half of the month of April respectively. There was no enhancement recorded in the population of natural enemies in the month of December, 2014, January, July and first half of the month of August, 2015.

From Locality Kayamganj: In this locality outfields, the highest population of selected predator, *Amblyseius cucumeris* was recorded 10 and 17 in the first and second half of the month of October, 2014. The maximum abundance of another selected thrips predator, *Macrotracheliella nigra* was recorded 7 and 8 in number in the first and second half of the month of October, 2014. While during the year 2015, the maximum abundance of the predator *Amblyseius cucumeris* was recorded 16 in the second half of the month of April and 16 in the second half of the month of May. In the same manner, the maximum count of another predator, *Macrotracheliella nigra* was recorded 11 and 7 in the first and second half of the month of April respectively. In this locality, the presence of parasitoids was also marked. *Thripobius semiluteus* was recorded 2 in the first half of the month of October and 3 in the second half of the month of October. Similarly, the presence of another parasitoid, *Ceranisus menes* was also marked. It was recorded 2 in the first half of the month of October. During regular research and sampling survey in the year 2015, it was recorded that the natural presence of parasitoids, *Thripobius semiluteus* was 5 in the second half of the month of March and 3 in the first half of the month of April. In the same manner, the samplings of another parasitoid, *Ceranisus menes* was done, and, it was recorded 2 in the first and second half of the month of March. There was no enhancement recorded in the population of natural enemies in the month of December, 2014, January, May, June, July and August, 2015.

For obtaining the pure yield production of chilli, we established a well developed net house in D.S. (P.G.) College, District Aligarh. The net house was set in such a manner that there was natural ventilation and regular effect of the climate. Weeds and grass were also unweeded at regular intervals. The supply of necessary minerals for the growth of Chilli plants was faciliated in a natural manner. Entire mineral capacity of the soil and natural environment was directed towards the suitable growth of the chilli plants. In order to facilitate our research, chilli, *Capsicum annuum* seedlings transplanted in sets of microplot. In total, twenty one experimental microplots of seven net houses were prepared for our controlled experiments. We collected certain sample plants from different nurseries and transplanted them in the experimental microplots in order to find out results over damage on chilli plants with the help of biological experiments.

The evaluation of biological management through natural enemies by the process of rearing and mass production performed continuously under research laboratory of Zoology Department, D.S. College, Aligarh. In this manner, some numbers of thrips were also collected from different selected chilli outfields of District Aligarh. On the other hand, few species of selected predators and parasitoids were collected during the rearing period of thrips under laboratory experiments, when they were emerging from the infected eggs and larvae of the sampled thrips. We were also obtained some live species of natural enemies of thrips from the Department of Zoology, Aligarh Muslim University, Indian Agricultural Research Institute, Bangalore. Thus, a maintained stock of thrips (all life stages) and their natural enemies were maintained for the experiment of biological control.

Experiments were performed biologically without any use of pesticides. Under 21 different sets / microplots of seven net houses. Each treated and replicated with three times release of predators and parasitoids. So, we covered all the microplots separately with the help of nylon nets.Each experiment was performed, till the population of thrips reached at significant level i.e. Economic Threshold Level (ETL). In this manner, we were recorded the remaining number of thrips population regularly for three weeks. In our experiments, the possibilities of use the native predators *Amblyseius cucumeris* and *Macrotracheliella nigra* and the parasitoids *Thripobius semiluteus* and *Ceranisus menes* for the biological management of chilli thrips on chilli plants under controlled conditions.

Control of *S. dorsalis* by Predators

Amblyseius cucumeris : The experiment of biological control during the year 2014-2015 against the infestation of chilli thrips with the help of predator, *Amblyseius cucumeris* was done. The first release of predators was 40, 45 and 50. At that time, the percentage reduction was recorded at 20.8%, 27.6% and 33.3% in the I, II and III experimental microplots.

After one week of the first release, the percentage reduction was recorded at 48.5%, 52.4% and 59.02% (for second release). In the same manner, after two weeks of the first release, the percentage reduction in the thrips population was recorded at 68.2%, 74.02% and 82.32%. During the year 2015-16, the release rate of predators was increased. It was made 50, 55 and 60 for first, second and third release. The percentage reduction after the first release was recorded at 17.8%, 24.8% and 30.2%. After the second release, it was recorded at 45.9%, 51.0% and 56.8%. At the time of the third release (after two weeks of the first release) it was recorded at 67.12%, 75.51% and 84.80%.

Macrotracheliella nigra : During the year 2014-15, the experiment was done to control the thrips population by the release of predator *Macrotracheliella nigra*. During the first release, the number of predators 45, 50 and 55 was released to control the thrips population. Thus, the percentage reduction was recorded at 25.2%, 32.9% and 41.1% in I, II and III experimental microplots. During the time second release, the gain in percent reduction in thrips population was recorded at 43.7%, 50.23% and 53.6%. In the same manner, at the time of the third release (after two weeks of the first release) the released number of predators was again the same, (45, 50 and 55). At that time, the reduction in thrips population was recorded at 63.01%, 77.04% and 85.8%.

During the year 2015-16, the significant reduction in thrips population was recorded. At the time of first release the release rate of predators was 55, 60 and 65. At that time, the percent reduction in thrips population was recorted at 32.4%, 35.6% and 39.9% in first, second and third experimental microplots. The second release of predators was same as the first release (55, 60 and 65). Thus, the reduction in thrips population was recorded at 41.8%, 49.4% and 59.04%. In the same manner, the third release of predators (after 2 weeks of the first release) was also done as same as the first and second release i.e. 55, 60 and 65. At that time, the reduction in thrips population 61.49%, 78.09% and 86.10%.

Amblyseius cucumeris and Macrotracheliella nigra : During the year 2014-15, the experiment was done to control the thrips population by the release of both the predators. During the first release, the number of predators (40+45), (45+50) and (50+55) was released to control the thrips population. After observation, the percentage reduction was recorded at 23.6%, 26.2% and 32.9% in I, II and III experimental microplots. During the second release (after one week) the number of predators was same as, the time of first release (40+45), (45+50) and (50+55). Thus, the gain in percent reduction in thrips population was recorded at 68.3%, 75.8% and 80.7%. In the same manner, at the time of the third release (after two weeks of the first release) the released number of predators was again the same, (40+45), (45+50) and (50+55). At that time, the reduction in thrips population was recorded at 88.7% and 92.0% in the I and II experimental microplots. At the time third release in the III microplot, the significant level in thrips population was recorded. Thus, there was no need of third release at that time.

During the year 2015-16, the first release of predators (50+55), (55+60) and (60+65) was done. At that time, the percent reduction in thrips population was recorded at 27.09%, 30.65% and 33.92%. The second release of predators was same as the first release (50+55), (55+60) and (60+65). Thus, the reduction in thrips population was recorded at 70.9%, 77.05% and 82.5%. In the same manner, the third release of predators (after 2 weeks of the first release) was also done as same as the first and second release i.e. (50+55), (55+60) and (60+65). At that time, the reduction in thrips population 86.78%, 90.8% of the I and II microplots. At the time third release in the III microplot, the significant level in thrips population was recorded. Thus, there was no need of third release at that time.

Control of *S. dorsalis* by Parasitoids

Thripobius semiluteus : The experiment of biological control during the year 2014-2015 against the infestation of chilli thrips with the help of parasitoid, *Thripobius semiluteus* was done. The first release of parasitoid was 15, 20 and 25. At that time, the percentage reduction was recorded at 10.6%, 36.68% and 32.16% in the I, II and III experimental microplots.

After one week of the first release, same number of parasitoids 15, 20 and 15 was released. At that time the percentage reduction was recorded at 43.23%, 46.7% and 55.6% (for second release). In the same manner, after two weeks of the first release the same number of parasitoids 15, 20 and 25 were released, and, thus, the percentage reduction in the thrips population was recorded at 71.4%, 80.7% and 85.09%. During the year 2015-16, the release rate of parasitoids was increased. It was 20, 25 and 30. At that time, the percentage reduction in thrips population was recorded at 10.56%, 17.6% and 28.4%. After the second release, it was recorded at 41.1%, 50.15% and 53.4%. At the time of the third release (after two weeks of the first release) it was recorded at 69.04%, 78.8% and 87.79%.

Ceranisus menes : The experiment of biological control during the year 2014-2015 against the infestation of chilli thrips with the help of parasitoid, *Ceranisus menes* was done. The first release of parasitoid was 20, 25 and 30. At that time, the percentage reduction was recorded at 11.9%, 21.02% and 27.93% in the I, II and III experimental microplots.

After one week of the first release, same number of parasitoids 20, 25 and 30 was released. At that time the percentage reduction was recorded at 52.1%, 55.3% and 64.8% (for second release). In the same manner, after two weeks of the first release the same number of parasitoids 20, 25 and 30 were released, and, thus, the percentage reduction in the thrips population was recorded at 68.4%, 75.01% and 78.75%. During the year 2015-16, the release rate of parasitoids was increased. The percentage reduction after the first release was recorded at 15.25%, 23.58% and 28.18%. After the second release, it was recorded at 45.21%, 54.11% and 67.5%. At the time of the third release (after two weeks of the first release) it was recorded at 65.64%, 77.3% and 83.1%.

Thripobius semiluteus and *Ceranisus menes* : During the year 2014-15, the experiment was done to control the thrips population by the release of both the parasitoids. During the first release, the number of parasitoids (15+20), (20+25) and (25+30) was released to control the thrips population. After observation, the percentage reduction was recorded at 21.7%, 33.11% and 43.7% in I, II and III experimental microplots. During the second release (after one week) the number of parasitoids was same as, the time of first release (15+20), (20+25) and (25+30). Thus, the gain in percent reduction in thrips population was recorded at 66.6%, 83.7% and 87.6%. In the same manner, at the time of the third release (after two weeks of the first release) the released number of parasitoids was again the same, (15+20), (20+25) and (25+30). At that time, the reduction in thrips population was recorded at 88.05% in the I experimental microplot. At the time second and third release in the II and III microplots, the significant level in thrips population was recorded. Thus, there was no need of second and third release at that time.

During the year 2015-16, the first release of parasitoids was increased (20+25), (25+30) and (30+35) was done. At that time, the percent reduction in thrips population was recorded at 19.6%, 33.72% and 39.70%. The second release of parasitoids was same as the first release (20+25), (25+30) and (30+35). Thus, the reduction in thrips population was recorded at 65.5%, 82.9% and 85.7%. In the same manner, the third release of parasitoids (after 2 weeks of the first release) was also done as same as the first and second release i.e. (20+25), (25+30) and (30+35). At that time, the reduction in thrips population 87.26% in the I microplot. At the time second and third release in the II and III microplots, the significant level in thrips population was recorded. Thus, there was no need of second and third release at that time.

Control of *S. dorsalis* by Predators and Parasitoids

Amblyseius cucumeris and *Macrotracheliella nigra* (Predators); *Thripobius semiluteus* and *Ceranisus menes* (Parasitoids) : During the year 2014-15, the experiment was done to control the thrips population by the release of both the predators (A.c. & M.n.), (40+45), (45+50) and (50+55) and the release of both the parasitioids (T.s. & C.m.), (15+20), (20+25) and (25+30) was done. After observation, the percentage reduction was recorded at 41.8%, 44.3% and 46.4% in I, II and III experimental microplots. After the second release, the gain in percent reduction in thrips population was recorded at 86.6%, 91.6% and 93.6%. The significant in thrips population was recorded. So, there was no need to release the natural enemies in the third release.

During the year 2015-16, the experiment was done to control the thrips population by the release of both the predators (A.c. & M.n.), (50+55), (55+60) and (60+65) and the release of both the parasitioids (T.s. & C.m.), (20+25), (25+30) and (30+35) was done. After observation, the percentage reduction in thrips population was recorded at 44.7%, 50.3% and 59.9% in I, II and III experimental microplots. After the second release, the gain in percent reduction in thrips population was recorded at 88.6%, 89.7% and 94.07%. The significant in thrips population was recorded. So, there was no need to release the natural enemies in the third release.

Data Analysis

In order to confirm the progressive data towards the desired research results. Statistical analysis of the data with T-test was done. On the basis of data analysis we found out the feasibility and viability of the biological agents i.e. Natural Enemies. Thus, we conclude statistically that the percentage gain in thrips reduction is affected by the release of predators, *Amblyseius cucumeris* and *Maracrotracheliella nigra* and the release of parastioids *Thripobius semiluteus* and *Ceranisus menes*, during two successive releases. We assumed that the difference between gain in percentage reduction of alive thrips during two successive releases is not significantly different and the gain in percentage reduction of the two samples is accepted at 5% level.

During our survey, the pathogencity of thrips was observed, in the outfields selected for regular monitoring it was found that the chilli plants were heavily infested with thrips breeding. As a result of the thick population of thrips, the proper growth of chilli plants at the flowering and fruit producing stage was affected. Either there were no fruits or at some places there were deformed fruits. Meristems, terminals, and other tender plant parts of the host plant were directly attacked by thrips. In our survey, on the plants above the soil surface we found that the young plant tissues were preferred by pests. Most of the mature plants were found unaffected. On the affected plants we marked the change of the colour from silvery to brown or black. It was also noticed that thrips began to rear their eggs from the early stage of seedling. At the flowering and fruiting stages thrips infestation became severe. It was found that the ventral surface of the chilli leaves was preferred by thrips. By their infestation at these places, they sucked the cell sap and the results was the loss of vitality and vigour of the leaves. Physically, the leaves curled and twisted. Both nymphs and adults sucked the cell sap from the twigs and stems. The affected twigs were also distorted due to loss of vitality. It was the direct damage to the plant. In the indirect damage we noticed the reduction of photosynthesis and finally there was loss to the plant. In our assessment we noticed that the thrips population grew at faster rate during favourable conditions and the net result was about 40% damage to the chilli production. In this manner, it can be concluded that thrips is a measure problem in the chilli production. Several losses in the form of reduction of the leaf size, shortening of photosynthesis activity and other such constraints carry the crop towards significant yield loss. So, thrips are major insect faunna causing big loss to chilli fields. It was also noticed that the chilli plants are highly sensitive to direct cosmetic damage caused by thrips. Moreover, yield loss has direct economic disadvantage.

Direct Damage : In our observations we have seen that plant tissues become the food for thrips and they suck them by their mouth parts. The cell contents of the host plant are easily sucked as the thrips pierce their frontal portion of their mouth parts into it. In this process of sucking and feeding by thrips direct damage can be seen on the host plant in the form of tissue scarification on the leaves. In this natural habit of the thrips, the natural growth of the plants is directly affected. The host plant becomes deficient in its vitality and it appears in a diseased form. It is also observed that this pest prefers to suck the tissues of the young plants. On mature tissues, thrips were found less frequent. There was heavy infestation of thrips on the young plants and it reduced the quality, quantity and value of the crop directly. It was on the leaves of the young plants that the thrips liked to sit frequently they used it as oviposition site as they could get food for their survival and vitality. The better health of thrips became the cause of the direct damage to the plant. It was visible in the deformation of the leaves, flowers and fruits. Heavy infestation of thrips could be seen as silver patches and flecking on the expanded leaves. Chilli thrips are not exclusive for chilli. They feed and survive on several other plant species also. But, their rapid growth and heavy infestation is found on chilli plants, where they get preferrable sites and safer places for their reproduction besides food and shelther. Large growth of thrips on chilli plants becomes the cause of direct damage to the host plants. Several symptoms of damage can be seen on the leaves, twigs and fruits of chilli. Silvery patches on the leaves raise alarm of immediate danger to the plant. Sometimes, leaf lamina is found thickened in linear manner. At another time, Brown Frass Markings are found on the infested leaves due to the occurence of senescence and abscission. Leaf distortion due to feeding scars is directly visible. Scarring of petals and distortion of flower buds and flowers are also visible as direct damage to the host plant. Feeding on chilli plant caused silvery or bronze streaks of spots on the chilli fruit. Thrips also fed on the fruit calyx, caused it to turn up and exposed the fruit, created ghost spotting. In this manner, heavy infestation reduced the ability of the host plant to photosynthesize. Thus, the infested plant became stunted or dwarfed, and leaves with petiole detached from the stem and caused defoliation in some plants. While feeding on the host plants thrips deposited their greenish black fecal on the leaves. It takes about 30 minutes for the thrips to suck the cell and tissue of the leaves for their food. After that, infection on the plant can be seen with naked eyes. On the basis of this observation, it can be understood that thrips is highly polyphagous pest. Its constant feeding may cause total defoliation in the host plant. Chilli thrips generally feed on succulent leaves and young fruits. As a result of this irregular bronzing or scarring on both upper and lower sides of the leaf is visible. Due to prolonged feeding on the host plants thrips could cause discolouration along the mid-rib and lateral leaf veins. As the population of thrips increases and feeding on the plant tissues continues, scattered patches between veins are also visible.

In our survey and keen observations, we found that thrips get thier vitality from the plants of their preferences. However, they are more frequent on Chilli plants because this plant is suitable for their food, protective habitat and even as oviposition site. The vitality of chilli thrips is always a danger to the host plant because their survival and growth reduces the vitality of the host plant as it diminishes the photosynthetic capacity of the host plant. Constant feeding on the tissues of the plant reduces physiobiological reaction of the host plant. A large number of tissues of the host plant is sucked by the increasing population of the thrips, its direct result changes the appearance of the leavs of the plant. It is called "Chilli Leaf Curl". In order to mark this effect, we concentrated our observation on newly flushed leaves of chilli plants by sampling and tapping methods. We wanted to see their impact till the young fruits appeared. It was in the high season of chilli crop that we preferred to conduct a survey in our biological monitoring. Leaves were seen and examined and the infestation of thrips was marked. We continued our monitoring from the time when the young fruits started coming and carried it till the fruits were full grown. Succulent leaves were examined with the help of magnifying glass lens. In our regular observation leaves were found sometimes in light green colour and at another time in reddish brown. We did not carry our examination on those leaves which were in dark green colour or in hardened character. During our examination & observation we also left those leaves which touched the fruit or were immediately lower to the flower. We concentrated our monitoring on the young fruits and pinched stem. Thereafter, we examined the entire fruit surface. In order to find out proper results we made use of sticky cards. In our survey, we found the presence of thrips on sticky traps. In order to get better results we preferred to use yellow sticky card.

Thrips are small insects and they are found in the average of 1mm length. Their habits and features may be found in diversity. It is true that these species of insects survive on the plant materials sucked out of the leaves, flowers and some other portions of the host plants. As the thrips grow in number, they become hazardous to the host plants. The rate of their proliferation directly affects the growth of the plants. It is also seen that the warm conditions provide healthy growth to this pest species. Cold is harmful to their growth. It is seen that their probability to survive becomes thinner when the temperature is below 4^0C. Wet conditions are also not suitable to them. It is in drought conditions and warmer climate that the thrips find abundant growth.

Chilli thrips are found in Terebrantia and related family suborder. The development of Thripidae; *Scirtothrips* can be found in several stages. There are two nymphal stages and two pupal stages. The physionomical features and structural apperances of thrips do not attract as much as several other insects do. It has the average size of 1mm and 1.2mm. Their mouthparts are designed in such a manner that they help them feed on plant tissues. From the egg position to the adult position we may find the growth at six stages. Obviously, egg is the first stage. Thereafter, its growth passes through two larval stages. Thereafter, we find non-feeding prepupal stage. From this stage they pass to pupal stage. The prepupal and pupal instars are recognized by their developing wing buds. Finally, they become adults either male or female. The adult thrips has a fully developed pair of fringed wings. All these developments grow within 7-10 days under favourable conditions. However, a few individuals may take two weeks' time under certain conditions. Needless to say, a little earlier or a little later, development in growth depends on an indviduals access to nourishing contents and climatic conditions. In growth in such a short time that it can be called the fastest spreading thrips. When the adult thrips fly in the air, it is very difficult to see them in our casual sights. But we can notice them with our naked eyes when we see them on paper or cards. The features of their body can be examined through microscope. We saw several distinctive features of their external bodies in our microscopic studies. Their long but narrow wings are supported by the two pairs of feathers. For their growth they need food of plant cells. Nature has provided them with the appropriate tools to suck the plant tissues by piercing. This piercing into the leaves causes direct damage to the plant. When thrips go feeding on the leaves, greenish black fecal is gathered on leaves. Initially, it appears a kind of decoration on the leaves but very soon it distorts the leaves. In the general opinion of the people this change in the leaves of the chilli plants due to thrips was considered as "Murda Disease." In the beginning farmers did not know the real cause of these changes in the leaves of the chilli plant. It was only at later stage that the real cause of this change in colour and forms of leaves was recognized with accuracy. It was confirmed that the heavy infestation of chilli thrips on the plant for a longer duration begins to change colours and forms of leaves and flowers. Flowers and fruits of chilli are turned into black from bronze. Feeding damage makes the plant material unmarketable. Feeding damage can reduce the sale value of the crop production. It is further noticeable that high density of feeding on leaves by the thrips can even cause dessication and death of the plant. However, the total result is heavy loss to the crop.

Luckily they can be controlled by a range of natural enemies. In this manner, we selected two species of predator to the measured control on thrips species. To evaluate the applied measures, we performed their taxonomy and biological study under laboratory conditions with the help of taxonomic experts.

Amblyseius cucumeris or *Neoseilus cucumeris*

A large genus of predatory mites is *Amblyseius*. It belongs to the family of phytoseiidae. Members of this genus take other mites as their food. Thrips also become their food. Among such predators *cucumeris* is better known for feeding on thrips and in this manner, it is very useful for thrips control. This predator looks hazy and tan in colour. Generally, this predator can be found on the underside of chilli leaves. It rests along the veins of the leaves. Sometimes, it is found inside mature flowers of chilli plant. It is always in search of thrips for its food. In this manner, it can prevent the proliferation of thrips. If we apply this predator in the early days of the growing season of chilli, it can check the growth of thrips to the extent that the damage level is lowered. The adult of this predator has four pairs of legs and its size in average is 0.30 - 0.5mm. On close examination it is also found that the females of this genus are comparatively bigger in size than their counterpart males. The body of this predator is pear-shaped and elongated. Unique feature of this mite is that it develops the colour of its body on the basis of what it eats. It may be dark, red and purple. It may also be in white colour tending towards light yellow. It is seen that their preying on thrips changes the colour of their body towards light orange. The body of this predator is shaped in such a manner that separate head portion is not visible. The whole body appears to be one conspicuous unit. It does not show segmentation or parts. In the chilli crop, this predator works very effectively. It is found in the layer of air tight space beneath the leaf surface. its generative growth is also not worthy as it lays 4-5 eggs per day with an average of 40 eggs in the entire life cycle. The shape of the eggs of this predator is oval and the appearance of the eggs is transparent. This predator prefers to deposit its eggs on the soft hairs of the underside of leaves. The eggs are hatched after three days. After hatching of the eggs several developmental stages occur in its rapid growth. The stages are as follows:

Egg → Larva → Protonymph → Deutonymph → Adult.

Macrotracheliella nigra

Predators are beneficial to the crop because they take the damage causing pests as their food. Anthocorids is one such beneficial predator. It is recognized as an item of economic importance to the farmers. It saves the crop from the harmful pests. This predator is phytophagous in nature. It has its unique features in its length. It is about 2.24mm and in its width 0.77mm.

Its head has distinction of its own and extents upto 0.45mm. The size of the female adult of this predator is more than the male one. In length the adult female is about 2.52 to 2.66mm. In width also female adult is larger than the male one. Its width is 0.91mm – 0.98mm. Females are more robust and bigger feeders on thrips. The colour of this predator is dark reddish brown. *Macrotracheliella nigra* is a known predator from the family of Anthocoridae under super family Cimicoidae. This predator grows among the population of chilli thrips in natural conditions. Thrips are their natural food. Different species of thrips are swallowed by them for their vitality. In this manner, thrips population is automatically reduced. The size of this predator varies according to the conditions of its growth. In length it is found from 1.07mm – 4.9mm. In its genral appearance Anthocorids has similarity to the mirids except that of ocelli. Anthocorids have ocelli where as mirids do not have. This predator is flattened and pubescent. The wings of this predator are also peculiar. They are brachypterous. The growth of this predator depends on the climatic conditions. Several developmental stages takes place either rapidly or slowly depending on weather conditions. It can grow only within a few minutes after hatching. There is no much difference between the nymphs and adults of this predator. Metamorphosis is not seen during growth after hatching. The process ecdysis is followed in their growth. This predator gladly feeds on the population of thrips. It likes to feed on the nymphs and adults of thrips all alike. However, it prefers to choose thrips on the basis of their size. It is seen that even at the nymphal stage a single individual of this predator could consume seveal hundred thrips. The most hardy meals comes for the phenomenon of ecdysis. It as also observed that at least one meal was necessary between two molts.

Parasitoids : Parasitoids also prey upon thrips. Many hymenopterans belonging to the family of Eulophidae feed upon thrips. It was observed that parasitoids were more effective in controlling the population growth of the thrips. Parasitoids are mostly host specific and they continue to feed on the specific mites and in this manner keep the damage causing pests at low densities. Once they begin to eat one kind of mite they have inabilty to switch on to another host. In this manner, their feeding on thrips yields better results for the crop. We studied the biology, reproductive behaviour of selected parasitoids. We tried to find out the bio-efficacy of parasitoids. On the basis of this, we would like to assess the role of parasitoids as biological agents for saving the chili crop from *Scirtothrips dorsalis*.

Hymenoptera is the dominant order among all the entomophagous insects. Over two-third cases of successful biological control of insect pest species have been achieved by hymenopterous parasites. Hymenopteroeus parasitoids of thrips belong to the superfamily Chalcidoidea. There are many extremely interesting biological adaptations among the Hymenopterans, in which one is the ovipositor (specialized organ or egg laying). It is composed of long interlocking chitinous stylets. Egg passes in Hymenopterans through this stylet. The ovipositor acts as a drill to pierce the host or the material surrounding the host. In many cases parasitoids serves as a hypotermic (inject paralyzing venom into the host). In some species it is used to secret and form a feeding tube. This feeding tube helps adults parasitoids suck the host body fluid in order to obtain protein for continued egg production. Most of

Hymenopterans are solitary endoparasitoids of larvae (Eulophidae). Ovipositor of Hymenopterans lacks muscles and it is supplied with nerves which extends its length to the tip. Ovipositor bears highly sensitive sense organs that cna discern by chemical stimuli whether a host is suitable or not. The well-known method of reproduction of female production is from fertilized eggs. But the production of male shows arrhenotoky (production from unfertilized eggs). Hymenopterans shows the characteristic of biparental. But this characteristic is supplemented in some species in which uniparental reproduction and parthenogenesis occurred. The phenomenon of parthenogenesis is called thelytoky. In this process, females give rise to females and amles are either lost or if they surrive they became capable of killing the pest.

Thripobius semiluteus

Thripobius semiluteus, a larval endoparasitoid belongs to the order Hymenoptera of the family Eulophidae. It is solitary in habit and shows a uniparental characteristic. In preference it parasites the first and early second larvae of *S. dorsalis*. Females of this species are 0.6mm in size. Head and thorax are purple black in colour and in the lateral view it seems strongly convex in shape. Their wings are almost transparent and are called hyaline. They have slightly curved subcubital vein. Their antennae are pale yellow in colour. The gaster of *T. semiluteus* has 1-2 small dark spots which is sublateral in position. The male of *T. semiluteus* was not known during our research examination. The gaster of *T. semiluteus* has 1-2 small dark spots which is sublateral in position. The male of *T. semiluteus* is unknown.

T. semiluteus has unique features in its genetics. Only females grow out of them. The female of this species always remains alert in search of the hosts. It walks slowly at the sideways of the leaf surface. Thrips larvae of first and early stages are taken for their food. In this manner, majority of thrips is eliminated before becoming adults. But, these parasitoids refrained from feeding upon the adults and second stage larvae of thrips. Due to want of food about half of the pupae could not grow into adulthood. This hardship of food material restrained their rapid growth for the successful elimination of thrips.

Ceranisus menes

Ceranisus menes belongs to the subfamily Entedoninae under the family Eulophidae. Hymenoptera is a large order of insects that is represented by several hundred thousand different species. Hymenoptera is a large order of insects that is represented by several hundred thousand different species. Head and mesosoma of adult female was observed as dark brown in colour. Females are 0.66mm till 1.06mm in size. Ovipositor sheath infuscata if viewed applically of *C. menes* is characterized as head and thorax. All legs are yellow in colour. The hyaline wings are a distnguished feature of this species. The antennae of male and female have marked differences and it is on this basis that they are distinguished.

It is a solitary internal parasitoid of thrips larvae. In our observations, we found that larvae of this species could find its growth within a few second. The larvae of *C. menes* made the entire body of the larval hosts as its food. They have their own behaviour in making search for their hosts. Their behaviour includes several parasitoids combinations. The incubation period of *C. menes* is from 1-3 days. Within 3-4 days, larvae grows into pupa. And from the pupa stage to adult it takes 5-6 days. In this manner, *C.menes* has the total life cycle of 18-22 days under normal temperature and favourable conditions.

6.2 Recommendations

Recommendations Towards Quantitatively and Qualitatively aspects of Chilli Healthy Production :

- Chilli : A Magic Article
- 'A measure constrained': By *S. Dorsalis*
- Eradication of *S. dorsalis* is compulsory
- Misidentification lead to misapplication
- Ambiguous identification
- Taxonomic characterization
- Agricultural Practices based on chemical pesticides
- Biological control vs. Sustainable Agriculture
- Natural Enemies : Have shown Promises
- Export Potential of Chilli Crop
- Regular Release of Predatory Mites
- Need of Native Vegetation
- Role of Government of India
- Plant Protection

Chilli : A Magic Article

    Chilli is a very useful vegetable as well as a magic article for several other uses for the health and wealth being of human beings. Its healthy production is a very necessary thing. Eating chilli on regular basis prevents the growth of cancer cells in body. Regular use of chilli has cleansing a effect on the skin. It also promotes fair complexion by stimulating perspiration. The hotness of any fruit is indicated on Scoville scale. Green chilli enhances the sensitivity of insulin. It is also a powerful anti-pyratic agent when applied over thumb. It can also be used as non-lethal self-defence spray. Therefore, we have to be careful towards its healthy production quantitatively as well as qualitatively.

A measure constrained by : *Scirtothrips dorsalis*

    In the healthy production of chilli *S. dorsalis* is a measure constrained. It is a threat to its proper growth. It causes much harm to the leaves, fruits and finally to the product of chilli.

Eradication is compulsory

It is very difficult to manage successful eradication of this disease causing thrips.

Recommendation: It can be eradicated only when it is detected early and the management practices are implemented immediately.

Misidentification : Leads to misapplication

In order to confirm the infestation of *S. dorsalis* we can use more susceptible host plants like roses and cotton.

Recommendation: If the symptoms of *S. dorsalis* are confirmed, thrips as samples require to be sent to appropriate laboratory for proper identification.

Identification of Thrips species is necessary

It is necessary because thrips species pest can lead to misidentification and in this manner misapplication of our management practice will lead to failure. It is wastage of money and resources.

Ambiguous identification of the target pest

Any kind of ambiguity in this regard i.e. the wrong identification of the target pest will dishearten the growers and farmers in the use of biological control methods.

Recommendation: So, an accurate and non-limiting method is necessary for the identification of thrips. It is necessary that species should be specifically recognized and their developmental stages should be known accurately.

Taxonomic Characterization : Still Challenge

Taxonomic characterization of thrips including *S. dorsalis* is always a challenge for the researchers due to its size and nature of behaviour. So, for bringing accuracy in identification we have to be much careful. For this purpose, we can use molecular techniques. We may bring in use all those modern techniques which may lead us to accuracy.

Recommendation: In this manner we may be saved from deception sometimes caused by morphological ambiguity.

Agricultural Practices : Based on Pesticides

Aggravating agricultural practices surmounted with the practices of using chemicals based pesticides have irritated the environment and it causes ecological imbalance. These things are causing serious problems to the human world in many ways.

Recommendation: Having found these dangers knocking the doors of humanity, the thinkers of the world are showing their awareness for promoting environmentally sustainable agricultural practices.

Biological Control vs. Sustainable Agriculture

Biological control management is now a technique which is recognized a fruitful strategy to promote agriculture in healthy way. Therefore, a thought provoking research is organized for the treatment of *S. dorsalis*, a measure threat to chilli crop.

Recommendation: Biological control through predators and parasitoids will not disturb the ecosystem. It will also promote agriculture and chilli production to the right direction. Inspite initial difficulties in the biological treatment of the pest, this treatment is recommended keeping in mind the long range perspectives in sustainable agriculture.

Natural Enemies : Have shown promises towards augmentative control

During the process of research experiments, natural enemies have shown promises for augmentative biological control of *S. dorsalis*. In the process of experiments it is also learnt that there is scope for the identification of the adopted and virulent strains of the natural predators, *Amblyseius cucumeris* and *Macrotracheliella nigra* have shown some potential in natural conditions.

Recommendation: But more research is needed and experiments need to be intensified on aspects like their mass production technology, field attract rate, predation potential, adaptation to physical and chemical stresses.

Regular Release of Predatory Mites

On the basis of the results obtained during our research it is learnt that regular releases of predatory mites is a very successful method to reduce the population of thrips and the resultant damage on agricultural crops of chilli.

Recommendation: We should release the mites weekly, fortnightly, monthly as the progress of the crops and the pressure of thrips require.

Promote Export Potential of Chilli crop

In order to promote export potential of the chilli crop for getting good price in return in the international market.

Recommendation: It is necessary to reduce unbriddled use of pesticides because in the developed countries measuring scales of the use of pesticides are very strict. They are more careful about the environmental pollution and human health hazards caused by the indiscriminate use of pesticides. In this regard the idea of biological control is very simply. We can manage the pest infestation by deliberate use of living organisms. This will promote ecosystem as the population of harmful pests will naturally be regulated.

Need of Native Vegetation : For Better Habitation

In order to get more profitable results in the use of biological methods we should grow native vegetation around the farm.

Recommendation: The beneficial insects naturally curbing the population of harmful pests may find better habitation and better growth. Plants bearing flowers with pollens are helpful in rearing the beneficial insects. So, such plants are recommend to be grown in pockets in and around the field decided for getting the crop of chilli.

Role of Government of India : Towards Biological Agents

Since, the biological methods for controlling the pest population are a bit costly in the beginning and therefore the Government of India should provide appropriate help to promote such activities through appropriate departments.

Recommendation: State Agricultural Universities and Other research organization should also be assisted financially for developing and producing more and more bio-control agents.

Minimum Qualifications in Plant Protection : For Health Group of Crop Product

At many forums, researchers and educators have expressed concerns on the nature of the distribution of pesticides among the users. They have noticed the irresponsible behaviour of the dealers and distributors of pesticides. Such dealers do not hesitate in guiding the farmers with their wasted interests and recommend even those pesticides which are banned at the international level. Our government agencies are not doing anything effectively in this regard.

Recommendation: So, it is recommended that minimum qualifications related to the knowledge of agriculture should be fixed as the minimum qualifications for a person seeking licence for the sale of pesticides.Such licence holders should also be trained by experts and scientists time-to-time for promoting and twisting the attention of the farmers towards the use of biological means for the healthy growth of the crop product.

6.3 Future Scope

Biological Management : Safeguard Points to be considered
- IPM reflects our immaturity
- Insecticides wise or not
- Obstacle in pest management
- Aim of Agro-industries
- Biological Management : Our Future Demands
- Role of Plant Protection Associations
- Hope for Future

IPM reflects our Immaturity

From the worldwide studies related to insects and pests it becomes clear that IPM is absolutely essential to the survival and future of the crop management. As the studies grow the previous systems of IPM reflect our immaturity and certain new actions with serious attention are promised to be taken. Examples show sophisticated multi-factor effects in our programmes and require further environmental planning to save our crops from different kinds of damages.

Our past attempts predict certain factors that will daily suppress or enhance pest density. One thing is certain that only IPM can not stand alone it must be integrated with crop production.

Insecticides : Wise or Not

It was during 1950s that awareness spread for insect control and the use of insecticides, whether it was wise or not, were used for pest control. It was in late 60s that real push in this regard came into existence. The term IPM was quickly adopted and readily applied in relation to any programme of pest control. Any quick progress in pest management meets further challenges and new problems arise.

Obstacle in Pest Management

The greatest Obstacle to the future success of pest management is the attitude of the people, growers and industry personnels on the one hand and entomologists and environmentalists on the other hand.

However, Biological Management is people oriented and its general practice meets successful teaching in its technology and behaviour. Soils and crops will have to be managed for the proper production of food quantitatively as well as qualitatively. It requires double responsibility – first to discover means more harmful to damaging insects & less harmful to human health and then to educate the users about its time-to-time management.

Aim of Agro-Industries

Agro industries are primarily businessmen and their sole aim is to get profits. They consider primarily the return on the investment. This kind of drive is stronger in the private-sector industries.

They always look for better opportunities for profits by increasing turnovers. In the public-sector certain principles regarding health in human life are seriously observed and the guidelines dictated by the government time-to-time are strictly followed.

Biological Management : Our Future Demands

In the recent years, the researches have indicated harmful effects of the insecticides in IPM. Therefore, our future demands Biological Management for pest control. In order to cope up with unavoidable pest problems a few biological methods are to be adopted.

In the beginning, it will be difficult to adopt this approach because it will directly affect and lower down profits in crop production. But, it is the need of the time that we have to switch over to Biological methods of pest control in order to safeguard our future.

Role of Plant Protection Associations : Safe and Judicious

In order to check the growing use of pesticides and to develop some other methods for plant protection several associations of awakened entomologists and agro-scientists have been working for the betterment of the future of the society. In India we find the following three associations:

1. ICPA : Indian Crop Protection Association
2. PAI : Pesticide Association of India
3. PAFAI : Pesticide Manufacturers and Formulators Association of India

These associations regularly remain in contact with pesticides regulatory and law-enforcing authorities. They assist law and work for the betterment of plant protection as well as for the improvement of healthy crop production. Such associations work for the better future of the human society. They keep in their minds, proper supply of food on the one hand, but on the other hand the proper health of the people also. With the greater national and international awareness, almost all such agencies have started working upon the biological management of pest control. Their aim is to recommend as well as to supervise the safe & judicious use of pesticides and to work upon the environmental protection.

They also work upon the methods to safeguard intellectual property rights. With the efforts of such associations farmers have also started realizing the importance of Biological management in the field of agriculture. In future they will understand the importance of such services more and more.

Hope for Future

It is hoped that in future the products grown with the excessive use of pesticides will be devalued in the market and the consumers will readily pay higher costs for those products which are obtained under Biological Pest Management. In this manner, the initial problem regarding the cost of production and market value which the new methods are facing will be overcome and more and more people will be attracted towards Biological Management in crop production. More and more participation of the farmers in the movement of biological management of pest control will yield significant benefits. Means of education and the electronic means of communication have spread awareness for getting healthy intake for sustenance. Particularly among the rich consumers there is growing awareness for food safety. The knowledge of chemical adulteration in food articles forces the consumers to discard such food articles. In the near future, the stronger awareness will bring the things in favour of the biological pest management.

6.4 Popularization : May Reach at Gross Root Level Need

To facilitate the commercialization and quick dissemination of biological management approaches through natural enemies. A thorough investigation has to be done in order to ascertain the intensity of the effectiveness of the predators and parasitoids used as natural enemies of the thrips species of Chilli. We have to find out the level of biological control so that our economic level in the product is not hurt. Having done completed our investigation with accuracy and certainty we need to popularize it among the farmers so that the benefit of the research may reach the gross root level. In order to facilitate commercialization and quick dissemination of biological control of thrips management by several ways of print media, electronic media and other techniques were used to popularize the results of the research for longer benefits.

Several extension activities also performed in order to popularize the biological control activities. Wide scale publicity of biocontrol technology was imparted through demonstrations, farmer meetings, lectures in colleges, mass media (newspapers) etc. in District Aligarh of Western Uttar Pradesh.

Biological Approaches : Give Fruitful Results

Through fruitful results of our research most of the farmers will be aware of the beneficial aspects of biological approaches and its management through collective actions. In order to get better results of Biological management appropriate policies should be devised by the people in local governance. Panchayats and NGOs can play a effective role.

Rural Unemployed and Educated Youths

The educated youths in want of employment should be encouraged to establish small scale biological management through natural enemies production units at gross root level.

Maintain Sustain Agriculture

To sustain agriculture towards its natural mode some solution are to be traced. The solution to replace pesticides is present in the implementation of Biological management. This method requires study of predators and parasitoids for successful control over harmful pests.

6.5 General Suggestions : Through Our Entire Research

Points to be considered
- Research Orientation
- Maintains P:D Ratio
- Sweep net and visual counts
- Collect infected plant parts
- Towards proper observation :
  - Plants
  - Pests
  - Defenders
- Natural Enemies may require
- Predators and Parasiotoids : gives better efficacy
- Prompt treatment
- Careful uses
- Our own experience
- Behaviour of Dispersal
- For Bugs
- Sanitation : A Process

Research Orientation

During the course of the present study our aim was to find out certain biological methods to control the damage causing thrips on chilli plants. This aim was set due to the current observance about the harmful effects of the excessive use of pesticides in controlling damage on plants. It is true that the use of pesticides gives good results in yield of the crop, but it causes several other harmful effects on human health.

Our research orientation was to find out the ways through which damage on the crop of chilli may be controlled and at the same time naturalness of the product may be caused to survive so as to maintain eco-friendly character of the crop product of Chilli. Certain natural enemies of the harmful thrips were found out in order to control the widespread growth of thrips on chill plants. In this process a few known predators and parasitoids were applied and results were estimated. But during the research it was also found out that some unknown predators also exists and their proper culture is also beneficial in the process of the Biological control of the damage by thrips. So, the entire course of our study leads to the following suggestions;

Maintains P:D Ratio (Pest : Defence Ratio)

Unknown predators should be collected in plastic containers with the help of camel hair brush and from the infected areas of the field they should be brought to the place earmarked for study. In order to know the correctness of its effect on the plant, each predator should be put inside a plastic bottle and a few parts of the plants alongwith a few known insects should also be placed inside. It is to be done for sometime in order to determine whether the test insect is a pest or a predator under the guidance of the known theories such as P:D (Pest : Defence) Ratio – whether it feeds on the plant or it feeds on other insects.

Sweep Net and Visual Counts

Sweep net and visual counts should be adopted in order to find out the number of pests and defenders (Natural Enemies).

Collect infected Plant Parts

For this purpose one should go to the field in groups (about 5 persons per group) in order to choose infected plants randomly. The persons in group should also walk across the field in order to pick up the infected plants and its infected parts.

Proper Observation

Thereafter, the choosen plants or the parts of the plants should be keenly observed and observation should be properly recorded. In order to know pests or defenders (Natural Enemies) in the following way;

For Plants : Observe the plant height, no. of branches, crop stages etc.

For Pests : Observe and count pests at different places on the plant.

For Defenders (Natural Enemies) : Observe and count the number of predators and parasitoids per plant.

For Thrips: We should count the number of nymphs and adults which are present on five terminal leaves per plant (Tapping method or sticky card method can be used to count the thrips population).

Natural Enemies May Require

- Pollen and nectar become food for adults.

- To a place of protection such as : overwintering locations, moderate micro climate are needed.
- Beneficial organisms may also require – alternate host when primary host are not present.
- Due to availability of nectar, pollen, fruits, insects etc., predatory natural enemies will also increased in number.

Predators and Parasitoids : Gives better efficacy

Biological management is counted the most important and ecofriendly components of Insect Pest Management. In recent times the predators and parasitoids should be preferred for better efficacy and to avoid ecological problems. It has been found that they control many crop pests by keeping it at low levels population.

Parasitoids - are of limited potential as biological control agents. Its use should be properly made in in-situ conservation strategies. Cost-effective mass-rearing and release systems should also be taken into account.

Prompt Treatment : Key of Dealing

Every crop faces a few problems – The solution of the problem lies in quick identification of the problem and prompt treatment with the available knowledge. It should be done in the following manner:

- **For Seedlings :** Poor growth in nethouses; Poor establishment after transplanting.
- **For Leaves :** Spots or marks on leaves; Older leaves yellow or drop; White cottony lumps on leaves and stems; Sparse white power under leaves; Patterned and distorted leaves; Holes in leaves.
- **For Plants :** Stunted plants; Plants collapse suddenly; Broken or dying branches.
- **For Fruits :** Spots on fruits; Sting marks on fruit; Small, distorted fruit; Holes in fruit; Brown, corky etches on fruit; Cracks at end of fruit; Fine cracking of skin; Skin changes colour; Postharvest problems.

Careful Uses

Amblyseius cucumeris will help to control damage caused by thrips if it is applied timely and preventively. So, it should be done in proper manner. *Amblyseius cucumeris* is small predatony mite and it feeds only on the infant thrips and their larvae. It cannot attack adult thrips. For best results, it will be a good strategy if we release as many predatory mites as possible because the smaller thrips and larvae will be prevented from becoming adult thrips and hence less damage to the crop.

Our Own Experience : Our Best Guide

In this manner, there will be less pressure of thrips. We can control thrips with less number of mites. Therefore, regular release of mites is also necessary. It should be remembered that we should use our discretion in choosing the number of mites and the time of release.

Behaviour of Dispersal

The packaging material such as bran, rice and hulls may be used to carry minute pirate bugs. Packaging material on to the plants should be properly tackled so that the bugs may readily disperse and locate their prey.

For Bugs

The Bugs should be released in early morning or at the time of dusk, so that they may not fly out of the greenhouse. It is necessary that we should keep the vents closed in order to prevent the bugs from runnig out of the greenhouse.

During the winter season after late October, it is necessary to facilitate regular releases in order to keep up their population. So proper attention should made in this regard yield better results in the growing season when the thrips develop as pupates in the soil.

Sanitation : A Process For Refuging

It should be properly arranged by removing older leaves, flowers and the stuff time-to-time that accumulate in odd corners which provide refuge for thrips.Demonstration should be useful of socio-economic benefits of Biological pest management. We should work for its horizontal spread on a large scale.

To practice these Biological management strategies in agriculture is the beneficial use of nature against the harms caused by nature itself. It is a cure of nature with nature itself. In this manner, it is an attempt to use the face of nature for better human health by producing commodities, food, fibre, horticultural crops and forest production.

6.6 Limitations : Over Restrain the Growth of Harmful Pests
Following points to be considered
- Limitations of Predatory Mites
- Limitations of Parasitoids
- Hyperparasitism : A Phenomnon
- Limitations of Minute Pirate Bugs
- Limitations towards Pesticides
- Limitations towards Biological Approach
- Limitations towards Economic Threshold Assessment (ETA)

In the field of agriculture of proper growth of the plants has to be managed in order to get good results in food production. For this purpose we have to save the plants from the harmful effects and damage caused by insects and pests. Easy and direct method to control the damage causing pets was to kill them by the use of the pesticides. For several years this method was applied. But in the long run it was realized and found out that the use of pesticides at large scale causes other harmful effects to human health when the crop production is consumed in human valley. So, biological control methods were tried later on. In this process biological agents are applied to restrain the growth of the harmful pests.

The real goal of biological methods is to preserve the population of the beneficial organisms and in this manner to reduce the pest activities causing damage. By using certain methods of biological control we have to check the growth of severe greenhouse pests such as aphids, Mites, whiteflies and thrips. But due to environmental variations biological agents may not be as effective as they would be expected in research conditions. So, there will always be further scope of research to improve the efficacy of the biological agents in a particular environment where the crop is grown.

Certain limitations in the use of biological agents for the pest control are observed in the following manner:

Limitations of Predatory Mites (*Amblyseius cucumeris*)

We should not use *A. cucumeris* at temperature that remains below 15^0C for longer duration. We should also not use it in dry areas.

During our intense research activity it is found that the predatory mites eat immature thrips and hardly attack adults. So, the adult thrips remain active and cause damage to the plant. We faced two major problems in the use of predatory mites in the greenhouse vegetables:

(a) Mites do not work better in winter

(b) Pollen resources as food for mites is less available in some crops.

In the course of the practical application of mites, it was found out that a pair of mites on average eats five instar larvae of chilli thrips each day. At the initial stage of the growth of thrips it is effective. But older life stages of thrips free from the attack due to their large size and more effective defensive behaviour. In this manner, certain population of thrips survives and continues to cause damage to the host plants.

Limitations of Parasitoids

In the use of paraditoids it was realized during the research that they are often (but not always) host specific. They attack only one species of pests. In this manner, the other species of pests which survive, continue to cause damage. Parasitoids are generally more delicate than predators and hence more vulnerable to pesticides.

Parasitoids spend most of their life cycle developing within their prey. They are less visible than predators and their performance may be underestimated in the result.

Hyperparasitism : A Phenomenon

Further, parasitoids can be parasitized by other parasitoids this is known as hyperparasitism – a natural and common occurrence which can reduce the effectiveness of some beneficial species.

During the research results it was found that some parasitoids species attack one stage but do not emerge from their host for long and the time of their effectiveness passes away.

During the research it was also found *semiluteus* could control the damage causing thrips but they were required several thousands per acre / per week.

Limitation: Our limitation is that it will affect things at commercial level. Commercially it may not available when we need it, it is one thing. And another thing is that it will be commercially heavy on the product.

Limitations of Minute Prate Bugs (Anthocorids)

During the process of research it was seen that Anthocorids played effective role in reducing thrips populations in chilli pepper fields. But limitation was seen in fluctuations in its effect due to change in environmental conditions. Away from the particular environment we found out reduction in their efficiency. In the quick variations of weather conditions in India and in particularly in our area of research its effectiveness will be questionable.

Minute pirate bugs can also be used in conjunction with predatory mites. But limitation is that the bugs will eat the mites if the food provided by thrips population is not sufficient to them.

Limitation towards Pesticidees

During our intense research survey it is found that there are more than 150 pesticides which have been registered for use in agriculture. But, the limitation is that there are many pesticides, though banned in the developed countries, are being freely supplied in Indian areas.

Limitation towards Biological Approach

Biological Management has been accepted as the very attractive approach for protection of agricultural crops with regard to destruction of pests.

But the implementation of Biological management at the farmers' level has been limited. The phenomenon of Biocontrol and its management keep thrips population away from explosion.

But they can be overwhelmed if the proper time of their outbreak is not seriously observed. Any kind of ignorance may prove results fruitless.

Limitations towards Economic Threshold Assessment (ETA)

Limitations may be marked at Economic threshold assessment also. ETA is taken into account that some plants can tolerate limited pest damage.

Sincere efforts in research have been made to determine the damage thresholds for a variety of crops and pests situations, but definite conclusions in this regard are still wanting.

Limitation: For the practical application of the laboratory results, it is also to be kept in mind whether the cost incurred in controlling the pests in justifiable in the cost of production. The research will be beneficial only when a proper ratio in the investment and yield is maintained.

6.7 The "Pros and Cons" of Biological Control
Points to be considered
- Proper Management in a Natural Way
- Natural Habit of Natural Enemies
- Manipulating our Natural System
- Save the Crop
- Biological Control : A 'green' alternative
- Self-prepetuating Nature & Behaviour of Natural Enemies
- Levels of Risks
- Some Disturbance in applied biological management
- Biological Control vs. Pesticides
- Food Chain of Predators

Proper Management in a Natural Way

There is growing awareness of the scientists plant pathologists, insect pathologists, and microbiologists towards biological control in the pest management. Excessive use of chemicals and pesticides in IPM has created a thoughtful situation for the thinkers and therefore, regular emphasis is being given to find out strategies of biological control methods. Entomologists have been trying to find out natural enemies of harmful pests so that proper pest management may be done in a natural way and the artificial methods of pest control may be avoided. In the modern concept of biological pest control the practice of using living natural enemies to control pest species is promoted.

Natural Habits of Natural Enemies

In this manner, we may understand that Biological control is the human use of selected living organism for the purpose of getting control over the harmful pests damaging plant for its parts. Among such organisms we may think about a predator, a parasite or a pathogen which have natural habit of preying upon the harmful species of pests.

Manipulating Our Natural System through Predators, Parasitoids and Pathogens

In this manner, predators, parasites and pathogens of pests which are used in biological control methods are very useful components in the area of biodiversity all over the world. Their natural habit of attacking and killing the harmful pests makes them natural enemies of the pests which cause damage to the plants. So, these natural enemies are of greater value to our agricultural system where we can control the damaging effects of pests without the use of pesticides. They are also useful and valuable in eliminating those factors which threaten our natural ecosystems.

Save the Crop : From the increasing damages

But, the use of biological control is not easy as the use of pesticides. It requires perfect background information about the genetic behaviour and ecology of pests under particular environmental conditions, it is to be assessed whether particular kind of natural enemies of the pests are effective. Close examination of the behaviour of natural enemies and the rate of their consumption of the pests has to be made in order to save the crop from the increasing damages. Proper laboratory system has to be applied in order to find total control over the pests by natural methods of biological control.

However, biological control has several advantages which are enumerated in the following manner:

Advantages of Biological Control : A green alternative

Biological control is a very particular and specific strategy. A selected predator will only control the population of those pests which are targeted. This approach makes a "green alternative" to chemical control methods.

By this approach, the long term management of the targeted pest is useful only time-to-time. This approach has limited side-effects also.

Biological Control form of manipulating Nature into Natural system

From what we have discussed above, it becomes clear that biological control is a form of turning the face of nature into natural system and thus, to get the desired effect of greater human use. Certain methods of biological control and different approaches in this regard have potential to promote agricultural philosophy in the right direction. These methods may bring us closer to nature and can reduce farmers' reliance on pesticides.

Natural Enemies : Known for their self-perpetuating Nature

Natural enemies brought to control thrips are capable of sustaining themselves. They are also known for their self-perpetuating behaviour and thus they continue to have their effect in lowering the growth rate of pest population. So, after the initial introduction of natural enemies, the system works automatically. They feed on the harmful pests and also multiply through reproduction

Biological Management can bring good results in the long run. In the beginning, it may be a bit costly but later on it will prove economically viable. But this tactic becomes less costly in the long run as the introduced new species goes on increasing because of its self-perpetuating biological nature.

Evaluation of different levels of risks

This system is also known for controlling the different levels of risks which have been identified and evaluated before the growth of pests.

Applied with Accuracy & Precision

Above all, biological control is very effective when it is applied with accuracy and precision. Even the large population of pests will be controlled, because the predators, as applied, will naturally feed on the pests. In this manner, the pest population will go on wind dwelling.

A Few Disadvantages to be kept in mind i.e. while switching over to biological control methods, we have to keep in our mind certain disadvantages:

- Disturbances

The natural enemies which we use to control the pest population are set loose in the environment and therefore they may cause some kind of disturbance in the ecosystem.

When we apply a predator with a pre-supposed target to kill a partiuclar pest, it may switch to a different target and may even decide to consume the crop.

While introducing a new species to environment, we may also face the risk of disturbance in the natural food chain.

Biological Control vs. Pesticides

Biological control is a slow process. It takes a lot of time and patience for the biological agents (natural enemies) to work. Pesticides work immediately and results are easily available at close hand. But the method of Biological control requires patience for results.

- **Destroying Food Chain of Predators**

Through Biological control we cannot completely wipe out the pest population. Predators can survive only when there is something for them to eat. Destroying pests means destroying the food chain of the predators. After the elimination of the pests there is the risk of the safety of the predators also. So, we can only reduce the number of harmful pests, but we can not totally eliminate them. In this manner, we have to bear limited harm caused by the remaining pest population.

> *"The result of natural biological control is that the earth is green and that plants can produce sufficient biomass to sustain other forms of life."*

Chapter 6: Summary and Conclusion
Number of Figures: 0
Number of Graphs: 0
Number of Tables: 0

Bibliography and Appendices

Bibliography

1. Acreman, S.J. and Dixon, A.F.G. 1989. The effect of temperature and host quality on the rate of increase of the grain aphid (*Stiobion avenae*) on wheat. *Annals of Applied Biology*, Vol. 115, pp. 3-9.
2. Affandi, affandi., rosa, Celia dela., Medina. 2012. Thrips (Scirtothrips dorsalis) Hood (Thysanoptera; Thripidae) Associate with MnagoAgroecosystem in East Java, Indoneshia. *Agrivita Journal of Agricultural Science,* Vol. 35: (3).
3. Ahmed, S.M., Gupta, M.R. and Mazumdar, S.K. 1987. Effect of Oleoresins and Powders from Some Dried Fruits and Rhizomes on the Residual Toxicity of Pyrethrins. *Botya Kaguku*, Vol. 41, pp. 135-138.
4. Aliakabarpour, Hamesh. , Salmah, cheMd.R. 2012. Seasonal abundance of Thripshawaiiensis (Morgan) and *Scirtothrips dorsalis* (Hood) (Thysanoptera: Thripidae) in mango orchard in Malaysia. *Pertanika Journal of Agricultural Science*, Vol. 35: (3), pp. 637-645, ISSN.
5. Anandam, R.J., Doraiswamy, Sabitha. 2002. Role of barrier crops in reducing the incidence of mosaic disease in chilli.
6. Ananthakrishnan, T.N. 1961. Studies on some Indian Thysanoptera VI. *Zoologischer Anzieger,* Vol. 167, pp. 259-271.
7. Ananthakrishnan, T.N. 1984. Bioecology of Thrips. *Indira Publishing house, Oak Park, Michigan,* p. 223.
8. Ananthakrishnan, T.N. 1993. Bionomics of thrips. *Annu. Rev. Entomol.* , Vol. 38, pp. 71-92.
9. Ananthakrishnan, T.N., and Annadurai, R.S. 2007. Thrips - tospovirus interactions: Biological and molecular implications. *Current Science*, Vol. 92: (8), pp. 1083-1086.
10. Anonymous. 1997. Agricultural Statistics at a Glance. Directorate of Economics and Statistics, Ministry of Agriculture, Government of India, New Delhi.

11. Asaf Ali, K., E. Abraham, Thirumurthi, S. and Subramaniam, T. 1973. Control of Scabthrips (*Scirtothrips dorsalis*) infesting grapevine (*Vitis vinifera*). *S. Indian Hortic*, Vol. 21, pp. 113-114.
12. Bagnall, R.S. 1919. Achalcid parasite on thrips (Thysanoptera). Rept. Brit. Assen. Adv. Sci. 1913, London : 531.
13. Boucek, Z. 1976. Taxonomic studies on some Eulophidae (Hym.) of economic interest, mainly from Africa. *Entomophaga*, Vol. 21(4): pp. 401-414.
14. Boucek, Z. 1977. "A Faunistic Review of the Yugoslavian Chalcidoidea (Parasitic Hym)." *Acta Entomol. Jugoslav. Suppl.*, Vol. 13, pp. 1-115.
15. Boucek, Z., Askew, R.R. 1968. Palearctic Eulophidae (excl. Tetrastichinae) (Hym. Chalcidoidea). Index of entomophagous insects 3. Le Francois, Paris, pp. 260.
16. Buckman, Rebacca. S, Mound, L.A., Whiting, Michael, F. 2013. Phylogeny of thrips (Insecta: Thysanoptera) based on five molecular loci. *Systematic Entomology*, Vol. 38, pp. 123-133.
17. Butani, D.K. 1976. Pests and Diseases of Chillies and Their Control. *Pesticides*, Vol. 10, pp. 38-41.
18. CABI. 2003. Crop Protection Compendium. Global Module. CAB International, Wallingford, UK.
19. CABI/EPPO. 1997. Quarantine Pests for Europe, 2nd edition. CAB International, Wallingford, UK. pp. 1440.
20. Chandra, Mahesh. , Verma, R.K., Prakesh, Rajesh. , Kumar, Manish., Verma, Deepmala., and Singh, D.K. 2010. Effect of Pyrethrin on Adult of *Thripstabaci* and *Scirtothrips dorsalis* (Thysanoptera: Thripidae). *Advances in Bioresearch*, Vol. 1, pp. 81-83. ISSN.
21. Chandra, Mahesh., Verma, R.K., Prakesh, Rajesh. , Kumar, Manish, Verma, Deepmala., and Singh, D.K. 2010. Effect of Malathion on Adult of *Scirtothrips dorsalis* and *Rhiphiphorothripscruentatus* (Thysanoptera: Thripidae). *Advances in Bioresearch*, Vol. 1, pp. 78-80, ISSN.
22. Charles, Francis and Garth, Youngberg. 1990. Sustainable Agriculture – An Overview. In : Sustainable AGriculture in Temperate

Zones (Eds. C.A. Francis, C.B. Flora and L.D. King), New York : Wiley.

23. Clausen, C.P. 1978. Thysanoptera. In : Bartlett, B.R., Clausen, C.P., DeBach, P. (Eds.) Introduced parasites and predators of arthropod pests and weeds: a world review. Agric. Handb., USDA, Wash., Vol. 480 (545), pp. 18-21.

24. Crawford, J.C. 1911. Two New Hymenoptera. Proc. Ent. Soc. Wash., Vol. 13, pp. 233-234.

25. D. Pimentel. 1971. Ecological Effects of Pesticides on Nontarget Species. Exec. Off. President. Off. Sci. Technol. Supplement Documents, U.S. Government Printing Office, No. 4106-0029, Washington, D.C., p.220.

26. De Bach, P. 1964. The Scope of Biological Control. P. 3-20. *In* Biological Control of Insect Pests and Weeds (P. DeBach, editor). Chapman and Hall Ltd., London, p. 844.

27. Densantis, L. 1961. Dos nuevos parasitos de tisanopteros de la Rebublica Argentina (Hymenoptera : Entedontinae). Notas Mus. La Plata, Zool., Vol. 20 (187), pp. 11-19.

28. Dent, D. 2000. Insect Pest Management. 2nd Edn., CABI Publishing, Massochusetts, U.S.A., ISBN : 0-85199-340-0, p. 366.

29. Dev, H.N. 1964. Preliminary studies on the biology of Assam thrips, *Scirtothrips dorsalis* Hood on tea. *Indian J. Entomol.* Vol. 26, pp. 184-194.

30. Dev, H.N. 1964. Preliminary syudies on the biology of Assam thrips, yellow tea thrips, strawberry thrips), *Scirtothrips dorsalis* Hood, *Provisional Management Guidelines.* Florida: University of Florida, EDIS: ENY 725.

31. Dhaliwal, G.S. and Ramesh Arora. 1996. Principles of Insect Pest Management. National Agricultural Technology Information Centre, Ludhiana.

32. Dogramaci, Mahmut, Arthurs, Steven, P., Chen, Jijanjun, Mckenzie, Cindy, Irrizary, Fabieli and Osborne, Lance. 2011. Management of Chilli Thrips; *Scirtothrips dorsalis* (Thysanoptera : Thripidae) on peppers by *Amblyseius swirskii* (Aeari : Phytoseiidae) and *Orius*

insidiosus (Hemiptera : Anthocoridae). *Biological Control,* Vol. 59, pp. 340-347, ISSN : 1049-9644.

33. Dogramaci, Mahmut. , Arthurs Steven P., Chen Jianjun, Osborne Lance. 2013. Silicon Applications have Minimal Effects on Scirtothrips dorsalis (Thysanoptera: Thripidae) Populations on Pepper Plant, Capsicum annum L. Vol. 96: (1).

34. Duraimurugan, P., Jagdish, A. 2011. Preliminary Studies on the Biology of *Scirtothrips dorsalis* (Thysanoptera; Thripidae) as a pest of rose in India. *Journal of Applied Zoological Research,* Vol. 69, pp. 37-45.

35. Ekram, Atakan. 2011. Population Densities and Distribution of the WFT (Thysanoptera : Thripidae) and its Predatory Bug, *Orius niger* (Hemiptera : Anthocoridae), in Strawberry. *International Journal of Agriculture and Biology,* Vol. 13(5), pp. 638-644, ISSN 1560-8530.

36. Eppo. 2005. EPPO standards Diagonastic protocols for regulated pests- *Scirtothripsaurantii, Scirtothripscitri, Scirtothrips dorsalis. OEPP/EPPO Bulletin,* Vol 35, pp. 353-356.

37. EPPO/CABI. 1996. *Scirtothripsaurantii, Scirtothripscitri.* In: Quarantine pests for Europe. Cab International, Wallingford, UK.

38. FAOSTAT; Food and Agriculture Organization of the United Nations.

39. Farris, R.E., Arce, R. Ruiz., Ciomperlik, M., Vasquez, J.D., and Delon, R. 2010. Development of a Ribosomal DNA ITS2 Marker for the identification of the Thrips, *Scirtothrips dorsalis. The Journal of Insect Science,* Vol. 10, pp. 26.

40. Ferriere, C. 1958. Un nouveau parasite de thrips en Europe central. (Hym. Euloph.) Mitt. Schweiz, ent. Ges., Vol. 31(3-4), pp. 320-324.

41. Ferriere, C., 1938. Descriptions of Some African Eulophidae. Bull. Ent. Res., Vol. 29(2), pp. 141-147.

42. Fery, Richard, L., Schalk, James, M. 1991. Resistance in pepper (*Capsicum annuum* L.) to Western Flower Thrips (*F. occidentalis*) (Pergande). *J. of Hort Science,* Vol. 26(8), pp. 1073-1074, ISSN 0018-5345.

43. Froud, K.J. and Stevens, P.S. 1997. Life Table Comparison between the Parasitoid *Thripobius semiluteus* and its host Greenhouse Thrips.

Horti Research, Private Bug. Vol. 92 (169), Aukland, pp. 1-3, ISSN : 1174-6947.

44. Fugro, P.A. 2000. Role of organic pesticides and manures in management of some important diseases. *Journal of mycology and plant pathology,* Vol. 30: (1), pp. 96-97.

45. Fukuoka, Masanobu. 1982. Japanese Farmer and Philosopher.

46. Gillespie, D.R. 1989. Biological Control of Thrips (Thysanoptera : Thripidae) on greenhouse Cucumber of *Amblyseius cucumeris. Entomophaga,* Vol. 34, pp. 185-192.

47. Gillespie, David, R. 1988. Life-History and Cold Storage of *Amblyseius cucumeris* (Acarina : Phytoseiidae). *Journal of the Entomological Society of British Columbia,* Vol. 85, pp. 1-6, ISSN : 0071-0733.

48. Giraddi, R.S., Mantur, S.M., Patil, R.K., Mallapur, C.P., Ashalatha, K.V. 2012. Population Dynamics and Extent Damage of Pests of *Capsicum* under Protected Cultivation. *Karnataka J. Agric.Sci.,* Vol. 25(1), pp. 150-151.

49. Girault, A.A. 1915. Australian Hymenoptera Chalcidoidea, IV. Mem. Qd. Mus., Vol. 3, pp. 211-216 (Eau. Emar).

50. Girault, A.A. 1917b. New Javanese Hymenoptera. Private Publ., Wash, 12pp. In : Gordh, G., Menke, A.S., Dahms, E.C. Hall, J.C., 1979. The Privately Printed Papers of A.A. Girault. Mem. Amer. Ent. Inst. Vol. 28, p. 400.

51. Gnanachandran, S., and Sivayoganathan, C. 1990. Adoption of non-chemical pest control and brinjal by Jaffana Farmers. *Tropical Agricultural Research,* Vol.2.

52. Gopal, K., Reddy, M. Krishna., Reddy, D.V.R., Muniyappa, V. 2009. Transmission of peanut yellow spot Virus (PYSV) by thrips, *Scirtothrips dorsalis* Hood in groundnut.

53. Greer, Lane and Diver, Steve/ 2000. Greenhouse IPM : Sustainable Thrips Control, *Pest Management Technical Notes ATTRA* – 800-346-9140, pp. 1-18.

54. Grimaldi, D.A., Shmakov, A. and Fraser, N. 2004. Mesozoic Thrips and Early Evolution of the Order Thysanoptera (Insecta). *Journal of Paleontology*, Vol. 78 (5), pp. 941-952.
55. Gurr, G.M., Wratten, S.D. and Barbosa, P. 2000. Success in Conservation Biological Control of Arthropods. In Measures of Success in Biological Control (eds. G. Gurr and S. Wratten). The Netherlands : Kluwer Academic Publishers, pp. 105-132.
56. Hans, Larsson. 2005. Aphids and Thrips : The Dynamics and Bio-Economics of Cereal Pests. Doctoral Thesis of Sciences, Swedish University of Agricultural, pp. 42.
57. Hoffman, M.P. and Frodsham, A.C. 1993. Natural Enemie of Vegetable Insect Pests. Cooperative Extension, Cornell University, Ithaca, NY, pp. 63.
58. Hoodle, M.S. and L.A., Mound. 2003. The genus Scirtothrips in Australia (Insecta, Thysanoptera, Thripidae). *Zootaxa*, Vol. 268, pp. 1-40.
59. Hoodle, M.S., Mound, L.A. and Paris, D.L. 2009. *Scirtothrips dorsalis*. Thrips of California, University of California, California, U.S.A. http://keys lucidcentral.org/keys/v3/thrips_of_california/data/key/ thysanoptera./ Media/Html/browse_species/scirtothrips_dorsalis.htm (Accessed 14th March, 2011).
60. Howard, R.J., Garland, J.A., Seaman, W.L. 1994. Diseases and Pests of Vegetable crops in Canada. The Canadian – Phytopathological Society and the Entomological Society of Canada, Ottawa.
61. International Biological Programme Synthesis Series, Vols. 1-24.
62. Ishii, T. 1933. Notes on two hymenopterous parasites of *thrips* in Japan. *Kontyu*, Vol. 7, pp. 13-16, 1 plate.
63. Jagdish, E.J., Purnima, A.P. 2011. Evaluation of selective botanicals and entomopathogens against *Scirtothrips dorsalis* Hood under polyhouse conditions on Rose. *Journal of Biopesticides,* Vol. 4, pp. 244.

64. Jagtap, P.P., Shingane, U.S. and Kulkarni, K.P. 2012. Economics of Chilli Production in Indian-African Journal of Basic and Applied Sciences, Vol. 4(5), pp. 161-164, ISSN : 2079-2034.
65. Jayaraj, J. 1993. Biopesticides and Integrated Pest Management and Sustainable Agriculture (Ed. N.K. Roy), New Delhi, APC Publications.
66. Jha, Vivek K., Seal, D.R., Schuster, David J., and Kakkar, Graima. 2009. Diel Flight pattern and Periodicity of Chili Thrips (Thysanoptera: Thripidae) on Selected Hosts in South Florida. Proc. Fla. State Hort. Soc., Vol. 122, pp. (267-271).
67. Jha, Vivek Kumar, Seal, D.R., Schuster, David, J., Kakkar, Garima. 2009. Diel Flight Pattern and Periodicity of Chilli Thrips (Thysanoptera : Thripidae) on selected Hosts in South Florida. *Proceedings of the Florida State Horticultural Society*, Vol. 122, pp. 267-271, ISSN : 0886-7283.
68. Jha, Vivek Kumar, Seal, Dakshina, R., Kaakar, Garima and Osborne, Lance. 2012. New Tropical fruit Hosts of *S. dorsalis* (Thysanoptera : Thripidae) and its Relative Abundance on them in South Florida. *Florida Entomologist*, Vol. 95(1), pp. 205-207, ISSN : 1938-5102.
69. Johnson, M.W. 2000. Nature and Scope of Biological Control. *Biological Control of Pests*, ENTO 675, UH-Manod, pp. 1-5.
70. Kakkar, Garima, Kumar, Vivek, Seal, D.R., Liburd, Oscar, E., and Stansly, Philip A. 2016. Predation by *Neoseilus cucumeris* and *Amblyseius swirskii* on *Thrips Palmi* and *Frankliniella schultzei* on cucumber. *Biological control,* Vol. 92, pp. 85-91, ISSN : 1049-9644.
71. Kaur, Sandeep, and Singh, Subash. 2013. Efficacy of some Insecticides and Botanicals against Sucking Pests on *Capsicum* under net house. *Agriculture for Sustainable Development,* Vol. 1: (1), pp. 25-29. ISSN.
72. Kethran, Rafiq Muhammad, Sun, Ying, Ying, Khan, Shahbaz, Baloeh, Sana Ullah, Wu, L.L., Lu, T.T., Yang Yang, Hu, Zhan, Salam, Abdul, Iqbal, Sohil, Ali, Sakhwat, and Bashir, Waseem. 2014. Effect of Different Sowing Dates on Insect Pest Population of Chillies (*Capsicum annuum* L.). *Journal of Biology, Agriculture and Healthcare*, Vol. 4(25), pp. 1-19, ISSN : 2224-3208.

73. Khader, H. Khan. 1996. Integrated Pest Management and Sustainable Agriculture. Farmers and Parliament, Vol. 30(2), pp. 15-17.

74. Kharbangar, Magdaline, Choudhury, S. and Hajong, S.R. 2014. Occurrence and Abundance of Thrips (Thysanoptera) Associated with Rice Crops from Meghalaya. *International Journal of Research Studies in Bio-Sciences (IJRSB)*, Vol. 2(5), pp. 1-7, ISSN-2349-0357.

75. Kolb, T.E., McCormick, L.H. and Shumway, D.L. 1991. Physiological responses of pea-thrips damaged sugar maples to light and water stress. *Tree Physiology*, Vol. 9, pp. 401-413.

76. Krishnamurthy, Rao, B.H. and Murthy, K.S.R.K. (eds.) 1983. Proceedings of National Seminar on Crop Losses due to Insect Pests, *Indian J. Ent.* (Special Issue), Hyderabad, Vols. I-II.

77. Kulkarni, G.S. 1922. The "Murda" disease of Chilli (*Capsicum*). *Agricultural Journal of India,* Vol. 22, pp. 51-54.

78. Kumar, Vivek. , Seal, D.R., Kakkar, Graima., Mckenzie Cindy L., and Osborne, Lance S. 2012. New Tropical Fruit Hosts of *Scirtothrips dorsalis*(Thysanoptera:Thripidae) and its Relative Abundance on them in South Florida. *Florida Entomologist*, Vol. 95: (1), pp. 205-207.

79. Kumar, Vivek. 2012. New Tropical Fruit Hosts of *Scirtothrips dorsalis* (Thysanoptera : Thripidae) and Its Relative Abundance on them in South Florida. *Florida Entomologist*, Vol. 95(1), pp. 205-207.

80. Kumar, Vivek. *Scirtothrips dorsalis* (Thysanoptera : Thripidae) : Scanning Electron Micrographs of Key taxonomic Traits & a Preliminary Morphametric Analysis of the General Morphology of Populations of Different Continents. Florida Entomological Society.

81. Landis, Doughlas A., Wratten, Stephen D. and Gurr, Geoff M. 2000. Habitat Management to Conserve Natural enemies of Arthropod Pests in Agriculture. *Ann. Rev. of Entomol.*, Vol. 45(1), pp. 175-201, ISSN : 0066-4170.

82. Lattin, J.D. 1999. Bionomics of the Anthocoridae. *Annu. Rev. Entomol.*, Vol. 44, pp. 207-231.

83. Lewis, T. 1973. Thrips : their biology, ecology and economic importance. Acad. Press, London, XV, p. 349.

84. Lewis, T. 1997. Pest thrips in perspective. In : Thrips as crop pests (Lewis, T., Ed.) – *CAB International*, Wallingford, U.K., pp. 1-13.
85. Lewis, T. 1997. Thrips as crop pests. *CAB International*, Oxon, G.B.
86. Loomans, A.J.M. 1991. Collection and first evaluation of Hymenopterous parasites of thrips as biological control agents of *Frankliniella occidentalis*. Bull. IOBC/SRWP, Vol. 14(5), pp. 73-82.
87. Loomans, A.J.M., Lenteren, J.C., Van. 1995. Biological control of thrips pests : A Review on thrips parasitoids. Research Gate, Wagningen Agricultural University Papers, Vol. 95(1), pp. 89-201.
88. Ludwig, Scott W. and Bogran, Carlos. 2007. Chilli Thrips : A New Pest in the Homeland Scape. Texas Cooperative Extension, pp. 1-4, EEE 00041 12/07.
89. Macleod, A., Collins, D. 2006. CSL Report, Pest Risk Analysis for *Scirtothrips dorsalis*. Central Science Laboratory, Sand Hilton, York, U.K.
90. Mandal, S.K. 2012. Field Evaluation of Alternate use of Insecticides against Chilli Thrips, *Scirtothrips dorsalis* (Hood). Annals of Plant Protection Sciences, Vol. 20: (1), pp. 59-62. ISSN.
91. Mandi, N. and Senapati, A.K. 2009. Integration of Chemical, Botanical and Microbial Insecticides for Control of thrips, *S. dorsalis* Hood infesting Chilli. *The Journal of Plant Protection Sciences,* Vol. 1(1), pp. 92-95, ISSN : 2249-7897.
92. Mandi, N., and Senapati, A.K. 2009. Integration of chemical botanical and microbial insecticides for control of thrips, *Scirtothrips dorsalis* Hood infesting chilli. *The Journal of Plant Protection Sciences*, Vol. 1, pp. 92-95.
93. Mari, J.M., Laghri, R.B., Mari, A.S. and Shahzadi, A.K. 2013. Eco-friendly Pest Management of Chilli Crop. *International Journal of Agricultural Technology*, Vol. 9(7), pp. 1981-1992, ISSN : 1686-9141.
94. Maris, P.C., Joosten, N.N., Peters, D., and Goldbach, R.W. 2003. Thrips Resistance in Pepper and its Consequences for the Acquisition and Inoculation of *Tomato spotted wilt virus* by the Western Flower Thirps. *The American Phytopathogical Society*, pub no., P-2002-1106-01R., Vol.93: (1), PP. 96-101.

95. Masui, S. 2007. Oviposition time of overwintered adults of yellow tea thrips, *Scirtothrips dorsalis* Hood (Thysanoptera:Thripidae). *Japanese Journal of Applied Entomology and Zoology,* Vol. 51: (4), pp. 289-291.

96. Masui, S. 2007a. Timing and distance of dispersal by flight of adult yellow tea thrips, *Scirtothrips dorsalis* Hood (Thysanoptera : Thripidae). Jpn. J. Appl. Entomol. Zool., Vol. 51, pp. 137-140.

97. Mehle, Natasa and Trdan, Stanislav. 2012. Traditional and Modern methods for the identification of Thrips (Thysanoptera) species. *Research Gate; Journal of Pest Science,* Vol. 85(2), pp. 179-190, ISSN : 1612-4758.

98. Millawithanachichi, M.C., Perera, A.L.T., Peiris, B.C.N., and Fonseka, H.M. 2004. Development of High Yielding Chilli Hybrids (*Capsicum annum* L.) based on Heterobeltiosis and Characterization of Parental germplasm for DNA polymorphisms. *Tropical Agricultural Research,* Vol. 18.

99. Miyazaki, M. & Kudo, I. 1986. Description of thrips larvae which are noteworthy on cultivation plants (Thysanoptera : Thripidae). I. Species occurring on Solanaceous and Cucurbitaceous crops. *A Kitu* Vol. 79, pp. 1-26.

100. Moorthy, Krishna, P.N., Saroja, S., Shivaramu, K. 2013. *Pest Management in Horticultural Ecosystems,* Vol. 19(2), pp. 1-3, ISSN : 0971-6831.

101. Moraes, G.J. de. Mcmurtry, J.A., Denmark, H.A. and Campos, C.B. 2004. A Revised Catalog of the Mite Family Phytoseiidae : Aukland, *Mangolia Press* (Zootaxa 434), pp. 494.

102. Moritz, G. 1997. Structure, growth and Development In T. Lewis (ed.), Thrips as Crop Pests. CAB International, New York, pp. 15-63.

103. Moritz, G.D, Morris, D.C. and Mound, L.A. 2001. Thrips ID : Pest thrips of the World. An interactive identification and information system. CD-ROM published by ACIAR, Australia. CSIRO Publishing, Melbourne, Australia.

104. Mound, L.A. 2005. Thysanoptera : Diversity and Interactions. *Annual Review of Entomology,* Vol. 50, pp. 247-269.

105. Mound, L.A., and Marullo, R. 1996. The thrips of Central and South America : an introduction. International memoirs on Entomology, Vol. 6, pp. 1-488.
106. Mound, L.A., Kibby, G. 1998. Thysanoptera – an identification guide. 2nd edn. CAB Intenational, Wallingford, p. 70.
107. Mound, L.A., Palmer, J.M. 1981. Identification, Distribution and Host Plants of the Pest Species of Scirtothrips (*Thysanoptera : Thripidae*). *Bulletin of Entmological Research*, Vol. 71(3), pp. 467-479.
108. Mound, L.A., Palmer, J.M. 1981. Identification, distribution and host plants of the pest species of *Scirtothrips* (Thysaoptera: Thripidae). *Bulletin of entomological Research*.Vol. 71, pp. 467-479.
109. Murai, T., and Toda, S. 2002. Variation of *Thrips tabaci* in colour and size, pp. 377-378. In Q. Marullo and L.A. Mound (eds.) Thrips and Tospoviruses : Proceedings of the 7th International Symposium on Thysanoptera. Australian National Insect Collection, canberra, Australia.
110. Nanskumar, J., Bandra, J.M.R.S., and Peiris, S.E. 1996. Establishment of Embryogenic Cell Suspension Culture of Chilli [Capsicum annum L. var. accuminatumFingerh] for Somatic Embryogenesis. *Tropical Agricultural Research*, Vol.8, pp. 203-213.
111. Newberger, S.J., McMurtry, J.A. 1992. Insectary Production of *Thripobius semiluteus* Boucek (Hymenoptera : Eulophidae), an imported parasitoid of the greenhouse thrips, *Heliothrips haemorrhoidalis* (Bouche) (in press).
112. Nietschke, Brett S., Borchert, D.M., Magarey, R.D., and Ciomperlik, M.A. 2008. Climatological Potential for *Scirtothrips dorsalis* (Thysanoptera; Thripidae) Establishment in the United States. Florida Entomologist, Vol. 91: (1), PP. 79-86.
113. OEEP / EPPO (European and Mediterranean Plant Protection Organization). *Aurantii, Scirtothrips citri, S. dorsalis.* 2005. Diagnostic protocols for regulated pests – *Scirtothrips*. Bulletin, Vol. 35, pp. 353-356.
114. Oetting, Ronald D. and Beshear, Ramona J. 1991. *Orius insidiosus* (Say) and Entomopathogens as Possible Biological Control Agents

for Thrips. General Technical Report (GTR); U.S. Dept. of Agriculture, Forest Service, North Eastern Forest Experiment Station, pp. 419-424.

115. Oltean, Ion. 2012. Research on Biology, Ecology and Integrated Management of Thysanoptera species in greenhouses. University of Agricultural Sciences and Veterinary Medicine, CIUJ NAPOCA Doctoral School Ph.D. Thesis, pp. 1-9.

116. Packiam, S. Maria., and Ignacimuthu, S. 2013. Effect of Botanical Pesticidal Formulations against The Chilli Thrips (*Scirtothrips dorsalis* Hood) on Peanut Ecosystem. *International Journal of Natural and Applied Science,* Vol. 2: (1), pp. 1-5.

117. Palmer, J.M., Mound, C.A., du, Heaume, G.J. 1989. Thysanoptera : CIE guides to insects of importance to man. *CAB International*, Wallingford, p. 73.

118. Paroda, R.S. 1998. Sustaining the Green Revolution : New Paradigms. Dr. B.P. Pal Memorial Lecture. In : Proceedings of the 2nd International Crop Science Congress, Oxford & I.B.H., New Delhi, pp. 79-110.

119. Peppers as Influenced by Farmer's Management Practices in UGANDA. *Journal of Plant Protection Research,* Vol. 53: (2), pp. 158-164.

120. Puttaswamy, Reddy, D.N.R. 1985. Pests Infecting Chilli (*Capsicum annuum* L.) in the Nursery. *Mysore Journal of Ag. Sciences,* Vol. 18(2), pp. 122-125, ISSN : 0047-8539.

121. Raizada, U. 1965. Life history of *Scirtothrips dorsalis* Hood with detailed external morphology of the immature stages. *Bulletin of Entomology,* Vol. 6, pp. 30-49.

122. Ramakers, P.M.J., Dissevelt, M. and Peeters, K. 1989. Large Scale Introduction of Phytoseiid Predators to Control Thrips on Cucumber Meded. Fac. Landbouww. Rijksuniv. Gent., Vol. 54 (3a), pp. 923-929.

123. Ramakrishna Ayyar, T.V. 1932. Bionomics of some thrips injurious to cultivated plants in South India. *Agriculture and Livestock of India,* Vol. 2, pp. 391-403.

124. Rao, R.D., A.S., Reddy, S.V., Reddy, Thirumala Devik K., Chander, Rao S., Manoj Kumar, V., Subranmaniam T.Y., Reddy, S.N., Reddy DVR. 2003. The hosts range of Tobacco Streak Virus in India and transmission by thrips. *Annals of Applied Biology*, Vol. 142, pp. 365-368.

125. Raspudic Emilijia, Ivezic, Marija, Brmez, Mirjana and Trdan Stanislav. 2009. Distribution of Thysanoptera species and their host plants in Croatia, *Acta agricultural Solvenica*, Vol. 93(3), pp. 275-283. ISSN. 1581-9175.

126. Reddy, D.N.R. 2010. Chilli. World Spice Congress, Feb. 4, New Delhi.

127. Reddy, D.N.R. and Puttaswamy. 1983. Pest-infesting Chilli (*Capsicum annuum* L.) In the transplanted crop. *Mysore J. Agric. Sci.*, Vol. 17, pp. 246-251.

128. Reddy, D.N.R. and Puttaswamy. 1996. Pests infesting chilli in the transplanted crop. *Mysore J. Agric. Sci.*, Vol. 17, pp. 122-125.

129. Reddy, S.G. Eswara., and Kumar, N.K. Krishna. 2006. A Comparison of Management of Thrips, *Scirtothrips dorsalis* Hood on Sweet Pepper Grown under Protected and Open Field Cultivation. Pest Management in Horticultural Ecosystems, Vol. 12(1), pp. 45-54.

130. Sakimura, K. 1969. A comment on the colour forms of *Frankliniella schultzei* (Thysanoptera : Thripidae) in relation to transmission of the *Tomato spotted Wilt virus*. Pacific Insects, Vol. 11, pp. 761-762.

131. Santharam, G., Kumar, K., Chandrasekaran, S., and Kuttalam, S. 2003. Bioefficacy and residues of imidacloprid in chillies used against chilli thrips. *Madras Agri. J.,* Vol. 90: (7-9), pp. 395-399.

132. Sardana, H.R., Bhatt, M.N. and Sehgal, Mukesh. 2012. Wide area Validation and Economic Analysis of Adoptable IPM – Technology in Bell pepper (*Capsicum annuum*). *Indian Journal of Agricultural Sciences,* Vol. 82(2), pp. 1-2, ISSN : 0367-8245.

133. Sathe, Tukaram Vithalrao, Pranoti, Mithari, Patil, Sampatrao Shivajirao and Desai, Abhjit Somanrau. 2015. Diversity, Economic-Importance and Control Strategies of Thrips (Thysanoptera) on Crop

Plants. *Indian Journal of Applied Research*, Vol. 5(8), pp. 761-763, ISSN :

134. Satpathy, S., Kumar, Akhilesh. , Shivalingaswamy, T.M., Rai, A.B., and Rai, Mathura. 2006. Field Efficacy of Methonyl against Thrips,*Scirtothrips dorsalis* (Hood) in Chilli. *Vegetable Science,* Vol. 33: (2), pp. 164-167.

135. Schuh, R.T., Slater, J..A. 1995. True bugs of the World (Htemiptera : Heteroptera). Classification and Natural History. Cornell University Press, Ithaca, New York, xii, p. 336.

136. Seal, D.R. and Klassen, W. 2005. Chilli Thrips (Castor thrips, Assam thrips, Yellow tea thrips, Strawberry thrips), *Scirtothrips dorsalis* Hood, Provisional Management Guidelines, Florida : University of Florida, EDIS ; ENY 725.

137. Seal, D.R., and Kumar, V. 2010. Biological responses of chilli thrips, *Scirtothrips dorsalis* Hood (Thysanoptera: Thripidae) to various regimes of chemical and biorational insecticides. *Crop Protection,* Vol. 39, pp. 1241-1247.

138. Seal, D.R., Klassen, W. and kumar, V. 2010. Biological parameters of *Scirtothrips dorsalis* (Thysanoptera: Thripidae) on selected hosts. *Envr. Ento.* , Vol. 39, pp. 1389-1398.

139. Seal, D.R., Klassen, W., Kumar, V., 2009a in Review. Biological parameters of chilli thrips, *Scirtothrips dorsalis* Hood, on selected hosts. *Environmental Entomology.*

140. Seal, D.R., Kumar, V., Klassen, W., Sabine, K. 2008. Response of Chilli thrips, *Scirtothrips dorsalis* and melan thrips, *Thrips palmi*, to some selected insecticides. Proceedings of the Caribbean Food Crops Society, pp. 44-578.

141. Shibao, Manabu. 1997. Effects of Insecticide Application on Population Density of the ChillieThrips, *Scirtothripsdorasalis*Hood (Thysanoptera: Thripidae), on Grape. *Applied Entomology zoology*, Vol. 32: (3), pp.512-514.

142. Shipp, J.L., Wang, K. and Binns, M.R. 2000. Economic Injury Levels of Western Flower Thrips (Thysanoptera : Thripidae) on green house cucumber. *J.Econ., Entomol.*, Vol. 93, pp. 1732-1740.

143. Shivaprasad, M., Chittapur, B.M., Mohankumar, H.D., Astaputre, S.A., Tatagar, M.H. and Mesta, R.K. 2010. Eco-friendly Approaches for the Management of Murda Complex in Chilli. *Academic Journal, Agricultural Reviews*, Vol. 31 (4), pp. 298. ISSN – 0253-1496.

144. Silagyi, A.J. and W.N. Dixon. 2006. Assessment of Chilli Thrips, *Scirtothrips dorsalis* Hood. Division of Plant Industry, Gaines Ville Florida, p. 9.

145. Sivinski, John. 2013. Augmentative Biological Control : Research and Methods to help make it work. *CAB International*, Vol. 8, pp. 1-11, ISSN : 1749-8848.

146. Skarlinsky, T.L. 2003. Survey of St. Vincent Pepper Fields for *Scirtothrips dorsalis* Hood, pp. 5. USDA –APHIS-PPQ. Miami, Florida.

147. Ssemwogerere, Charles., Ssemakula, Mildred Kathrym Nyaburu Ochwo., Kovach, Joe. , Kyamanywa, Samuel. , Karungi, Jeninah. 2013. Species Composition and Occurrence of Thrips on Tomato and

148. Stern, V.M., Smith, R.F., Bosch, R., Van Den and Hogen, K.S. 1959. The Integrated Control Concept, Hilgardia, Vol. 29, pp. 81-101.

149. Stout, Rex. Todhunter, an american writer, Noblesville, Indiana, United States.

150. Swaminathan, M.S. 1995. Agrochemicals in sustainable Agriculture. In : Natural and Manmade Chemicals in Sustainable Agriculture in Asia (Ed. B.S. Parmar).

151. Tachikawa, T. 1986. A note on *Ceranisus brui* (Vuillet) in Japan (Hymenoptera : Chalcidoidea, Eulophidae). Trans. Shikoku Entomol. Soc., Vol. 17(4), pp. 267-269.

152. Tatagar, M.H., Awaknavar, J.S., Giraddi, R.S., Mohan Kumar, H.D., Mallapur, C.P., and Katarki, P.A. 2011. Role of border crop for the management of chilli laef curl caused/ due to thrips *Scirtothrips dorsalis* (Hood) and mites, *Polyphagotarsonemus latus* (Banks). *Journal of Agricultural Science,* Vol.24, pp. 294-299.

153. Tatara, A. 1994. Effect of temperature and host plant on the development, fertility, and longevity of *Scirtothrips dorsalis* Hood

(Thysanoptera : Thripidae). *Applied Entomology and Zoology*, Vol. 29, pp. 31-37.

154. Tatara, A., Furuhashi, K. 1992. Analytical study on damage to SatusmaMandarian fruit by *Scirtothrips dorsalis*, with particular reference to pest density. *Japanese Journal of Applied Entomology and Zoology*, Vol. 36, pp. 217-223.

155. Toda, Sathoshi, Hirose, Takuya, Kakiuchi, Kanako, Kadama, Hirosato., Kijima, Keisuke., Mochizuki, Masatoshi. 2013. Occurrence of a novel strain of *Scirtothrips dorsalis* (Thysanoptera: Thripidae) in Japan and development of its molecular diagonastics. *Applied Entomology and Zoology*, Vol. 49: (2), pp. 231-239.

156. Umar, Muhammad, Akram, Waseem. , Ali, Baboo., Tariq, Muhammad., Shah, Nazar Ali. 2003. Description of Three Genera (Thripidae: Thysanoptera) from Azad Jammu and Kashmir (Pakistan). *Online Journal of Biological Sciences*, Vol. 3: (5), pp. 524-534. ISSN.

157. Unlu, Levent, Ogur, Ekrem and Celik, Yusuf. 2012. The Importance of Integrated Pest Management for Sustainable Agriculture. *Journal of Selcuk University Natural and Applied Science*, Vol. 1(3), pp. 1-7.

158. Van emden, Meiracker R.A.F. and Ramakers, P.M.J. 1989. Biological Control of the Western Flower thrips *Frankliniella occidentalis* in Sweet pepper, with the anthocorid predator *Orius insidiosus*. *Mededelingen Faculteit Landbouwwetenschappen, Rijksuniversiteit Gent.*, Vol. 56(2a), pp. 241-249.

159. Van, Lanteren, J.C. 2006. Assessing risks of releasing exotic biological control agents of arthropod pests. *Annu. Rev. Entomol.*, Vol. 51, pp. 609-634.

160. Van, Nouhuys, S., Singer, M.C. and Nieminen, M. 2003. Spatial and Temporal pattern of Caterpillar performance and the Suitability of two host plant species. *Ecol. Entomol.*, Vol. 28, pp. 193-202.

161. Varghese., Sara, Thania, Mathew. , Biju, Thomas. 2013. Bioefficacy and Safety Evaluation of Newer Insecticides and acaricides against chilli thrips and mites, *Journal of Tropical Agriculture*, Vol. 51: (1/2), pp. 111.

162. Venette, R.C. Davis, E.E. 2004. Chilli thrips/ yellow thrips, *Scirtothrips dorsalis* Hood. (Thysaoptera: Thripidae). *Mini Pest Risk Assessment.* University of Minnesota, St. Paul. MN, USA, pp. 31.

163. Venette, R.C., Davis, E.E. 2004. Chilli thrips/ yellow thrips, *Scirtothrips dorsalis* Hood (Thysanoptera: Thripidae). Mini pest Risk Asseseement, University of Minnesota, St. Paul, MN, USA.pp.31.

164. Venkanna, Yasa. , Ranga, Rao GV. , Dharma, Reddy, K. 2010. Bioefficacy of neonicotinoid insecticides against Thrips, *Scirtothrips dorsalis and Leafhoppers, Empoascakerri* in groundnut. *Indian Journal of Plant Protection,* Vol. 38; (2), pp. 134-138, ISSN.

165. Wackers, F.L. 2003. The Parasitoids' Weed for Sweets : Sugars in Mass Rearing and Biological Control. In Quality Control and Production of Biological Control Agents : Theory and Testing Procedures (Ed. J.C. Van Canterren) Wallingford, U.K. *CAB International,* pp. 59-72.

166. Walker, F. 1839. Monographia Chalciditun, London, England, pp. 17-18.

167. Welter, S.C., Rosenheim, J.A., Johnson, M.W., Mav and Gusukuma-Minuto, L.R. 1990. Effects of Melon thrips, *Thrips palmi,* and Western Flower Thrips, *Frankliniella occidentalis* (Thysanoptera : Thripidae) on the yield, growth, and carbon allocation pattern in cucumbers. *J. Econ. Entomol.,* Vol. 83, pp. 2092-2101.

168. Wennergren, U. and Landin, J. 1993. Population growth and Structure in a variable environment I. Aphids and temperature variation. *Oecologica,* Vol. 93, pp. 394-405.

169. Winkler, K. 2005. Assessing the Risks and Benefits of Flowering Field Edges Strategic Use of Nectar Sources to Boost Biological Control. Ph.D. Thesis, Wageningen University.

170. Zamora, Gonzalez J.E. and Mari, Garcia F. 2003. Efficiency of several sampling methods for *F. occidentalis* (Thysanoptera : Thripidae) in Strawberry flowers. *Journal of Applied Entomology,* Vol. 127, pp. 516-521, ISSN : 0931-2048.

Appendix-I
Glossary for Terms

1. **Arrhenotokous:** Arrhenotoky, also known as arrhenotokous parthenogenesis in which unfertilized eggs develop into males.
2. **Coxae:** The first joint of the leg, connecting the rest of the joints of the leg to the body at the thorax.
3. **Ecdysis:** The process of shedding the old skin or casting off the outer cuticle (in insects and other arthropods).
4. **Ectoparasitoids:** An ectoparasitoid is a parasite that lives outside in another animal and ultimately kills it.
5. **Endoparasitoid:** An endoparasitoid is a parasite that lives inside another animal and ultimately kills it.
6. **Entomophagous parasites:** Entomophagous parasites are insects that are parasitic on other insects.
7. **Fecundity:** Fecundity is the actual reproductive rate of an organism or population, measured by the number of gametes (eggs). Fecundity is similar to fertility, the natural capability to produce offspring.
8. **Gaster:** The enlarged part of the abdomen behind the pedicel in hymenopterous insects.
9. **Hyaline:** Transparent, like glass.
10. **Larva:** A larva is a distinct juvenile form many animals, undergo before metamorphosis into adults.
11. **Metastome:** A median elevation behind the mouth in the arthropods.
12. **Moulting:** Moulting, also known as sloughing, shedding. In which an animal routinely casts off a part of its body, either at specific times of the year, or at specific points in its life cycle.
13. **Multivoltine:** A multivoltine species is a species that has two or more broods of offspring per year.
14. **Mummify:** Embalming the body.
15. **Nymph:** A nymph is the immature form of some invertebrates, particularly insects, which undergoes gradual metamorphosis (hemimetabolism) before reaching its adult stage.

16. **Ovipositor:** A tubular organ through which a female insect deposits eggs.
17. **Phytophagous:** (especially of an insect or other invertebrate) feeding on plants.
18. **Proboscis:** A probascis is an elongated appendage from the head of an animal.
19. **Protelean parasites:** Protelean organisms are those that beging the growing phase of their lives as parasites, and in particular, typically as internal parasites.
20. **Pterygotes:** The pterygota are a subclass of insects that includes the winged insects.
21. **Pupae:** An insect in its inactive immature form between larva and adult.
22. **Solitary:** Alone, without companions, unattended.
23. **Thelytokous:** Thelytoky is a type of parthenogenesis in which females are produced from unfertilized eggs.
24. **Uniparental:** Having, involving, or derived from a single parent.
25. **Univoltine:** A univoltine species is a species that has one brood of offspring per year.

Appendix-II

Sublethal Effects of Pesticides on Natural Enemies

Behavioural and physiological effects
- Communication
- Developmental time
- Locomotion
- Voracity / Consumption / Parasitism
- Repellency
- Deformation
- Longevity
- Vigour
- Reproduction
    - Fecundity
    - Fertility
- Foraging / Parasitism
    - Searching
    - Oviposition
    - Handling time

Appendix-III
Natural Enemies

The following genera and species of Predators and Parasitoids have been evaluated, on chilli crop in District Aligarh of Western Uttar Pradesh. This is by no means on exhaustive list but it does contain some of the most common natural enemies likely to be encountered.

Order	Family	Example
Hemiptera	Anthocoridae	*Orius spp.;* *Macrotracheliella nigra;* *Amblyseius cucumeris*
Neuroptera	Chrysopidae	Green lacewings
Diptera	Phytoseiidae	Tachinid flies
Mantodea	Mantidae	Praying mantids
Hymenoptera	Eulophidae	*Ceranisus menes;* *Thripobius semiluteus*
Acari	Tetranychidae	Mites
Arachnida	Phytoseiidae	Spiders spp.

Appendix-IV
Other Sources Referred for our Research Work

1. Annals of the Entomological Society of America, 1908 present - United States, ISSN 0013-8746, Entomological Society of America.
2. Annual Review of Entomology, 1956 present United States, ISSN 0066-4170.
3. Applied Entomology and Zoology, 1966, present – Japan – ISSN 0003-6862, Japanese Society of Applied Entomology, Zoology.
4. Biocontrol (Formerly Entomophaga), 1956 – present, ISSN 1386-6141, Internation Organization for Biological Control.
5. Biocontrol Science and Technology, 1991 – present – ISSN 0958-3157, Taylor and Francis.
6. Bulletin of Entomological Research, 1910 – present – ISSN 0007-4853, Cambridge Journals.
7. Entomologia Experimenttalis et Applicata, 1958 – present Netherlands, ISSN 0013-8703. The Neterhlands Entomological Society.
8. Entomological News, 1890, present – United States, ISSN 0013-872X, American Entomological society.
9. Environmental Entomology, 1972, present – United States, ISSN 0046-225X. Entomological Society of America.
10. Florida Entomologist, 1917, present – United States, ISSN 0015-4040, Florida Entomological Society.
11. Japanese Journal of Applied Entomology and Zoology, 1957, present – Japan, ISSN 0021-4914, Japanese Society of Applied Entomology and Zoology.
12. Journal of Economic Entomology, 1908, present – United States, ISSN 0022-0493, Entomological Society of America.

13. Journal of Entomology, 1862-1866, United Kingdom, ISSN 1812-5670, Academic Journals.
14. Journal of Hymenoptera Research, 1992, present – ISSN 1070-9428, International Society of Hymenopterists.
15. Journal of Insect Science, 2001, present – United States, ISSN 1536-2442, Entomological Society of America.
16. European Journal of Entomology : Book Review : Thrips as Crop Pests. www.eje.cz BY jpelikan-2013. Eur. J. Entomol. 95(3): 454, 1998. ISSN 1210-5759 – open access.

Appendix-V
List of Publications

Papers published during the course work:
1. Paper published in **'NATURE & ENVIRONMENT'** on 'Infestation and Damage Level of chilli thrips, *Scirtothrips dorsalis* on chilli *Capsicum annuum* crop', Volume-19, Issue-2, 2014, pp. 245-246, ISSN: 2321-810X (Print); 2321-8738 (Online), www.natureandenvironment.com
2. Paper published in **'ANNALS OF NATURAL SCIENCES'** on 'Release of Parasitoid, *Ceranisus menes* as Biological Control Agent of Chilli Thrips, *Scirtothrips dorsalis* in Experimental Net House', Volume-1, Issue-1, December 2015, pp.1-2, ISSN: 2455-667X, www.crsdindia.com/ans.html
3. Paper published in **'NATURE & ENVIRONMENT'** on 'Effect of some species of thrips, Thysanoptera: Thripidae on chilli crop of District Aligarh', Volume-20, Issue-2, 2015, pp. 59-60, ISSN: 2321-810X (Print); 2321-8738 (Online), www.natureandenvironment.com

Actively Participated in National and International Conferences
1. **INTERNATIONAL CONFERENCE AT JHUNJHUNU (RAJ.):** As on 10th – 11th May, 2014 with the Theme of **"Perspective for India's Development: A Commercial, Linguistic and Technological Approach"**.
2. **CONFERENCE AT AGRA (U.P.):** As on 18th – 19th October, 2014 with the Theme of **"Transformation of Indian Educational Structure"**.
3. **CONFERENCE AT BHARATPUR (RAJ.):** As on 11th – 12th December, 2015 with the participated paper of **"Biological Control of Chilli Thrips, *Scirtothrips dorsalis* with the predatory mite, *Amblyseius cucumeris*."**.
4. **CONFERENCE AT MAINPURI (U.P.):** As on 28th – 29th February, 2016 with the poster paper of **"Role of Science and Technology in Socio-Economic Development"**.
5. **CONFERENCE AT KANPUR (U.P.):** As on 29th – 30th September, 2016 with the oral presentation entitled **"Control Agent in Experimental Net House"**.
6. **CONFERENCE AT SHAHJAHANPUR (U.P.):** As on 26th – 27th November, 2016 with the poster entitled **"Chilli Thrips; *Scirtothrips dorsalis* As an Applied Problem on Chilli Crops in the field of Agriculture in Western Uttar Pradesh"**.
7. **WORKSHOP AT FIROZABAD (U.P.):** As on 28th – 30th April, 2016 with the Discussed Topic : **"Natural Enemies (Predators and Parasitoids) Play an Enormous Role for Sustainable Development."**

Appendix-VI

Reviews References followed in our Research Work

A	
Sr. No. 1	
Name of the Author	Hans Larsson
Name of the Article	Aphids and Thrips: The Dynamics and Bio-Economics of Cereal Pests
Pages	42
Year of Publication	2005
Journal Name	Doctoral Thesis Sciences, Swedish University of Agricultural

Sr. No. 87	
Name of the Author	S. Gnanachandran and C. Sivayoganathan
Name of the Article	Adoption of non-chemical pest control and Brinjal by Jaffana Farmers
Pages	8, Vol. 2
Year of Publication	1990
Journal Name	Tropical Agricultural Research

Sr. No. 39	
Name of the Author	Ali Hosseini – Gharalari, Ali Mohammadipour and Nazanin Koupi
Name of the Article	A New Method for Analyzing Sticky-Card Data in Entomology
Pages	7
Year of Publication	2009
Journal Name	The Weta 50 : 48-54

Sr. No. 45	
Name of the Author	Alexandre Pires Aguiar
Name of the Article	A technique to dry Mount Hymenaptera (Hexapoda) from alcohol a few seconds and its application to other Insect orders
Pages	9
Year of Publication	2012
Journal Name	Zootaxa

Sr. No. 86	
Name of the Author	A. Tatara and K. Furuhashi
Name of the Article	Analytical Study on Damage to Satusma Mandarian Fruit by *Scirtothrips dorsalis,* with particular reference to Pest Density
Pages	7, Vol. 36
Year of Publication	1992
Journal Name	Japanese Journal of Applied Entomology and Zoology

Sr. No. 28	
Name of the Author	John Sivinski
Name of the Article	Augmentative Biological Control : Research and Methods to help make it work.
Pages	11
Year of Publication	2013
Journal Name	CAB International, Vol. 8, ISSN 1749-8848

B	
Sr. No. 8	
Name of the Author	P.N. Krishna Moorthy, S. Saroja and K. Shivaramu
Name of the Article	Bio-efficacy of neem product and essential oils against thrips (*S. dorsalis* Hood) in *Capsicum*.
Pages	3
Year of Publication	2013
Journal Name	Pest Management in Hortiucultural Ecosystems, Vol. 19(2), 2013, ISSN 0971-6831

Sr. No. 54	
Name of the Author	Sara Varghese, Biju Mathew and Thomas
Name of the Article	Bioefficacy and safety evaluation of Newer Insecticides and Acaricides Against Chilli Thrips and Mites
Pages	1, Vol. 51 (1/2)
Year of Publication	2013
Journal Name	Journal of Tropical Agriculture

Sr. No. 80	
Name of the Author	G. Santharam, K. Kumar, S. Chandrasekaran and S. Kuttalam
Name of the Article	Bio efficacy and Residues of Imidalcoprid in Chillies used Against Chilli thrips
Pages	5, Vol. 90 (7-9)
Year of Publication	2003
Journal Name	Madras Agri. J.

Sr. No. 42	
Name of the Author	J.S. Bale, J.C. Van Lenteren and F. Bigler
Name of the Article	Biological Control and Sustainable Food Production
Pages	17
Year of Publication	2008
Journal Name	*Phil. Trans. R. Soc.* Philosophical Transactions of the Royal Society

Sr. No. 37	
Name of the Author	Dionysio Pedikis, Eleftheria Kapaxidi and Georgios Papadoulis
Name of the Article	Biological Control of Insect and Mite Pests in Greenhouse Solanaceous Crops
Pages	20
Year of Publication	2008
Journal Name	The European Journal of Plant Science and Biotechnology

Sr. No. 49	
Name of the Author	Yuonne M. Van Houten, Mai Linn Ostile, Hans Hoogerbrugge and Karel Bokkmans
Name of the Article	Biological Control of Western Flower Thrips on Sweet Pepper using the Predatory Mites, *Amblyseius cucumeris, Iphiseius degenerans*
Pages	8
Year of Publication	2005
Journal Name	The European Journal of Plant Science and Biotechnology

Sr. No. 94	
Name of the Author	D.J. Greathead
Name of the Article	Biological Control of Insect Pests by Insect Parasitoids and Predators : The BIOCAT Database
Pages	6
Year of Publication	1986
Journal Name	Biocontrol News and Information

Sr. No. 40	
Name of the Author	U. Bernardo, G. Viggani and R. Sasso
Name of the Article	Biological Parameters of *Thripobius semiluteus* Boucek (Hymenoptera: Eulophidae), a larvae Endoparasitoid of *Heliothrips haemorrhoidalis* (Bouche) (Thysanoptera: Thripidae).
Pages	8
Year of Publication	2005
Journal Name	Journal of Applied Entomology, Vol.129 (5), pp. 250-257.

Sr. No. 60	
Name of the Author	D.R. Seal and V. Kumar
Name of the Article	Biological Responses of Chilli Thrips, *S. dorsalis* Hood (Thysanoptera : Thripidae) to various Regimes of Chemical and Biorational Insecticides
Pages	7, Vol. 39
Year of Publication	2012
Journal Name	Crop Protection

Sr. No. 68	
Name of the Author	Yasa Venkanna, G.V. Rao and K. Reddy
Name of the Article	Bio-Efficacy of Neonicotinoid Insecticides Against Thrips, S. dorsalis and Leafhoppers, Empoasca Kerri in groundnut
Pages	5, Vol. 38(2)
Year of Publication	2010
Journal Name	Indian Journal of Plant Protection

C	
Sr. No. 11	
Name of the Author	Scott W. Ludwig and Carlos Bogran
Name of the Article	Chilli Thrips : A New Pest in the Home Land Scape
Pages	4
Year of Publication	2007
Journal Name	Texas Cooperative Extension EEE-00041 12/07

Sr. No. 46	
Name of the Author	D. Galazzi, S. Maini and A.J.M. Lomans
Name of the Article	Ceranisus menes (Walker) (Hymenoptera Eulophidae) : Collection and Initial Rearing on Frankliniella occidentalis (Pergande) (Thysanoptera : Thripidae)
Pages	5
Year of Publication	1991-92
Journal Name	Bulletin of Insectology

Sr. No. 73	
Name of the Author	S. Brett Nietschke, D.M. Borchert, R.D. Magarey and M.A. Ciomperlik
Name of the Article	Climatological Potential for *S. dorsalis* (Thysanoptera : Thripidae) Establishment in the United States
Pages	8, Vol. 91 (1)
Year of Publication	2008
Journal Name	Florida Entomologist

Sr. No. 44	
Name of the Author	T. Sankaran
Name of the Article	Current Status and Future Projections for Biological Control of Insect Pests in India
Pages	9
Year of Publication	1986
Journal Name	Proc. Indian natn. Sci. Acad.

D	
Sr. No. 9	
Name of the Author	Vivek K. Jha, Dakshina R. Seal, David J. Schuster and Garima Kakkar
Name of the Article	Diet Flight Pattern and Periodicity of Chilli thrips (Thysanoptera : Thripdae) on Selected Hosts in South Florida
Pages	4
Year of Publication	2009
Journal Name	Proceedings of the Florida State Horticultural Society ISSN 0886-7283. Vol. 122 (2009), pp. 267-271, 2009

Sr. No. 79	
Name of the Author	Muhammad Umar, Waseem Akram, Baboo Ali, Muhammad Tariq and Nazar Ali Shah
Name of the Article	Description of Three Genera (Thripidae : Thysanoptera) from Azad Jammu & Kashmir (Pakistan)
Pages	10, Vol. 3(5)
Year of Publication	2003
Journal Name	Online Journal of Biological Sciences

Sr. No. 78	
Name of the Author	M.C. Millawithananchichi, A.L.T. Perera, B.C.N. Peiris and H.M. Fonseka
Name of the Article	Development of High Yielding Chilli Hybrids (*Capsicum annuum* L.) based on Heterobeltiosis and Characterization of Parental Germplasm for DNA polymorphisms
Pages	7, Vol. 18
Year of Publication	2004
Journal Name	Tropical Agricultural Research

Sr. No. 65	
Name of the Author	R.E. Farris, R. Ruiz Arce, M. Ciomperlik, J.D. Vasquez and R. Delon
Name of the Article	Development of a Ribosomal DNATI S2 Marker for the Identification of the Thrips, S. dorsalis
Pages	26, Vol. 10
Year of Publication	2010
Journal Name	The Journal of Insect Science

Sr. No. 5	
Name of the Author	Emilija Raspudic, Marija Ivezic, Mirjana Brmez and Stanislav Trdan
Name of the Article	Distribution of Thysanoptera Species and Their Host Plants in Croatia
Pages	8
Year of Publication	2009
Journal Name	Acta agriculturae Solvenica, 93-3, Sept., 2009, pp. 275-283, ISSN 1581-9175

Sr. No. 30	
Name of the Author	Dr. T.V. Sathe, Mithari Pranoti, Dr. S.S. Patil and A.S. Desai
Name of the Article	Diversity, Economic-Importance and Control Strategies of Thrips (Thysanoptera) on crop plants
Pages	3
Year of Publication	2015
Journal Name	Indian Journal of Applied Research, Vol. 5(8), pp. 761-763

E	
Sr. No. 2	
Name of the Author	M. Shivaprasad, B.M. Chittapur, H.D. Mohankumar, S.A. Astaputre, M.H. Tatagar and R.K. Mesta
Name of the Article	Eco-Friendly Approaches for the Management of Murda Complex in Chilli
Pages	1
Year of Publication	2010
Journal Name	Academic Journal; Agricultural Reviews; 2010, Vol. 31(4), p. 298, ISSN 0253-1496

Sr. No. 22	
Name of the Author	J.M. Mari, R.B. Laghri, A.S. Mari and A.K. Shahzadi
Name of the Article	Eco-Friendly Pest Management of Chilli Crop
Pages	12
Year of Publication	2013
Journal Name	International Journal of Agricultral Technology Vol.9(5), pp. 1981-1992

Sr. No. 21	
Name of the Author	P.P. Jagtap, U.S. Shingane and K.P. Kulkarni
Name of the Article	Economics of Chilli Production in India.
Pages	4
Year of Publication	2012
Journal Name	African Journal of Basic & Applied Sciences Vol. 4(5), pp. 161-164, 2012, ISSN 2079-2034

Sr. No. 52	
Name of the Author	Sandeep Kaur and Subhash Singh
Name of the Article	Efficacy of Some Insecticides and Botanicals against Sucking Pests on *Capsicum* under net house
Pages	5, Vol. 1 (1)
Year of Publication	2013
Journal Name	Agriculture for Sustainable Development

Sr. No. 56	
Name of the Author	S. Maria Packiam and Ignacimuthu

Name of the Article	Effect of Botanical Pesticidal Formulations Against The Chilli Thrips (*S. dorsalis*) Hood on Peanut Ecosystem
Pages	5, Vol. 2(1)
Year of Publication	2013
Journal Name	International Journal of Natural and Applied Science

Sr. No. 93	
Name of the Author	C.M. Mannion, Andrew I. Derksan, D.R. Seal, Lance S.s Osborne and G. Ciff Martin
Name of the Article	Effects of Rose Cultivaors and Fertilization Rates on Populations of *Scirtothrips dorsalis* (Thysanoptera :Thripidae) in Southern Florida
Pages	5
Year of Publication	2013
Journal Name	Florida Entomdogist – Bioone

Sr. No. 25	
Name of the Author	Muhammad Rafiq Kethran, Ying Ying Sun, Shahbaz Khan, Sana Ullah Baloch, L.L. Wu, T.T. Lu, Yang Yang, Zhan Hu, Abdul Salam, Sohil Iqbal, Sakhwat Ali and Waseem Bashir
Name of the Article	Effect of Different Sowing Dates on Insect Pest Population of Chillies (*Capsicum annuum* L.)
Pages	19
Year of Publication	2014
Journal Name	Journal of Biology Agriculture and Healthcare, Vol. 4(25), 2014, ISSN 2224-3208

Sr. No. 35	
Name of the Author	Virendra Kumar and Manika Gupta

Name of the Article	Effect of some species of thrips, Thysanoptera : Thripidae on chilli crop of District Aligarh
Pages	2
Year of Publication	2015
Journal Name	Nature and Environment, Vol. 20(2), 2015: 59-60 ISSN : 2321-810X

Sr. No. 84	
Name of the Author	Manaku Shibao
Name of the Article	Effects of Insecticide Application on Population Density of the Chilli Thrips, S. dorsalis Hood (Thysanoptera : Thripidae), on Grape
Pages	3, Vol. 32(3)
Year of Publication	1997
Journal Name	Applied Entomology Zoology

Sr. No. 66	
Name of the Author	Mahesh Chandra, R.. Verma, Rajesh Prakash, Manish Kumar, Deepmala Verma and D.K. Singh
Name of the Article	Effect of Malathion on Adult of S. dorsalis and Rhiphiphorothrips cruenateis (Thysanoptera : Thripidae)
Pages	3, Vol. 1
Year of Publication	2010
Journal Name	Advances in Bioresearch

Sr. No. 67	
Name of the Author	Mahesh Chandra, R.K. Verma, Rajesh Prakash, Manish Kumar, Deepmala Verma and D.K. Singh
Name of the Article	Effect of Pyrethrin on Adult of Thrips tabaci and S. dorsalis (Thysanoptera : Thripidae)

Pages	3, Vol. 1
Year of Publication	2010
Journal Name	Advances in Bioresearch

Sr. No. 15	
Name of the Author	J.E. Gonzalez – Zamora and F. Garcia-Mari
Name of the Article	Efficiency of Several Sampling Methods for *F. occidentalis* (Thysanoptera : Thripidae) in strawberry flowers
Pages	5
Year of Publication	2003
Journal Name	Journal of Applied Entomology, 2003, pp. 516-521 ISSN 0931-2048

Sr. No. 85	
Name of the Author	J. Nandkumar, J.M.R.S. Bandra and S.E. Peiris
Name of the Article	Establishment of Embryogenic Cell Suspension Culture of Chilli *Capsicum annuum* (Var. accuminatum Fingerh. for Somatic Embryogenesis
Pages	10, Vol. 8
Year of Publication	1996
Journal Name	Tropical Agricultural Research

Sr. No. 33	
Name of the Author	Steven Arthurs, Cindy L. McKenzie, Jianjun Chen, Mahmut Dogramaci and Mary Brennan, Katherine Houben and Lance Osborne
Name of the Article	Evaluation of *Neoseiulus cucumeris* and *Amblyuseius swirskii* (Acari : Phytoseiidae) as Biological Control Agents of Chilli Thrips, *Scirtothrips dorsalis* (Thysanoptera : Thripidae) on Pepper

Pages	6
Year of Publication	2009
Journal Name	Biological control, Vol. 49, 2009, pp. 91-96

Sr. No. 62	
Name of the Author	C.J. Jagdish and A.P. Purnima
Name of the Article	Evaluation of Selective Botanicals and Entomopathogens Against *S. dorsalis* Hood under Polyhouse Conditions on Rose.
Pages	1, Vol. 4
Year of Publication	2011
Journal Name	Journal of Biopesticides

F

Sr. No. 61	
Name of the Author	S.K. Mandal
Name of the Article	Field Evaluation of Alternate use of Insecticides Against Chilli Thrips, *Scirtothrips dorsalis* (Hood.)
Pages	4, Vol. 20(1)
Year of Publication	2012
Journal Name	Annals of Plant Protection Sciences

Sr. No. 77	
Name of the Author	S. Satpathy, Akhilesh Kumar, T.M. Shivalingaswamy, A.B. Rai and Mathura Rai
Name of the Article	Field Efficacy of Metonyl Against Thrips, *S. dorsalis* (Hood) in Chilli
Pages	4, Vol. 33(2)

Year of Publication	2006
Journal Name	Vegetable Science

G	
Sr. No. 19	
Name of the Author	Lane Greer and Steve Diver
Name of the Article	Greenhouse IPM : Sustainable Thrips Control
Pages	18
Year of Publication	2000
Journal Name	Pest Management Technical Notes ATTRA – 800-346-9140

H	
Sr. No. 16	
Name of the Author	Douglas A. Landis, Stephen D. Wratten and Geoff M. Gurr.
Name of the Article	Habitat Management to Conserve Natural Enemies of Arthropod Pests in Agriculture
Pages	30
Year of Publication	2000
Journal Name	Ann. Rev. of Entomol. Vol. 45(1), pp. 175-201, ISSN 0066-4170

I	
Sr. No. 34	
Name of the Author	Manika Gupta and Virendra Kumar
Name of the Article	Infestation and Damage level of Chilli thrips, *S. dorsalis* on Chilli, *Capsicum annuum* crop
Pages	2

Year of Publication	2014
Journal Name	Nature and Environment, Vol. 19 (2), 2014, pp. 245-246

Sr. No. 10	
Name of the Author	N. Mandi and A.K. Senapati
Name of the Article	Integration of Chemical Botnaical and Microbial Insecticides for control of thrips, *S. dorsalis* Hood infesting chilli.
Pages	4
Year of Publication	2009
Journal Name	The Journal of Plant Protection Sciences 1(1): 92-95, 2009, ISSN 2249-7897

L	
Sr. No. 18	
Name of the Author	K.J. Froud and P.S. Stevens
Name of the Article	Life Table Comparison between the Parasitoid *Thripobius semiluteus* and its Host Greenhouse Thrips
Pages	3
Year of Publication	1997
Journal Name	Horti Research, Private Bug Vol. 92 (169), Aukland, ISSN 1174-6947

Sr. No. 29	
Name of the Author	David R. Gillespie
Name of the Article	Life-History and Cold Storage of *Amblyseius cucumeris* (Acarina : Phytoseiidae)
Pages	6

Year of Publication	1988
Journal Name	Journal of the Entomological Society of British Columbia, Vol. 85, 1988, ISSN 0071-0733

M	
Sr. No. 31	
Name of the Author	Mahmut Dogramaci, Steven P. Arthurs, Jijanjun Chen, Cindy McKenzie, Fabieli Irrizary and Lance Osborne
Name of the Article	Management of Chilli thrips *Scirtothrips dorsalis* (Thysanoptera : Thripidae) on Peppers by *Amblyseius swirskii* (Acari : Phytoseiidae) and *Orius insidiosus* (Hemiptera : Anthocoridae)
Pages	7
Year of Publication	2011
Journal Name	Biological control 59, 2011, pp. 340-347

N	
Sr. No. 17	
Name of the Author	M.W. Johnson
Name of the Article	Nature and Scope of Biological Control
Pages	5
Year of Publication	2000
Journal Name	Bio-control of Pests, ENTO 675, UH-Manoa

Sr. No. 7	
Name of the Author	Vivek Kumar, Dakshina R. Seal, Garima Kakkar and Lance Osborne
Name of the Article	New Tropical fruit Hosts of *S. dorsalis* (Thysanoptera : Thripidae) and its Relative Abundance on them in South Florida.

Pages	2
Year of Publication	2012
Journal Name	Florida Entomologist 95(1): 205-207. March 2012, ISSN 1938-5102

O

Sr. No. 3	
Name of the Author	Magdaline Kharbangar, S. Choudhury and S.R. Hajong
Name of the Article	Occurrence and Abundance of Thrips (Thysanoptera) Associated with Rice Crops from Meghalaya
Pages	7
Year of Publication	2014
Journal Name	International Journal of Research Studies in Bio-Sciences (IJRSB), Vol. 2(5), June 2014, pp. 1-7, ISSN 2349-0357

Sr. No. 13	
Name of the Author	Ronald D. Oetting and Ramona J. Beshear
Name of the Article	*Orius insidiosus* (Say) and Entomopathogens as Possible Biological Control Agents for Thrips
Pages	6
Date of Published	1991
Publication Information	General Technical Report (GTR) : U.S. Dept. of Agriculture, Forest Service, North Eastern Forest exp. Station , pp. 419-424

Sr. No. 55	
Name of the Author	Santoshi Toda, Fakuya Hirose, Kanako Kakiuchi, Hirosato Kadana, Keisuke Kijima and Masatoshi Mochizuki

Name of the Article	Occurrence of a novel strain of *S. dorsalis* (Thysanoptera : Thripidae) in Japan and Development of its Molecular Diagnostics
Pages	9, Vol. 49(2)
Year of Publication	2013
Journal Name	Applied Entomology and Zoology

P	
Sr. No. 43	
Name of the Author	A. Bonet
Name of the Article	Parasitoid Wasps; natural Enemies of Insects
Pages	7
Year of Publication	2009
Journal Name	Tropical Biology and Conservation Management, Vol. VII

Sr. No. 26	
Name of the Author	D.N.R. Reddy, Puttaswamy
Name of the Article	Pests infesting Chilli (*Capsicum annuum* L.) in the Nursery
Pages	4
Year of Publication	1985
Journal Name	Mysore Journal of Ag. Sciences Pub. 1985, Vol 18(2), pp. 122-125, ISSN 0047-8539

Sr. No. 6	
Name of the Author	Ekram Atakan
Name of the Article	Population Densities and Distribution of the WFT (Thysanoptera : Thripidae) and its Predatory Bug,

	Orius niger (Hemiptera : Anthocoridae), in Strawberry.
Pages	6
Year of Publication	2011
Journal Name	International Journal of Agriculture and Biology 13(5): 638-644, January 2011, ISSN 1560-8530

Sr. No. 96	
Name of the Author	N. Nandini
Name of the Article	Population Dynamics and Extent Damage of Pests of *Capsicum* under protected cultivation
Pages	1
Year of Publication	2012
Journal Name	Karnataka J. Agricultural Science

Sr. No. 20	
Name of the Author	R.S. Giraddi, S.M. Mantur, R.K. Patel, C.P. Mallapur, K.V. Ashalatha
Name of the Article	Population Dynamics and Extent Damage of Pests of *Capsicum* under Protected Cultivation
Pages	2
Year of Publication	2012
Journal Name	Karnataka J. Agric. Sci. 25 (1), pp. 150-151

Sr. No. 53	
Name of the Author	S. Rebacca Buckman, L.A. Mound, F. Michael Whiting
Name of the Article	Phylogeny of thrips (Insecta : Thysanoptera) based on five molecular loci

Pages	11, Vol. 38
Year of Publication	2013
Journal Name	Systematic Entomology

Sr. No. 32	
Name of the Author	Anais Chailleux, Philippe Bearez, Jearnnine Pizzol, Edwige Amiens-Desneux, Ricardo Ramirez-Romero and Nicolas Desneux.
Name of the Article	Potential for Combined Use of Parasitoids and Generalist Predators for Biological Control of the Key Invasive Tomato Pest *Tuta absoluta*.
Pages	8
Year of Publication	2013
Journal Name	J. Pest Sci. (Springer) Vol. 86(3), pp. 533-541

Sr. No. 23	
Name of the Author	Garima Kakkar, Vivek Kumar, Dakshina R. Seal, Oscar E. Liburd and Philip A. Stansly
Name of the Article	Predation by *Neoseilus cucumeris* and *Amblyseius swirskii* on *Thrips palmi* and *Franklineilla schultzei* on cucumber
Pages	6
Year of Publication	2016
Journal Name	Biological Conrol Vol. 92, 2016, pp. 85-91, ISSN 1049-9644

Sr. No. 62	
Name of the Author	P. Duraimurugan and A. Jagdish
Name of the Article	Preliminary studies on the Biology of *Scirtothrips dorsalis* (Thysanoptera : Thripidae) as a pest of rose in India

Pages	9, Vol. 69
Year of Publication	2011
Journal Name	Journal of Applied Zoological Research

R	
Sr. No. 36	
Name of the Author	Manika Gupta and Virendra Kumar
Name of the Article	Release of Parasitoid, *Ceranisus menes* as Biological Control Agent of Chilli Thrips, *Scirtothrips dorsalis* in Experimental Net House
Pages	2
Year of Publication	2015
Journal Name	Annals of Natural Sciences, Vol. 1(1), December 2015: 1-2

Sr. No. 12	
Name of the Author	Ion Oltean
Name of the Article	Research on Biology, Ecology and Integrated Management of Thysanoptera species in greenhouse
Pages	9
Year of Publication	2012
Journal Name	University of Agricultural Sciences & Veterinary Medicine CIUJ NAPOCA Doctoral School Ph.D. Thesis

Sr. No. 4	
Name of the Author	Richard L. Fery and James M. Schalk
Name of the Article	Resistance in Pepper (*Capsicum annuum* L.) to Western Flower Thrips (*Frankliniella occidentalis*) (Pergande)

Pages	1
Year of Publication	1991
Journal Name	Hort Science Vol. 26(8), pp. 1073-1074, August 1991

Sr. No. 63	
Name of the Author	M.H. Tatagar, J.S. Awaknavar, R.S. Giraddi, Mohan Kumar, H.D. Mallapur and P.A. Kataraki
Name of the Article	Role of border crop for the Management of Chili Leaf cure caused due to thrips *S. dorsalis* (Hood) and mites, *Polyphagotarsonemus latus* (Banks)
Pages	6, Vol. 24
Year of Publication	2011
Journal Name	Journal of Agricultural Science

Sr. No. 83	
Name of the Author	P.A. Fugro
Name of the Article	Role of Organic Pesticides and Manures in Management of Some Important Diseases
Pages	2, Vol. 30(1)
Year of Publication	2000
Journal Name	Journal of Mycology and Plant Pathology

Sr. No. 81	
Name of the Author	R.J. Anandan and Sabitha Doraiswamy
Name of the Article	Role of Barrier Crops in Reducing the Incidence of Mosaic disease in Chilli
Pages	4
Year of Publication	2002
Journal Name	Journal of Plant Diseases & Protection

S

Sr. No. 57	
Name of the Author	Charles Ssemwogerere, Ssemakula,a Mildred Kathrym Nyaburu Oehwo, Joe Kovach, Samuel Kyamanywa and Jeninah Karungi
Name of the Article	Species Composition and Occurrence of Thrips on tomato and Pepper as influenced by Farmer's Management Practices in UGANDA
Pages	10, Vol. 39
Year of Publication	2010
Journal Name	Journal of Plant Protection Research

Sr. No. 59	
Name of the Author	Hamesh Aliakabarpour and R. Md. Che Salmah
Name of the Article	Seasonal abundance of *Thrips hawaiiensis* (Morgan) and *S. dorsalis* Hood (Thysanoptera : Thripidae) in Mango Orchard in Malaysia
Pages	9, Vol. 35(3)
Year of Publication	2012
Journal Name	Pertanika Journal of Agricultural Science

Sr. No. 58	
Name of the Author	Mahmut Dogramaci, P. Steven Arthurs, Jianjun Chen & Lance Osborne
Name of the Article	Silicon Applications have Minimal Effects on *Scirtothrips dorsalis* (Thysanoptera : Thripidae) Populations on Pepper Plant, *Capsicum annuum* L.
Pages	7, Vol. 96(1)
Year of Publication	2013
Journal Name	Florida Entomologist, Bioone

T	
Sr. No. 27	
Name of the Author	Levent Unlu, Ekrem Ogur, Yusuf Celik
Name of the Article	The Importance of Integrated Pest Management for Sustainable Agriculture
Pages	7
Year of Publication	2012
Journal Name	Journal of Selcuk University Natural and Applied Science, Vol. 1(3)

Sr. No. 51	
Name of the Author	Affandi, Affandi, Rosacelia and Medina
Name of the Article	Thrips (*Scirtothrips dorsalis*) Hood (Thysanoptera : Thripidae) Associate with Mango Agro ecosystem in East Java, Indonesia.
Pages	3
Year of Publication	2013
Journal Name	Agrivita Journal of Agricultural Science

Sr. No. 82	
Name of the Author	P.C. Maris, N.N. Joosten, D. Peters and R.W. Goldbach
Name of the Article	Thrips Resistance in Pepper and its consequences for the Acquisition and Inoculation of Tomato spotted wilt virus by the Western Flower Thrips
Pages	6, Vol. 93(1)
Year of Publication	2003
Journal Name	The American Phytopathological Society

Sr. No. 75	
Name of the Author	T.N. Ananthakrishnan and R.S. Annadurai
Name of the Article	Thrips – Tospovirus Interactions : Biological and Molecular Implications
Pages	4, Vol. 92 (8)
Year of Publication	2007
Journal Name	Current Science

Sr. No. 14	
Name of the Author	Natasa Mehle and Stanislav Trdan
Name of the Article	Traditional and Modern Methods for the Identification of Thrips (Thysanoptera) species
Pages	13
Year of Publication	2012
Journal Name	(Research Gate), Journal of Pest Science, June 2012, Vol. 85(2), pp. 179-190

Sr. No. 71	
Name of the Author	K. Gopal, M. Krishna Reddy, D.V.R. Reddy and V. Muniyappa
Name of the Article	Transmission of Peanut Yellow Spot Virus (PYSV) by Thrips, *S. dorsalis* Hood in Groundnut
Pages	9, Vol. 43(5)
Year of Publication	2009
Journal Name	Archives of Phytopathology & Plant Protection

Sr. No. 41	
Name of the Author	K. Kavitha and K. Dharma Reddy
Name of the Article	Tritrophy - A New Dimension in IPM – A Review
Pages	9
Year of Publication	2014
Journal Name	Agri. Review, 35(3): 207-215

U

Sr. No. 38	
Name of the Author	Stephanie Williamson
Name of the Article	Understanding Natural Enemies; A Review of Training and Information in the Practical Use of Biological Control.
Pages	10
Year of Publication	1998
Journal Name	Bio-control News and Information, Vol. 19(4)

W

Sr. No. 24	
Name of the Author	H.R. Sardana, M.N. Bhatt and Mukesh Sehgal
Name of the Article	Wide Area Validation and Economic Analysis of Adoptable IPM Technology in Bell Pepper (*Capsicum annuum*)
Pages	3
Year of Publication	2012
Journal Name	Indian Journal of Agricultural Sciences 82(2), pp. 186-189, Feb. 2012